Physics of Comets

2nd Edition

WORLD SCIENTIFIC SERIES IN ASTRONOMY AND ASTROPHYSICS

Editor: Jayant V. Narlikar
Inter-University Centre for Astronomy and Astrophysics, Pune, India

Volume 1:	Lectures on Cosmology and Action at a Distance Electrodynamics *F. Hoyle and J. V. Narlikar*
Volume 2:	Physics of Comets (2nd Ed.) *K. S. Krishna Swamy*
Volume 3:	Catastrophes and Comets *V. Clube and B. Napier*
Volume 4:	From Black Clouds to Black Holes (2nd Ed.) *J. V. Narlikar*
Volume 5:	Solar and Interplanetary Disturbances *S. K. Alurkar*
Volume 6:	The Universe: From Inflation to Structure Formation *V. Sahni and A. A. Starobinsky*

WORLD SCIENTIFIC SERIES IN
ASTRONOMY AND ASTROPHYSICS — Vol. 2

Series Editor: **Jayant V Narlikar**

Physics of Comets

2nd Edition

K S Krishna Swamy
Tata Institute of Fundamental Research,
Bombay, India

World Scientific
Singapore • New Jersey • London • Hong Kong

Published by

World Scientific Publishing Co. Pte. Ltd.
P O Box 128, Farrer Road, Singapore 912805
USA office: Suite 1B, 1060 Main Street, River Edge, NJ 07661
UK office: 57 Shelton Street, Covent Garden, London WC2H 9HE

Library of Congress Cataloging-in-Publication Data
Krishna Swamy, K. S.
   Physics of comets / K. S. Krishna Swamy, -- 2nd ed.
      p.   cm. -- (World Scientific series in astronomy and astrophysics ; vol. 2)
   Includes bibliographical references and index.
   ISBN 9810226322
   1. Comets. I. Title. II. Series.
   QB721.K74   1997
   523.6--dc21   97-7970
                                                                                                                               CIP

British Library Cataloguing-in-Publication Data
A catalogue record for this book is available from the British Library.

Copyright © 1997 by World Scientific Publishing Co. Pte. Ltd.

All rights reserved. This book, or parts thereof, may not be reproduced in any form or by any means, electronic or mechanical, including photocopying, recording or any information storage and retrieval system now known or to be invented, without written permission from the Publisher.

For photocopying of material in this volume, please pay a copying fee through the Copyright Clearance Center, Inc., 222 Rosewood Drive, Danvers, MA 01923, USA. In this case permission to photocopy is not required from the publisher.

Printed in Singapore.

FOREWORD TO THE REVISED EDITION

Only in recent years has the true nature of comets become evident. With this knowledge has come the realization that comets are almost certainly debris left over from the building of the outermost planets. This material must be typical of the gas and dust in the interstellar cloud from which the Sun and planetary system evolved. Thus comets appear to offer us an opportunity to study primitive matter involved in the origin of the solar system, material that has been stored in deep freeze for 4.6×10^9 years. At the same time, the comets provide a link with interstellar solids, of which they are probably largely composed.

The physical and chemical study of comets has now replaced classical celestial mechanics as the major focus of observational and theoretical research. Familiar areas of astronomical spectroscopy are approached from a somewhat unusual point of view in cometary studies, while comets provide an entirely new laboratory for an increased understanding of magneto-hydrodynamics. On the other extreme, the physics of amorphous ices and other solids can be applied to probe the internal structure, origin and activity of the comet nuclei.

In summation, the study of comets has now become a manifold discipline with fascinating potential, particularly as the space age provides new observational input over the entire electromagnetic spectrum, coupled with the expectation of direct *in situ* studies and even the eventual return of samples from cometary nuclei.

The marked progress in this direction is represented by the new results described in this book's revised edition. The several missions to Halley's comet have filled in many lacunae of comet knowledge.

Fred L. Whipple

PREFACE

The study of Comet Halley in 1986 was a tremendous success giving a big boost to cometary science. The greatest adventure was when six spacecrafts passed through Comet Halley in March 1986 as close as 600 km from the nucleus and made the *in situ* measurements of various kinds. The combined effort of these space missions to Comet Halley and that of ICE spacecraft to Comet Giacobini-Zinner, with studies, both ground-based and above atmosphere, has been a remarkable success and has increased our knowledge of cometary science in a dramatic way. Therefore, there is a lot of new and exciting material on comets that has come out after the publication of the first edition of the book *Physics of Comets* in 1986. Hence, it is now an appropriate time to revise the book. In addition to several chapters being rewritten, extensive revisions have been carried out in all the other chapters. A significant amount of new material is also incorporated in the chapters. However, the general arrangement, purpose and level of the book do not differ from the old one.

I am grateful to Dr S Ramadurai for taking a deep interest in the critical reading of the revised chapters of the book. I am thankful to Dr N C Rana for reading the chapter on Dynamics and making helpful comments. A thorough reading of the entire book by my daughter Sujata is well appreciated. I wish to thank her as well as my wife Shyamala and son Suresh for their patience, encouragement and the moral support extended to me during the revision of the book. I am also indebted to my parents for their encouragement.

I wish to thank all the publishers and authors who have given permission for reproduction of the figures. The help of Mr P Joseph and Ms Flory Fernandes in the preparation of the manuscript and assistance in other matters is greatly appreciated. I would also like to thank the

Drawing, the Photography and the Xerox Sections of the Institute for their wholehearted cooperation. I am also grateful to the Director, Professor V Singh and the Institute for its excellent research facilities and conducive atmosphere, which have been helpful for bringing this revised edition to fruition.

K S Krishna Swamy

Bombay
April 1995

CONTENTS

Foreword	v
Preface	vii
1. General Introduction	**1**
1.1. Historical Perspective	1
1.2. Encounter with Comet Halley	4
1.3. Discovery	6
1.4. Appearance	7
1.5. Statistics	8
1.6. Importance	10
1.7. Brightness	11
1.8. Main Characteristics	15
1.9. An Overall View	23
Problems	26
References	27
2. Dynamics	**28**
2.1. Orbital Elements	28
2.2. Orbit in Space	30
2.2.1. Relevant equations	30
2.2.2. Orbital elements from position and velocity	35
2.2.3. Orbital elements from observations	37
Problems	40
References	41
3. Physical Aspects	**42**
3.1. Black Body Radiation	42
3.2. Perfect Gas Law	43
3.3. Dissociative Equilibrium	44
3.4. Doppler Shift	45

3.5. Spectroscopy	45
3.5.1. Atomic spectroscopy	45
3.5.2. Molecular spectroscopy	48
3.6. Isotope Effect	52
3.7. Franck-Condon Factors	52
3.8. Intensity of Emitted Lines	54
3.9. Boltzmann Distribution	57
3.10. Λ-Doubling	57
3.11. Solar Radiation	58
3.12. Solar Wind	61
Problems	62
References	63
4. Spectra	**64**
4.1. Main Characteristics	64
4.2. Forbidden Transitions	78
4.3. Line-to-Continuum Ratio	79
Problems	80
References	81
5. Spectra of Coma	**82**
5.1. Fluorescence Process	82
5.1.1. Rotational structure	85
5.1.2. Vibrational structure	88
5.1.3. Comparison with observations	90
5.1.4. Case of C_2 molecule	95
5.1.5. Molecules other than diatomic	101
5.1.6. OH radio lines	102
5.1.7. Oxygen lines	106
5.1.8. Forbidden transitions	106
5.1.9. Molecular band polarization	107
5.2. Excitation Temperature	109
5.2.1. Rotational temperature	109
5.2.2. Vibrational temperature	111
5.3. Abundances of Heavy Elements	113
5.4. Isotopic Abundances	115
Problems	121
References	121
6. Gas-Production Rates in Coma	**123**
6.1. Theoretical Models	123

6.1.1. From the total luminosity		123
6.1.2. From surface brightness distribution		128
6.1.3. From number densities		132
6.1.4. Semi-empirical photometric theory		134
6.2. Results		136
6.2.1. OH and H		136
6.2.2. CN, C_2, C_3, CH, NH_2, CO		141
6.2.3. CS, S_2		146
6.2.4. Ions		146
6.2.5. Complex molecules		149
6.2.6. O, C, N, S		151
6.3. Analysis of Hydrogen Observations		157
6.3.1. Analysis of $Ly\alpha$ measurements		157
6.3.2. Analysis of $H\alpha$ observations		165
6.4. Gas-Phase Chemistry in the Coma		167
6.4.1. In situ mass-spectrometer for ions		173
6.5. Temperature and Velocities of the Coma Gas		177
6.6. Parent Molecules		184
6.7. Summary		189
Problems		190
References		191
7. Dust Tail		**193**
7.1. Dynamics		193
7.2. Anti-Tail		200
7.3. Dust Features		202
7.4. Icy-Halo		205
Problems		206
References		206
8. Light Scattering Theory		**207**
8.1. Mie Scattering Theory		207
8.1.1. Efficiency factors		207
8.1.2. Albedo		210
8.1.3. Scattered intensity		211
8.1.4. Polarization		212
8.2. Approximate Expressions		212
8.3. Computation of Cross Sections		213
8.4. Results		214
8.5. Particles of Other Types		218

	8.6. Optical Constants	223
	Problems	226
	References	227
9.	**The Nature of Dust Particles**	**228**
	9.1. Visible Continuum	228
	9.1.1. Phase function	235
	9.1.2. Dust production rate from continuum	236
	9.2. Polarization	238
	9.3. Infrared Measurements	242
	9.3.1. Dust production from infrared observation	250
	9.3.2. Anti-tail	251
	9.4. Spectral Features	254
	9.4.1. Silicate signature	254
	9.4.2. CH stretch feature	258
	9.4.3. Ice signature	260
	9.4.4. Possible new features	260
	9.5. Properties Derived from *in situ* Measurements	261
	9.6. Albedo of the Particles	262
	9.7. Continuum Emission in the Radio Region	263
	9.8. Radiation-Pressure Effects	264
	9.9. Summary	264
	Problems	267
	References	268
10.	**Ion Tails**	**270**
	10.1. Evidence for the Solar Wind	270
	10.2. Dynamical Aberration	271
	10.3. Theoretical Considerations	275
	10.3.1. Comparison with observations	280
	10.4. Instabilities and Waves	283
	10.5. Acceleration of Cometary Ions	286
	10.6. Large Scale Structures	288
	10.6.1. Tail rays or streamers	288
	10.6.2. Knots or condensations	289
	10.6.3. Oscillatory structure	290
	10.6.4. 'Swan-like' feature	291
	10.6.5. Bend in the tail	291
	10.6.6. Disconnection events	293

	Problems	295
	References	295
11.	**Nucleus**	**297**
	11.1. Theory of Vaporization	297
	11.2. Outbursts	303
	11.3. Albedo and Radius	304
	11.4. Rotation	307
	11.5. Density	309
	11.6. Chemical Composition	310
	11.7. Mass Loss	314
	11.8. Structure	314
	11.9. Non-Gravitational Forces	316
	Problems	322
	References	323
12.	**Origin**	**324**
	12.1. Evidence for the Oort Cloud	324
	12.2. Evolution and Properties of Oort Cloud	327
	12.2.1. Short period comets	333
	12.3. Origin of the Oort Cloud	334
	Problems	339
	References	339
13.	**Relation to Other Solar System**	**340**
	13.1. Asteroids	340
	13.2. Meteorites	343
	13.3. Meteor Streams	345
	13.4. Particles Collected at High Altitudes	349
	13.5. Primordial Material	350
	13.6. Chemical Evolution	351
	13.7. Overview	353
	Problems	359
	References	359
14.	**Problems and Prospects**	**361**
	14.1. Epilogue	361
	14.2. Future Studies	364
	14.3. Postscript	366
	Problems	369
	References	369
	Index	371

CHAPTER 1

GENERAL INTRODUCTION

1.1. Historical Perspective

Among the various objects of the solar system, comets have attracted and fascinated the common man to a large extent for the last two thousand years or so. This information comes from the ancient records of paintings or drawings of comets on caves, clothes, etc. as well as from the observations of early writers. It was not until the sixteenth century that comets were demonstrated to be celestial objects. This came from the work of Tycho Brahe who observed the bright comet of 1577 AD with accurate instruments and from various locations in Europe. This really revolutionized the ideas about comets and from then on, observers took a serious view of comets and started making position measurements.

The complete credit for the discovery that comets are part of the solar system goes to Edmond Halley. Halley, using Newtonian mechanics, showed that the comets which had appeared in 1531, 1607 and 1682 are the one and the same with a period of about $75\frac{1}{2}$ years. He also noticed that the time interval between the successive perihelion passages was not the same. He concluded rightly that this could be due to the perturbation of the cometary orbit produced by the planets Jupiter and Saturn. Following these successes, he predicted that the same comet would return in 1758. As predicted, the comet did appear in 1758, though Halley, dead by then, was not there to witness the glorious triumph of his prediction. This comet is therefore named after him. In recent years this comet has been traced backwards through many centuries by several investigators through orbit calculations. Through this work, it has been possible to identify every appearance of the comet as shown by ancient records until about 240 BC. So far, it seems to have made about twenty-eight appearances.

Since early times, the appearance of a comet has been associated with disasters, calamities, tragedies and so on. One beneficial result of such wrong notions and ideas is that the appearances of most comets are recorded. These observations have proved very valuable to modern astronomers. Although there were many bright comets which have been recorded since early times, somehow the Comet Halley seems to have attracted much more attention than the others. The comet which has been depicted in the Bayeux tapestry is the Comet Halley which appeared in 1066 AD (Fig. 1.1). The comet in the tapestry can be seen to hover above the English King Harold who is being told of the bad omen. Such types of association of Comet Halley with the occurrence of bad things on Earth have also been made for many other apparitions. The last apparition of Comet Halley in 1910 drew wide publicity (Fig. 1.2). It was very bright and enormous in extent. Actually the Earth passed through the tenuous gas of this comet's tail.

There was also the fear that a comet might collide with the Earth and bring disastrous consequences. There are several indirect observational evidences, such as the presence of Cretaceous-Tertiary boundary, Tunguska event and so on, which show that such events must have happened on the

Fig. 1.1. A portion of the Bayeux tapestry showing the 1066 apparition of Comet Halley. The tapestry depicts the people pointing at the comet with fear for its effect on King Harold of England (Report of the Science Working group. The International Halley, Watch, NASA, July 1980).

BARNARD PICTURES OF HALLEY'S COMET

Taken at Yerkes Observatory May 4, They Tally with Observation from Times Tower May 5.

VIEWED BY MISS PROCTOR

Negatives Show the Tail Extending 20 Degrees, Equivalent to 24,000,000 Miles in Length.

IN COMET'S TAIL ON WEDNESDAY

European and American Astronomers Agree the Earth Will Not Suffer in the Passage.

TELL THE TIMES ABOUT IT

And of Proposed Observations— Yerkes Observatory to Use Balloons if the Weather's Cloudy.

TAIL 46,000,000 MILES LONG?

Scarfed in a Filmy Bit of It, We'll Whirl On In Our Dance Through Space, Unharmed, and, Most of Us, Unheeding.

SIX HOURS TO-NIGHT IN THE COMET'S TAIL

Few New Yorkers Likely to Know It by Ocular Demonstration, for It May Be Cloudy.

OUR MILLION-MILE JOURNEY

Takes Us Through 48 Trillion Cubic Miles of the Tail, Weighing All Told Half an Ounce!

BALLOON TRIP TO VIEW COMET.

Aeronaut Harmon Invites College Deans to Join Him in Ascension.

MAY SEE COMET TO-DAY.

Harvard Observers Think It May Be Visible in Afternoon.

MAY BE METEORIC SHOWERS.

Prof. Hall Doubts This, Though, but There's No Danger, Anyway.

YERKES OBSERVATORY READY.

Experts and a Battery of Cameras and Telescopes Already Prepared.

CHICAGO IS TERRIFIED.

Women Are Stopping Up Doors and Windows to Keep Out Cyanogen.

Fig. 1.2. Some of the newspaper headlines which appeared in The New York Times during Comet Halley's appearance in 1910 (Report of the Science Working Group, *loc, cit.*).

Earth in the past. The collision of Comet Shoemaker-Levy 9 with Jupiter during July 1994 has given supporting evidence, although the probability of such an event happening on the Earth is very small. With the passage of time some of the fears have been erased from the people's minds. Today the appearance of a bright comet in the sky like that of the Comet Ikeya-Seki of 1965 or the Comet West in 1975 is welcomed both by the scientists and the public at large. Scientists look forward to observing and studying these objects and understanding their nature. The public look forward to viewing a spectacular and colorful event in the sky.

During early times, comets were studied more from the point of their dynamics. This was made possible through the efforts of many pioneers in celestial mechanics. These extensive dynamical studies of various comets have shown, for the first time, the existence of some important physical effects like the presence of non-gravitational forces in comets. With the passage of time, cometary research has evolved from the study of dynamics to the study of these objects *per se*. Specifically, in the last two or three decades, emphasis has been laid more in understanding the origin, physics and chemistry of these objects. The presence of complex organic molecules in comets, which may have some relation to the existence of life on Earth, has interested biologists too, in the study of comets.

1.2. Encounter with Comet Halley

The general nature of comets has been revealed mainly through indirect means from the studies of ground based observations, rockets, aeroplanes and satellites. In order to make *in situ* measurements of a comet and to have direct access to the nucleus, several spacecrafts were sent to Comet Halley (1986 III) in 1986. Even prior to Comet Halley observations, *in situ* measurements of particles, fields and waves were carried out on Comet Giacobini-Zinner (1985 XIII) by the ICE Satellite on September 11, 1985. This spacecraft, which was earlier called the Third International Sun-Earth Explorer (ISEE-3), was launched in 1978 for the purpose of studying the solar-wind interaction with the Earth's magnetosphere. This spacecraft which was renamed the International Cometary Explorer (ICE) passed through the plasma tail of Comet Giacobini-Zinner.

During March 1986, around the time of the closest approach of Comet Halley to the Sun, six spacecrafts from various space agencies made detailed and extensive *in situ* measurements of various kinds in the coma of Comet Halley (Fig. 1.3). All the encounters took place on the Sunward side of the Comet. The closest approach to the nucleus was made by the European Space Agency's (ESA) Spacecraft Giotto, which passed at a distance of approximately 600 km from the nucleus. The Russian spacecrafts Vega 1 and 2 passed at distances of around 8000 km from the nucleus. The distances of the closest approach of the Japanese spacecrafts Suisei and Sakigake were around 1.5×10^5 km and 7.6×10^6 km respectively. The ICE spacecraft also passed through at a distance of around 0.2 au upstream of Comet Halley later in March 1986. Another space encounter with a comet took place on July 10, 1992 when the Giotto Spacecraft, which was successfully redirected in July 1990, passed through the Comet P/Grigg-Skjerllerup. The spacecrafts to Comet Halley carried a variety of scientific experiments, which performed a wide range of *in-situ* measurements, some of which complemented each other and others which overlapped.

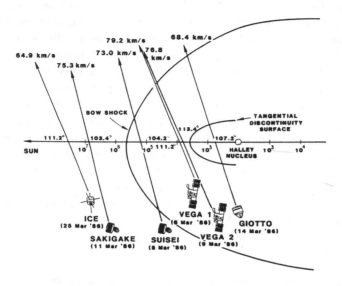

Fig. 1.3. The geometry of the six spacecraft flybys to Comet Halley. The distances are marked in logarithmic scale and the Sun is to the left of the Comet. The flyby dates for each mission are given at the bottom, flyby phase angle in the centre and flyby speeds at the top (Mendis, D. A. 1988. Ann. Rev. Astron. Astrophys. **26**, 11).

The sampled Comets Giacobini-Zinner and Halley belonged to nearly two different types of Comets. The Comet Halley, with a period of about 76 yrs, and aphelion distance of 35 au, has an inclination of $162°$ to the ecliptic plane and moves in a retrograde motion with respect to the planets. The Comet Giacobini-Zinner, on the other hand, is a short period Comet with a period of about 6.5 yrs and aphelion distance of only about Jupiter. Due to the retrograde motion of Comet Halley with respect to the Earth and the Spacecrafts, the relative encounter velocity was very high, ~ 68 km/sec. Since the gas and dust velocities in the coma are ~ 1 km/sec, the Giotto Spacecraft essentially saw the static situation while it passed through the coma as particles hit the spacecraft from the forward direction. The scientific payload on the spacecrafts, which passed through Comet Halley, had experiments for the study of flux and composition of neutrals, ions, electrons and dust, magnetic field and waves, imaging the nucleus, infrared spectra and ultraviolet images among others. These missions have met with tremendous success and have provided an enormous amount of information, which provides new insights into the nucleus, coma and the tails. These observations were supplemented by extensive observations of various kinds carried out with worldwide ground based telescopes, satellites, rockets and Kuiper Airborne Observatory, with sophisticated instrumentation. All the observations made worldwide by professionals and amateur astronomers were co-ordinated in a systematic way by the International Halley Watch (IHW) to obtain as much coverage as possible, over the orbit of the comet. The main motivation for such a co-ordinated venture was to make sure that maximum information is revealed out of these efforts. This co-ordinated effort of an unprecedented nature has provided an enormous amount of new information on the nature of comets. They have not only confirmed our knowledge about comets, theories and hypothesis that existed before these measurements were made, but have also provided new and unexpected insight into the cometary phenomena.

1.3. Discovery

Many comets are discovered by amateur astronomers who just scan the sky with a low-power telescope. They are called 'comet seekers'. The comets are usually named after their discoverers. If two or even more observers find the same comet nearly simultaneously, all the names are attached to that comet. For example, the Comet Ikeya-Seki (1965 VIII) which was visible to the naked eye in 1965, was discovered by two Japanese amateurs, Ikeya and Seki. However, not all comets are found by amateurs. Many are being discovered nowadays by astronomers in their photographic plates taken from some other scientific study. A typical example of this class is Comet Kohoutek (1973 XII) discovered in 1973. In addition to the names of the discoverers, comets are also assigned temporary designations, indicating the year of their discovery followed by a small letter denoting the order of their discovery in that year. For example, the first two comets found in the year say 1968, are designated as 1968a and 1968b respectively. Later on when the orbits of all the comets discovered in that particular year are well determined, permanent designations are given. This consists of the year in which the comet passed nearest to the Sun i.e., perihelion, followed by a Roman numeral which indicates the order of perihelion passage during that year. Hence, the comets mentioned above will be given the permanent designations 1968I and 1968II and so on. If the comet is periodic, one also attaches P to the name of the comet. Thus the periodic Comet Encke is written as P/Encke.

However, in the present system of naming of comets, there is the difficulty sometimes in deciding whether a particular object is a comet or a minor planet. e.g. 2060 Chiron, 1990 UL3=1990P=1990XVI. In view of this, a change in the cometary designation will be followed effective January 1, 1995. In the new system, the present year/letter and year/Roman numeral systems will be replaced by a single system resembling that for minor planets, with objects recorded by the halfmonth. As for example, the third comet reported as discovered during the half of February 1995 would be designated as 1995D3. If there is an indication of the nature of the object it could be expressed by preceding the designation with C/ (for comet), P/ (as now, for periodic comet), etc. In a process similar to the numbering of minor planets, sequential numbers will be defined for comets whose periodicity has been well established and the numbers should immediately precede the P/ notation. Routine recoveries of periodic comets will not in future receive additional designations. The new scheme also proposes to retain in general terms the tradition of naming comets after their discoverers.

1.4. Appearance

Comets spend almost all their time at great distances from the Sun. The cometary activity starts showing up only when it approaches the Sun. At far-off distances from the Sun, it appears as a faint fuzzy patch of light. The fuzzy patch of light is a cloud of gas and dust called *coma*. The coma grows in size and brightens as it nears the Sun. In addition to the brightening of the coma, the *tail* starts developing and reaches its maximum extent at about the closest approach to the Sun. After its perihelion passage, the reverse process takes place in the sense that the comet starts fading away as it recedes from the Sun. These effects can be seen clearly in the time sequence photographs of Comet Halley, as shown in Fig. 1.4. Cometary activity is therefore transient in nature. These observations clearly show that the material composed of gas and dust must have come from a central compact solid source called the *nucleus* of comet. The diameter of the nucleus is extremely small and is estimated to be about 1 to 10 km. This size is so small that it appears as a point source and cannot be resolved even with the largest telescope. On the other hand, the diameter of the coma is much larger and lies in the range of about 10^4 to 10^5 km. The nucleus and the coma forms the *head* of a comet. The most characteristic feature of a comet is the tail which may extend up to about 10^7 to 10^8 km.

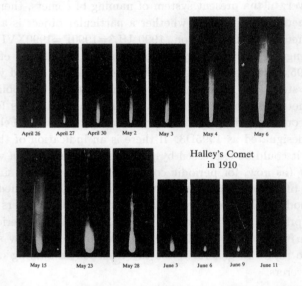

Fig. 1.4. Shows the time sequence photographs of Comet Halley in 1910 which brings out the transient nature of comets (Report of the Science Working Group, *loc. cit.*).

1.5. Statistics

The rate of discovery of comets has increased steadily since the beginning of the century. On an average, around 20 to 25 comets are seen every year. Of these, the newly discovered comets are around 12 to 15 per year and the recoverable ones are around 6 to 9 per year. The general convention which has been followed is that *short-period* comets are those which have periods less than 200 years. Those which have periods greater than 200 years are called *long-period* comets. Among the newly discovered comets, about 8-10 are long period comets and 5 are short period ones. Most of these comets are generally faint. The bright comets and, in particular, the Sun-grazing comets occur occasionally. Comets have been seen as close as 0.01 au from the Earth. Comets have also been classified as 'old' and 'new' based purely on their orbital characteristics. Comets which have made several perihelion passages around the Sun are generally termed 'old' and those which are entering for the first time are called 'new'. If the direction of motion of the comet is the same as that of the Earth's motion, in its orbit, it is said to have a *direct* orbit. If they are in opposite directions, the comet is said to have a *retrograde* orbit.

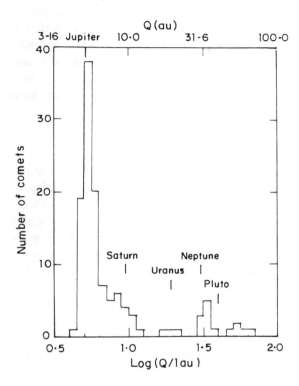

Fig. 1.5. Shows the distribution of the number of comets as a function of the aphelion distance for short period comets. The peak around Jupiter's distance can clearly be seen. Smaller peaks at the distances of other planets also appears to be present (Bailey, M. E., Clube, S. V. M. and Napier, W. M. 1990 *The origin of comets*, Pergamon press).

The total number of comets catalogued upto now is around 1000. The ratio of long-period to short-period comets is $\approx$ 5 : 1. In short-period comets, the period of a few years has been seen as in the case of Comet Encke which has a period of 3.3 years. Among the long period comets, most of them seem to have parabolic and osculating elliptical orbits. It is of interest to know how far the comets reach away from the Sun in their orbit, which is called the *aphelion* distance. Figure 1.5 shows a histogram of the number of comets versus aphelion distance. A peak in the distribution occurs around 5 au, which corresponds roughly to the distance of Jupiter. These comets are generally classified as belonging to the Jupiter family. There is also a slight indication of Comets peaking around Saturn, Uranus and Neptune, the other major planets. On the other hand, the long-period comets seem to peak around 4×10^4 au (Fig. 12.1). It is also found that for most of the comets the closest approach to the Sun, called the *perihelion* distance, is around 0.6 to 1.5 au. The short period comets generally have direct orbits and relatively small inclinations to the ecliptic plane, $i \lesssim 30^0$. However, for long-period comets, the inclinations are randomly distributed. They also approach the Sun more or less isotropically.

1.6. Importance

The study of comets is important from several points of view. Cometary activity arises basically from the solar heating of the nucleus, releasing the gas and dust, which finally are lost into the solar system. The time which the comet spends near the Sun is a very small fraction of its total period. So only a thin layer of the material of the nucleus is abrased at every perihelion passage. The inner core of the comet may thus represent the composition of the original material at the time of its formation. Therefore it is hoped that a systematic study of the material of the nucleus of comets can give information with regard to the nature of the material present at the early phase of the solar nebula, 4 to 5 billion years ago, even before the formation of the Earth and the solar system. The recent findings of the isotopic anomalies in certain meteorites indicate the possibility of the existence of interstellar grains in the cometary nucleus material. The influx of the cometary material into the solar system may have some effect on the atmosphere of planets. In addition the highly complex molecules and organic compounds seen in comets can finally find their way on to the Earth. These might have played a key role in the complex scenario of chemical evolution finally leading to life on the Earth. The tails pointing away from the Sun arise primarily due to the interaction of the dust and gas of the cometary material with the solar radiation and solar wind. Therefore the study of cometary tails may throw light on the physical conditions of the interplanetary medium as well as of the solar wind and the solar activity. It is also of great interest from the point of view of Plasma Physics for the study of interactions, generation of instabilities and waves and so on, many of which cannot be produced under the laboratory conditions. In fact, the existence of the solar wind, i.e. the flow of high velocity charged particles from the Sun, was predicted by Biermann in the 1950s from the study of the ion tails of comets. After a large number of revolutions around the Sun, the cometary activity may die out completely leading finally to a residual solid nucleus, which possibly may lead to an asteroid. They are also believed to be the sources of meteors and interplanetary dust. It is generally believed that the origin of comets is intimately related to the origin of the solar system, a problem of great current interest. Therefore, the study of comets can provide clues which may help in understanding the origin of the solar system. In addition to these possible interrelationships, the comets themselves are interesting objects to study, as their nature and origin are still not well understood.

1.7. Brightness

One of the uncertain facts about a comet is its brightness. The comet shines mostly due to the reflected sunlight at far-off distances from the Sun. The brightness depends upon three factors: (i) the distance r from the Sun to the comet: (ii) the nature of the comet and (iii) the distance Δ from the comet to the earth. The brightness depends upon the nature of the comet as it is the one which is producing the observed radiation.

The expected brightness of a comet I, can be written as

$$I = \frac{I_o}{r^2 \Delta^2} \phi(\alpha) \tag{1.1}$$

where $\phi(\alpha)$ is the appropriate phase function which is not important for the total brightness, I_o is the constant of proportionality, usually taken to be the brightness of the comet at $r = \Delta = 1$ au. It has been found that the brightness of comets rarely follow a simple relation of the above type. Figure 1.6 shows results for a few comets. Mostly the power of r is greater than 2. One usually writes a modified form of Eq. (1.1) as

$$I = \frac{I_o}{r^n \Delta^2}. \tag{1.2}$$

The above equation can be written in terms of magnitudes as

$$m = m_0 + 5 \log \Delta + 2.5 n \log r \tag{1.3}$$

where m refers to the total apparent magnitude and m_0 the absolute magnitude formally corresponding to $r = \Delta = 1$ au. The study of a large number

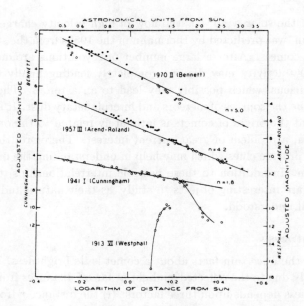

Fig. 1.6. The magnitude of the comet reduced to a standard distance of 1 au from the Earth is plotted as a function of log of the solar distance in au. Circles and dots refer to observations made before and after perihelion respectively (Jacchia, L. G. 1974. *Sky and Telescope* **47**, 216; courtesy of *Sky and Telescope*).

of comets has given a pretty good idea as to the variation of brightness with heliocentric distance r as well as the mean value of n.

Quite often it is found that it is not possible to find a single value of n covering the whole range of the Sun-comet distance. In addition, the variation of the observed brightness, before and after perihelion passage, seems to require different values of n. Because n varies from comet to comet, a mean value for n has been derived, with the proper weightage given to the observations. The results of such a study based on carefully analyzed photometric data for more than 100 comets grouped into four classes are shown in Table 1.1. The range of parameters of the classes selected are,

I (new) $1/a \times 10^6 \leq 50 (au)^{-1}$; $(P) > 2.8 \times 10^6 yr$
II $5 \times 10^{-5} < 1/a < 0.00215 (au)^{-1}$; $10^4 < P < 2.8 \times 10^6 yr$
III $0.00215 < 1/a < 0.01 (au)^{-1}$; $10^3 < P < 10^4 yr$
IV $0.01 < 1/a < 0.117 (au)^{-1}$; $25 < P < 10^3 yr$

where P is the period. Table 1.1 give weighted mean perihelion distance

Table 1.1. Values of $<n>$, $<q>$ and $\sigma(n)$ for various comets.

Orbit Class	Mean Values of $\bar{n}$			
	I	II	III	IV
Maximum period (yr)	∞	2.8×10^6	10^4	10^3
Minimum period (yr)	2.8×10^6	10^4	10^3	25
	Pre-Perihelion Dominated			
Number of Comets	10	5	5	6
$<q>$ (au)	0.43	0.90	1.17	0.71
$<n>$	2.45	3.11	3.32	3.83
$\sigma(n)$	±0.35	±0.71	±0.52	±0.49
	Post-Perihelion Dominated			
Number of Comets	15	16	9	17
$<q>$ (au)	1.20	1.54	0.88	0.80
$<n>$	3.16	3.87	4.48	4.94
$\sigma(n)$	±0.26	±0.58	±0.53	±0.79

(Whipple, F.L. 1991, In *Comets in the Post-Halley Era*, Eds. R.L. Newburn, Jr. et al., Kluwer Academic Pubilshers, P. 1259.)

$<q>$, the weighted mean $<n>$ and $\sigma(n)$ the mean error of $<n>$. The mean value of n shows a variation with the comet class and with the age of the comet (i.e. inversely with the period). This is indicated for both pre-and post - q data. The post -q observations give a considerably larger value of $<n>$ than the pre -q observations. In general, the large value of n required for comets arises due to the fact that the brightness is the sum total of the reflection component and the emission of gases from the coma. For the case of pure reflection the value of n is 2. Therefore the expected brightness of a comet is a complicated function of the physical condition of the coma. However, in the absence of any knowledge or data on the comet, one can get an idea of the average behaviour of its brightness from Eq. (1.3) with a suitable value of n.

As remarked earlier the larger value of n arises due to the contribution of the emission component into the total light that is observed. The amount of emission depends upon the total number of the molecules present in the coma. In principle, it is possible to predict the brightness of a comet by relating it to the evaporation of the gases from the nucleus. This is a difficult practical problem. However, the problem is simplified if the comet has already been observed once before, for which the visual light curve is usually available. For such cases, one can use a simplified approach for predicting the expected brightness for its next apparition. In the visual spectral region, the emission is mainly due to the C_2 molecule. Therefore, to a first approximation one can assume the visible light is mainly due to the fluorescence process of the C_2 molecule, which depends upon the production rate of the C_2 molecule. This could in turn be related to the total hydrogen production rate which is very well studied in various comets. From such a procedure and using the observed light curve, it is possible to evaluate the unknown constants occurring in the photometric equation. Knowing the photometric equation, the calculation of the expected brightness of a comet as a function of r and Δ is quite simple. (Sec. 6.1.4).

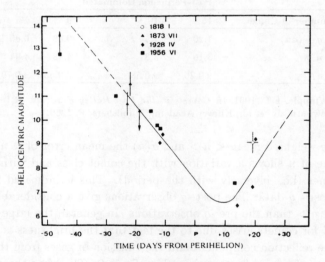

Fig. 1.7. A plot of the heliocentric magnitude of the Comet Crommelin as a function of time (equivalent to perihelion distance) from perihelion for four different appearances (Festou, M. 1983. International Halley Watch News Letter No. 3, p. 4).

There is still a simpler way than the above method. Since in Eq. (1.3) the value of n is uncertain, it can be written in a simplified form as

$$m = 5\log \Delta + m(r) \qquad (1.4)$$

where

$$m(r) = m_0 + 2.5n \log r.$$

From the observed light curve, the value of $m(r)$ can be calculated from Eq. (1.4) as a function of the time from the perihelion passage ($\equiv r$). Figure 1.7 shows the results of such calculations for Comet Crommelin (1984 IV) based on the last four apparitions. Based on Eq. (1.4) and the curve of Fig. 1.7, the expected brightness can easily be predicted.

1.8. Main Characteristics

Some of the main observed features in comets are the following:

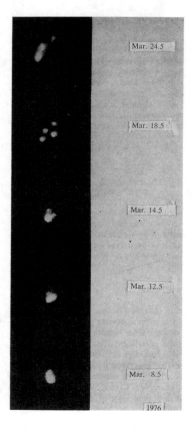

Fig. 1.8. Photograph shows the splitting of the nucleus of the Comet West 1976 VI into four parts in a time period of 10 days. The separation of the various components with time can easily be seen (Whipple, F. L. 1978. In *Cosmic Dust*, ed. J. A. M. McDonnell, New York: John Wiley and Sons. p. 1).

Fig. 1.9. Image of Shoemaker-Levy 9 taken with the Hubble Space Telescope on July 1, 1993. The comets heliocentric distance was 5.46 au and the geocentric distance was 5.4 au. The various fragments of the comet can clearly be seen. (Courtesy of H. A. Weaver and collaborators).

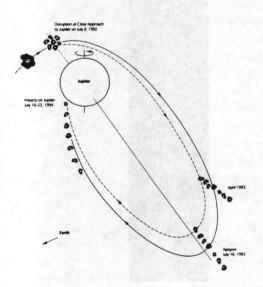

Fig. 1.10. The orbit of the Comet Shoemaker-Levy 9 around Jupiter is shown schematically. The comet which appears to have come close to Jupiter around July 8, 1992 was disintegrated due to Jupiters tidal force. This comet discovered on March 25, 1993 had as much as 21 components. These components penetrated Jupiters atmosphere during July 16 to 22, 1994 (Courtesy of Sekanina, Z., Chodas, P. W. and Yeomans, D. K.).

Most of the comets seem to deviate from their predicted orbits. They are known to arrive into the solar vicinity earlier or later than the predictions based on Newtonian gravitation. The classic example for which the data exist for the last two centuries is Comet Encke, which has a period of 3.3 years. It arrives earlier every time by about $2\frac{1}{2}$ hours. The splitting of the nuclei of comets into two or more fragments has been seen in many comets. The best example is Comet West (1976VI) in which the nuclei split up into four components in a time span of a few days. This can be seen from Fig. 1.8. There appears to be some correlation between the time of fragmentation and the increase in brightness of the comet. This might mean that a spurt of cometary activity leads to fragmentation. For sungrazing or planet grazing comets the splitting might also take place because of the tidal forces. The classic example is the nucleus of P/Shoemaker-Levy 9 which is believed to have fragmented into several pieces primarily due to the tidal forces exerted by Jupiter during its closest approach in July 1992. The spectacular feature of such an event is that as many as 21 nuclei all in a line, were discovered in mid July 1993. (Fig. 1.9). The subsequent orbit of these fragments and their collision with Jupiter in July 1994 is schematically shown in Fig. 1.10.

Comets which come very close to the Sun can be completely destroyed. An example of this type is shown in Fig. 1.11 which was discovered accidentally in the satellite observation and shows the time sequence photograph of a comet. After its closest approach to the Sun, the comet was not visible at all, most probably due to complete evaporation or falling into Sun.

Many comets show a sudden increase in brightness of one to two magnitudes in a short time scale usually called *outbursts* or *flares*. These outbursts are not periodic in character. An outburst may mean a sudden release of material from the nucleus. A classic example in which the flares have been seen very frequently is Comet Schwassmann-Wachmann 1. Figure 1.12 shows one such event for this comet. This comet has a period of 16.5 years and has been seen to brighten as much as 8 magnitudes. The flaring activity appears to be a general property of comets and is not associated with any particular type of comet. A strong outbursts was seen in Comet Halley even at a heliocentric distance of around 14 au.

In many comets successive haloes coming out of the coma have also been seen. Comet Donati (1858VI) is an example of this class where the successive haloes can distinctly be seen (Fig. 1.13). In many comets a broad fan-shaped coma coming out of the central condensation has also been seen.

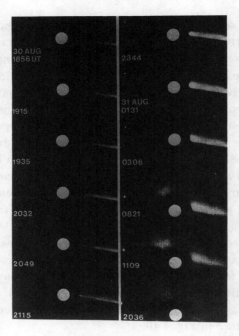

Fig. 1.11. Time sequence photographs taken between August 30, $18^h\ 56^m$ to August 31, $20^h\ 36^m$ by Solwind satellite show the disappearance of the comet. The comet enters from the right. The photographs are taken by Howard, R. A., Koomen, M. J. and Michels D. I. official US Navy photograph. (Courtesy of Howard, R. A. and collaborators).

1961   Oct. 12   Oct. 18   Nov. 3

Fig. 1.12. An outburst seen in Comet Schwassmann-Wachmann 1. The position of the comet on October 12, is shown by lines in the margin. The comet is very bright on October 18, but has become faint by November 3 (Roemer, E. 1996. In *Nature and Origin of Comets*. Memoirs of the Society Royale de Sciences of Liege **12**, 15; Courtesy of E. Roemer, official US Navy photographs).

Fig. 1.13. Successive haloes around Comet Donati (1858 VI) as observed visually by Bond on October 1858 (Whipple, F. L. 1981. In *Comets and Origin of Life*, ed. C. Ponnamperuma, Dordrecht: D. Reidel Publishing Company, p. 1).

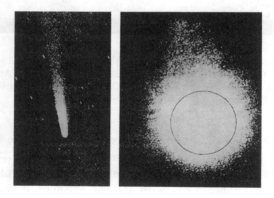

Fig. 1.14. Comet Kohoutek (1973 XII) as seen in visible (left) and in ultraviolet region taken in Lyman α line (right). Circle represents the apparent size of the Sun at the same distance. The vast extent of the hydrogen halo can clearly be seen. (Whipple, F. L. 1981, *loc. cit.*).

The satellite observations in the Lyman α line of hydrogen at 1216 Å led to the discovery of the enormous extent of the hydrogen envelope ($\sim 10^7 km$) around the visible coma of about 10^4 to 10^5 km. The size of the hydrogen halo was found to be larger than the size of the Sun at the same distance. This can clearly be seen in the observations of Comet Kohoutek which is shown in Fig. 1.14.

The most characteristic feature of a comet is, of course, the presence of two tails. One is the *dust tail* which is curved, also called *Type II tail*. The other is the *plasma tail* which is straight, also called *Type I tail* or *ion tail*.

Fig. 1.15. Photograph of Comet Mrkos (1957 V) taken on 27 August 1957 which shows the characteristic feature of the two tails, ion tail and the dust tail. (Arpigny, C. 1977, Proceedings of the Robert A. Welch Foundation Conferences on Chemical Research XXI, Cosmochemistry, Houston).

Figure 1.15 shows these two well-developed tails for the Comet Mrkos (1957V). Both the tails point to the direction away from the Sun. The nature of these two tails can be seen clearly in the colour photographs in which the dust tail appears yellowish and the plasma tail bluish in colour. Quite often a third short tail has also been seen in the direction towards the Sun. This is called the *anti-tail* of the comet. Figure 1.16 shows an example of this class for Comet Arend-Roland (1957 III).

The dust tail is generally very smooth and structureless. But this is

Fig. 1.16. Photograph of Comet Arend-Roland (1957 III) which shows the anti-tail (sharp ray towards the left). Taken on 24 April 1957 (Whipple, F. L. 1981, *loc. cit.*).

Fig. 1.17. Comet Kohoutek showing the helical structure. Photograph was taken on 13 January 1974. The oscillations in the tail at far off distances from the head can clearly be seen. (Brandt, J. C. and Chapman, R. C. 1981. *Introduction to Comets*, Cambridge: Cambridge University Press, illustrations credited to Joint Observatory for Cometary Research (JOCR), NASA.)

not so in the case of the plasma tail. Large scale structures of various kinds have been seen in the plasma tail of comets. As the name indicates, this tail is composed mainly of ions. Therefore many of the features seen in

Fig. 1.18. Photograph showing a big knot in the tail of Comet Kohoutek. This is generally called as 'Swan-like' feature. The photograph was taken on 11 January 1974 (Brandt, J. C. and Chapman, R. C. 1981, *loc. cit.* JOCR photograph).

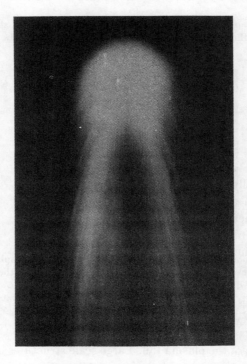

Fig. 1.19. Photograph of Comet Halley showing well developed streamers. Taken on 8 May 1910. (Report of the Science Working Group, *loc. cit.*).

the laboratory plasma arising out of various physical processes as well as others, which cannot be seen in the laboratory plasma, have also been seen in the plasma tail of comets. For example features like oscillations, kinks, helices, knots, filaments, rays etc., have been seen in many comets. These indicate clearly the presence of complex interactions of the magnetic field with the tail plasma. Some typical large scale observed features in comets are shown in Figs. 1.17 to 1.19.

1.9. An Overall View

The three major parts of a comet are the nucleus, the coma and the tail (Fig. 1.20). Most of the information about these three components

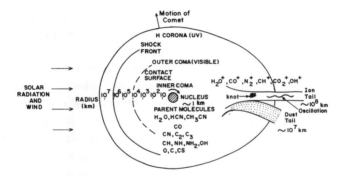

Fig. 1.20. Descriptive sketch of a comet.

has come basically from the study of spectra. Almost all the observed activities seen in a comet should be related directly or indirectly to the nucleus of a comet. A reasonable working model for the nucleus which is the *icy-conglomerate model* was first proposed by Whipple in the 1950's. In this model the nucleus was believed to be a discrete rotating body consisting of frozen water, complex molecules formed out of abundant elements H,C,N and O, and dust. All the subsequent observations of comets for the last three decades had supported this model by indirect means, basically through the observation of the dissociated products of H_2O (i.e. OH, H and O) and H_2O^+. The first actual detection of H_2O in a comet came from the observation of Comet Halley in 1986, when well resolved rotational lines of the $2.7\mu m$ band of H_2O were detected with observations carried out with the Kuiper Airborne Observatory. The nucleus contains around 80% of

H_2O -ice and the rest is made up of other constituents. The single body nature of the nucleus, in contrast to that of 'loosely bound system', was confirmed by the photographs taken of the nucleus of Comet Halley by Giotto Spacecraft when it was at a distance of 600 km from the nucleus, supporting the general concept of the Whipple model. The nucleus was observed to be irregular in shape. The presence of a few active areas on the nucleus was also confirmed. The rotation of the nucleus of Comet Halley was seen by various means, which confirmed that the nucleus of comets rotates as envisaged earlier. The inferred bulk density from Comet Halley observations is around 0.2 to 0.6 gm/cm^3. This indicates the nucleus to be fragile. With dimension of the nucleus of around 1 to 10 km and for a density 1 gm/cm^3, the total mass of a comet turns out to be in the range of about 10^{16} to 10^{18} gm.

When the comet is far off from the Sun, the continuum spectrum seen is simply that of the reflected sunlight. As it approaches the Sun, the gas, mostly made up of complex molecules and the dust are released from the nucleus, due to solar heating. This then expands outwards into the vacuum at about 0.5 km/sec giving rise to the observed coma. The dimension of the visible coma is around 10^4 to 10^5 km, while the ultraviolet coma extends up to about 10^6 to 10^7 km. As the gas expands, it is subjected to various physical processes like dissociation, ionization, gas-phase reactions, etc. Therefore, the gaseous material in the coma is modified to a large extent. The complex molecules released by the nucleus, generally known as *parent molecules*, ultimately give rise to simple molecules and radicals like CN, C_2, OH, CH, NH_2, etc., which are seen in the comet's emission spectra. The emission lines of various elements like Na, Si, Ca, Cr, etc., show up in sun-grazing comets. Through the study of the microwave region, several molecules like H_2CO, HCN, H_2S, etc., possibly the parent molecules of some of the observed species, have been identified.

The dust coming out of the nucleus is dragged outwards by the gas accompanying it. These dust particles are subjected to the radiation pressure of the Sun which pushes them in the direction away from the Sun. Since the dust particles lag behind the coma as they stream away from the sun, they take a curved path. This gives rise to the observed curved nature of the dust tail. The tail that is usually seen in the sky is the dust tail, which is made visible through the scattering of the solar radiation by the dust particles in the tail. The tail extends up to about 10^7 km or so.

The understanding of the nature and composition of the dust particles came mostly from the interpretation of continuum, polarization and infrared radiation from comets. That the grain material could be some form of silicate, was inferred from the detection of broad emission features occurring around 10 and $20\mu m$ in comets, which are characteristic features of all silicate materials. The silicate material could be of the olivine type which can be further inferred from the detection of a feature at $11.3\mu m$ in the high resolution spectral observations of Comet Halley. There is another component of the grains in comets came from the discovery of CHON particles from the *in situ* measurements of Comet Halley. As the name indicates, these particles are mostly made up of H,C, N and O. This was also supported from the ground-based observations of Comet Halley and other comets by the detection of a broad emission feature around 3.4 μm attributed to C-H stretching bond of hydrocarbons. Therefore there appears to be two major components of the grains in comets, namely silicate and some form of carbon. The presence of large numbers of small-sized grains, $\leq 0.1\mu m$, which cannot be detected through observations made in the visible region, was found to exist in Comet Halley in abundance. The grains in the coma, like the CHON particles, can also act as a source of observed molecules which comprise the gas. This can be inferred from the *in situ* measurements of Comet Halley.

The solar radiation also breaks up the original molecules released by the nucleus and ionizes them. The ionized gas in the coma is swept outwards by a stream of charged particles present in the solar wind. These two are coupled through the interplanetary magnetic field. This gives rise to the plasma tail which extends up to about 10^7 to 10^8 km. Therefore, the structure and the dynamics of plasma tail are basically due to the interaction of the cometary plasma with the solar wind. The theoretical modeling of the interaction between the solar wind plasma and the cometary ions had predicted the existence of several gross features like bow shock, ionopause etc., which were proved to be correct based on the *in situ* measurements carried out on Comet Halley and Giacobini-Zinner by the spacecrafts. Therefore the multi-layered structure and variations seen in comets arise due to complex interactions taking place in the plasma between electrons, heavy cometary ions, cometary protons and solar wind protons as they are being thermalised at different positions in the coma. The spectrum of the plasma tail shows mainly emissions from ions like CO^+, CO_2^+, N_2^+, etc; of these the emission due to CO^+ is the dominant one. Since this emission spectrum lies in the blue spectral region, the ion tail appears blue in colour photographs.

The comets are believed to be members of the solar system, which inference is deduced from the observed orbital characteristics of comets. The study based on the isotopic ratio of $^{12}C/^{13}C$ in many comets gives a value ~ 90 which is the same as the solar system value. Several other isotopic ratios seen in Comet Halley are also consistent with the solar system values. These results seem to suggest that the cometary material and the solar system materials are similar in nature. However, the question of the origin of comets is still an open one, although the widely accepted hypothesis is that of Oort. He pointed out that there appears to be a cometary reservoir whose aphelia is about 50,000 au or more from the Sun. This is usually called the *Oort cloud*. It is estimated that there may be around 10^{11} comets in the Oort cloud. Many of the comets leaving this cloud are finally brought into the solar system due to stellar perturbations. This accounts for the steady influx of comets into the solar system. However, the long period comets coming from the Oort cloud faces serious problems in explaining the number and inclination of the observed short period comets. This has led to the hypothesis that the short period comets come from another population of comets from a region which is believed to contain around 10^8 to 10^{10} comets in the ecliptic plane beyond the orbit of Neptune between 30 and 50 au. This region is generally called the *Kuiper belt*. The origin of comets therefore is related to the origin of the Oort cloud and the Kuiper belt. Several hypotheses have been put forward to explain the origin of these clouds. Each hypothesis has its own difficulties. Therefore, at the present time the origin of comets is still not understood.

In the succeeding chapters, we would like to elaborate on some of these aspects. Before going into the actual subject matter, we would like to give a brief account of the physical background required for the interpretation of various observed phenomena.

Problems

1. What is the basis of the assumption that a comet possesses a nucleus at its centre? Since it is hard to see the nucleus of a comet directly, what is the best way to locate it?
2. Suppose a Comet A has $n = 3$ [Eq. (1.2)] in its 50th orbit around the Sun while Comet B has $n = 6$ in its first orbit around the Sun. Does

one expect the same brightness variation of Comets A and B in their next orbits? Give reasons. Explain why the two values of n could be vastly different.
3. Is it possible for the comets to be in orbit around the Sun, but not seen from the Earth?
4. Discuss with examples that impact of comets on solar system objects is a natural phenomena.
5. Discuss the consequences of a comet hitting the Earth.
6. Compare the energy released by a 1 km size comet moving at 60 km/sec and suddenly coming to a stop with that of two 3500 lb cars colliding head-on at 50 km/hr.

References

1. Bailey, M. E., Clube, S. V. M. and Napier, W. M. 1990. *The origin of comets.* Pergamon Press, Oxford.
2. Brandt, J. C. and Chapman, R. D. 1981 *Introduction to comets.* Cambridge University Press, Cambridge.
3. Festou, M. C., Rickman, H. and West, R. M. 1993, *Astr. Ap.* Review, **4**, 363; **5**, 37.
4. Grewing, M., Praderie, F. and Reinhard, R. (eds). 1987. *Exploration of Halleys Comet.* Springer-Verlag, Berlin.
5. Huebner, W. F. (ed.) 1990. *Physics and Chemistry of comets.* Springer-Verlag, Berlin.
6. Mason, J. (ed.). 1990, *Comet Halley: Investigations, Results, Interpretations*, Vol, 1 and 2. Ellis Horwood, New York.
7. Nature. 1986 *Encounters with Halley.* **321**, 259.
8. Newburn, Jr., R. L., Neugebauer, M. and Rahe, J. (eds.). 1991 *Comets in the Post-Halley Era.* Vols. 1 and 2. Kluwer Publishers, Dordrecht.
9. Wilkening, L. (ed.). 1982 *Comets.* The University of Arizona Press, Tucson.
10. Whipple, F. L. 1985. *The Mystery of Comets*, Cambridge University Press, Cambridge.
11. The method of new designation and names of comets is given in IAU Circular No. 6076, September 10, 1994.

CHAPTER 2

DYNAMICS

2.1. Orbital Elements

The objects in space are generally specified with respect to the ecliptic or to the equatorial system of coordinates. In the former the Earth's orbit around the Sun, i.e., the *ecliptic plane* is the reference frame, while in the latter it is the plane of the Earth's equator. The position of an object is specified by the *longitude* and the *latitude* in the ecliptic system and by the right ascension (α) and the declination (δ) in the equatorial system. The right ascension is measured from the vernal equinox, which is the point where the ecliptic plane cuts the Earth's equator. The declination is the angular distance from the north to the south of the celestial equator. The two systems can be transformed from one to the other through trigonometric relations.

The orbit of a body around the Sun is generally a conic section. In general, the conics are the *ellipse, parabola* and *hyperbola*. The *major axis* refers to the maximum diameter of the ellipse and it determines the size of the ellipse (Fig. 2.1). The *eccentricity e* of the ellipse is defined as the ratio of the distance between the center and a focus to the length of the semi major axis. An ellipse is completely defined by the eccentricity and the major axis. The value of *e* varies between 0 and 1. A *circle* is a special case of the ellipse when the eccentricity is zero. The parabola has an eccentricity equal to unity while a hyperbola has an eccentricity greater than one. In an elliptical orbit the closest and the farthest distance of the object from the Sun which is stationed at one of the foci is known as the *perihelion* and *aphelion* distances respectively.

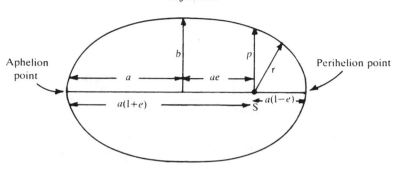

Fig. 2.1. Various parameters defined for an elliptical orbit. S denotes the position of the Sun occupying one of the foci.

The position of a comet in the sky is generally referred to the ecliptic system of coordinates. To define completely an orbit in space, six quantities usually termed as *elements* of the orbit are to be specified (Fig. 2.2). They are the following:

a = length of the semi major axis;
e = eccentricity;
i = angle between the orbital plane and the plane of the ecliptic;
Ω = longitude of the ascending node. This is the angle from the vernal equinox along the ecliptic plane to the point of intersection of orbital plane with the ecliptic plane;
ω = argument of the perihelion, which is the angle measured from the ascending node to the perihelion point.

The first two parameters specify the size and the shape of the orbit, while the other three define the orientation of the orbit with respect to the ecliptic plane.

The sixth element is the *time parameter* which defines the position of the body in its orbit at that time. This is taken to be T, the time of perihelion passage. This gives a reference time to fix the body at other times in its orbit.

Therefore, the quantities, a, e, T and the angle i, ω, Ω define completely the position of the body and its orbit at any given time. For a parabolic orbit, the semi major axis which is infinite is replaced by the perihelion distance q, which defines the size of the parabola.

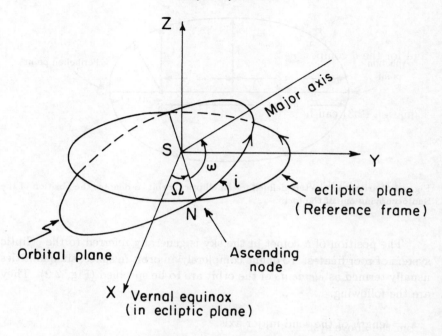

Fig. 2.2. The orbital elements required for specifying an orbit.

2.2. Orbit in Space

2.2.1. *Relevant equations*

The orbit of a comet is a conic section about the Sun and it can be defined under the Newtonian Gravitation.

The equation of motion of a comet of mass m around the Sun can be represented as

$$\left(\frac{d^2\mathbf{r}}{dt^2}\right) = -\left(\frac{G(M_\odot + m)}{r^3}\right)\mathbf{r} \equiv \frac{\mu \mathbf{r}}{r^3} \qquad (2.1)$$

where $M_\odot$ is the mass of the Sun and G is the gravitational constant. Since $m \ll M_\odot$, $\mu = GM_\odot$ and $\mathbf{r}$ is the position vector of the comet relative to the Sun, the use of plane polar coordinates r and θ in the orbital plane of motion allows one to separate the Eq. (2.1) into r and θ components, giving, for the r component

$$\ddot{r} - r\dot{\theta}^2 = -\frac{\mu}{r^2} \tag{2.2}$$

and for the θ component

$$r\ddot{\theta} + 2\dot{r}\dot{\theta} = 0 \tag{2.3}$$

Equation (2.3) can be written as

$$\frac{1}{r}\frac{d}{dt}(r^2\dot{\theta}) = 0. \tag{2.4}$$

The integration of the above equation gives the specific (means per unit mass) angular momentum integral

$$r^2\dot{\theta} = \text{constant} = h. \tag{2.5}$$

Specific energy integral

$$E' = \frac{1}{2}(\dot{r}^2 + r^2\dot{\theta}^2) - \frac{\mu}{r}. \tag{2.5a}$$

From the above system of equations, one can derive the polar equation for the conic section for an angle θ as

$$r = \frac{p}{1 + e\cos(\theta - \omega)} \tag{2.6}$$

where ω is constant, $p = h^2/\mu$ is the semi latus rectum and $e = \sqrt{1 + \frac{2E'h^2}{\mu^2}}$ is the eccentricity of the orbit and therefore the length of the semi major axis $a = -\frac{\mu}{2E'}$. ω is actually the angle that the major axis makes with the axis $\theta = 0$. At the perihelion point $\theta = \omega = 0$ and therefore the perihelion distance is given by $h^2/\mu(1 + e)$. The perihelion distance is also equal to $a(1 - e)$. Hence

$$h^2 = \mu a(1 - e^2). \tag{2.6a}$$

Equation (2.6) can be expressed as

$$r = \frac{a(1 - e^2)}{1 + e\cos(\theta - \omega)} = \frac{a(1 - e^2)}{1 + e\cos f} \tag{2.7}$$

where a is the semimajor axis and the angle f is known as the *true anomaly* (Fig. 2.3).

The other quantity of interest is the velocity of the object in its orbit.

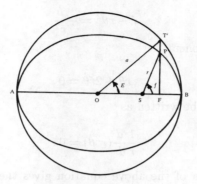

Fig. 2.3. Definition of eccentric anomaly E. f represents the true anomaly.

From Eqs. (2.1) and (2.4), the expression for the orbital speed of the particle can be obtained and it is given by

$$v^2 = \mu \left[\frac{2}{r} - \frac{1}{a} \right]. \tag{2.8}$$

The quantity $1/a$ is $+ve$, zero or $-ve$ depending on whether the orbit is an ellipse, a parabola or a hyperbola. The velocity at the perihelion point for an ellipse is given by

$$v_p^2 = \frac{\mu}{a} \left(\frac{1+e}{1-e} \right); \tag{2.9}$$

similarly at the aphelion point, the expression for the velocity is given by

$$v_a^2 = \frac{\mu}{a} \left(\frac{1-e}{1+e} \right). \tag{2.10}$$

From Eq. (2.7) it is clear that the value of $\frac{1}{r}$ monotonically decreases from perihelion to aphelion, Eq. (2.8) would then suggest that the velocity of the object is maximum at perihelion and minimum at aphelion and varies along its orbit in between these, two limits.

Another quantity of interest is the heliocentric radial velocity of the object. For elliptical orbits, it can be shown from Eq. (2.7) by taking its time derivative that

$$\dot{r} = \mp \mu^{1/2} \left[\frac{a^2 e^2 - (a-r)^2}{ar^2} \right]^{1/2}. \tag{2.11}$$

The minus and positive signs refer to preperihelion and postperihelion radial velocities.

The *mean angular motion n* of the body in its orbit with the period p is by definition

$$n = \frac{2\pi}{P}. \tag{2.12}$$

Since Eq. (2.5) suggest that $h = r^2\dot{\theta} = $ constant and $\frac{1}{2}r^2\dot{\theta}$ is the areal velocity, the orbital period

$$P = \frac{Area\ of\ the\ ellipse}{areal\ velocity}$$

$$= \frac{2\pi a^2 \sqrt{1-e^2}}{h} \tag{2.12a}$$

$$P = 2\pi \sqrt{\frac{a^3}{\mu}}. \tag{2.13}$$

Hence

$$n = \mu^{1/2} a^{-3/2}. \tag{2.14}$$

If T represents the time of perihelion passage, the angle swept by the radius vector in a time interval $(t - T)$ is called the *mean anomaly M* and is defined as

$$M = n(t - T). \tag{2.15}$$

If a circle is drawn with OB as radius, then the line FP referring to the ellipse when extended perpendicular to the major axis cuts the circle at T' (Fig. 2.3). The angle T' OB is called by definition *eccentric anomaly*, generally denoted as E. Since PF/T'F=$\sqrt{1-e^2}$ for any point P on the orbit, it can be shown that

$$SF = x = a\cos E - ae$$

$$PF = y = a\sqrt{1-e^2}\ \sin E$$

Therefore

$$r = \sqrt{x^2 + y^2} = a(1 - e\cos E) \tag{2.16}$$

Further

$$\cos f = 1 - 2\sin^2 \frac{f}{2} \tag{2.16a}$$

or
$$2r \sin^2 \frac{f}{2} = r(1 - \cos f) \tag{2.16b}$$
However
$$\cos f = \frac{SF}{r} = \frac{a \cos E - ae}{a(1 - e \cos E)} \tag{2.16c}$$
and
$$\sin f = \frac{PF}{r} = \frac{a(1 - e^2)^{1/2} \sin E}{a(1 - e \cos E)} \tag{2.16d}$$
Therefore the Eq. (2.16b) can be written as
$$2r \sin^2 f/2 = a(1 + e)(1 - \cos E) \tag{2.16e}$$
similarly
$$2r \cos^2 f/2 = a(1 - e)(1 + \cos E) \tag{2.16f}$$
Therefore
$$\tan f/2 = \left(\frac{1+e}{1-e}\right)^{1/2} \tan \frac{E}{2} \tag{2.17}$$
This is the relation which connects the eccentric anomaly E with the true anomaly f.

There is also a relation connecting the mean anomaly M and the eccentric anomaly E. This is generally referred to as *Kepler's* equation which is derived as follows:

The quantity $(t - T)/P$ represents the fractional area of the ellipse swept by SP with respect to the point B, where P is the orbital period. This is also equal to $n(t-T)/2\pi$. From the law of areas and the properties of the auxiliary circle, it follows that

$$\frac{n(t-T)}{2\pi} = \frac{M}{2\pi} = \frac{area\ BSP}{area\ of\ ellipse} = \frac{area\ BST'}{area\ of\ circle}$$

But $Area\ BST'$ = area of circular sector BOT'- area SOT'
$$= \frac{a^2 E}{2} - \frac{a}{2} ae \sin E$$

Therefore
$$\frac{M}{2\pi} = \frac{a^2}{2} \frac{(E - e \sin E)}{\pi a^2}$$
or
$$E - e \sin E = M$$
which is the Kepler's equation.

The above relation can also be written as

$$E - e \sin E = n(t - T). \qquad (2.19)$$

The corresponding relations for the parabolic case are

$$r = \frac{2q}{1 + \cos f} \qquad (2.20)$$

$$v^2 = \frac{2\mu}{r} \qquad (2.21)$$

$$\left(\frac{\mu}{2q^3}\right)^{1/2}(t - T) = \tan\frac{f}{2} + \frac{1}{3}\tan^3\frac{f}{2} \qquad (2.22)$$

where $q = a(1 - e)$, and for the hyperbolic case,

$$r = \frac{a(e^2 - 1)}{1 + e \cos f} \qquad (2.23)$$

$$r = a(e \cosh F - 1) \qquad (2.24)$$

$$\tanh\frac{F}{2} = \left(\frac{e-1}{e+1}\right)^{1/2}\tan\frac{f}{2} \qquad (2.25)$$

$$v^2 = \mu\left(\frac{2}{r} + \frac{1}{a}\right) \qquad (2.26)$$

$$e \sinh F - F = M = \left(\frac{\mu}{a^3}\right)^{1/2}(t - T). \qquad (2.27)$$

2.2.2. Orbital elements from position and velocity

The position and the velocity of an object along its orbit can be obtained from the solution of the above equations. The reverse problem which is often of interest in cometary studies is to determine the elements of the orbit from a given set of position, velocity and time.

Let the position of the body in the heliocentric coordinate system at a time t be (x, y, z) and the velocity components $(\dot{x}, \dot{y}, \dot{z})$. Then

$$r^2 = x^2 + y^2 + z^2 \qquad (2.28)$$

and

$$v^2 = \dot{x}^2 + \dot{y}^2 + \dot{z}^2. \qquad (2.29)$$

Since r and v are known, the semimajor axis can be calculated from Eq. (2.8).

The areal constant h may be regarded as the vector product of $\mathbf{r}$ and the orbital velocity $\mathbf{v}$ of the comet relative to the Sun. If the components of $\mathbf{h}$ along x, y and z directions are represented as h_x, h_y and h_z then,

$$h_x = y\dot{z} - z\dot{y}$$
$$h_y = z\dot{x} - x\dot{z} \qquad (2.30)$$
$$h_z = x\dot{y} - y\dot{x}.$$

and

$$h^2 = h_x^2 + h_y^2 + h_z^2 = \mu p.$$

Therefore the value of p the length of the semilatus rectum can be calculated. From a knowledge of p and a, e can be obtained from the relation

$$p = a(1 - e^2). \qquad (2.31)$$

The projection of $\mathbf{h}$ onto the three planes yz, zx and xy gives the following expressions

$$h \sin i \sin \Omega = \pm h_x$$
$$h \sin i \cos \Omega = \mp h_y \qquad (2.32)$$
$$h \cos i = h_z.$$

Hence, the above equations give i and Ω. The upper sign and lower sign refer to the cases when i is less than or greater than $90°$.

The value of $(\omega + f)$ can be derived from the following relations which relate the position of the point (x, y, z) in terms of the angles of the orbit.

$$\sin(\omega + f) = \frac{z}{r} \csc i \qquad (2.33)$$

and

$$\cos(\omega + f) = \frac{1}{r}(x \cos \Omega + y \sin \Omega).$$

The value of f can be obtained from the relation

$$r = \frac{(h^2/\mu)}{1 + e \cos f}. \qquad (2.34)$$

Hence ω can be determined. The only other quantity remaining to be determined is the time of the perihelion passage T, which depends upon the conic section. For an elliptical orbit the eccentric anomaly E can be obtained from the Eqs. (2.16) or (2.17). knowing E, e, n and f, the value of T can be calculated from the Eq. (2.19).

Therefore all the elements a, e, i, ω, Ω and T can be determined. The procedure for the other two types of orbits are also similar. For parabolic and hyperbolic orbits, the Eqs. (2.22) and (2.27) may be used to evaluate T.

2.2.3. Orbital elements from observations

Since in general six elements are required to specify completely the orbit in space, it follows that six independent quantities must be obtained by the observations. A single observation gives only two quantities say, in terms of the angular coordinates α and δ of the body. Therefore in all, three different sets of observations are required to define its orbit.

Let the heliocentric equatorial rectangular coordinates of the comet and the Earth at any given time be denoted by (x, y, z) and (X, Y, Z) respectively with respect to the plane of the celestial equator. Let their heliocentric distances be r and R which are given by

$$r^2 = x^2 + y^2 + z^2$$

and

$$R^2 = X^2 + Y^2 + Z^2. \tag{2.35}$$

Then neglecting comet's mass,

$$\ddot{x} = -\frac{GMx}{r^3} \tag{2.36}$$

and for the Earth

$$\ddot{X} = -\frac{G(M + m_e)X}{R^3} \tag{2.37}$$

where M and m_e are the mass of the Sun and the Earth respectively.

Let the geocentric direction cosines of the comet be l, m and n and its geocentric distance be ρ. Then

$$x = X + l\rho$$

$$y = Y + m\rho$$

and

$$z = Z + n\rho. \tag{2.38}$$

The direction cosines l, m and n are given in terms of the observed position of the comet, right ascension α and declination δ, as

$$l = \cos\alpha \cos\delta$$

$$m = \cos\delta \sin\alpha$$

and

$$n = \sin\delta. \quad (2.39)$$

Equations (2.36), (2.37) and (2.38) give

$$\ddot{\rho}l + 2\dot{\rho}\dot{l} + \rho\ddot{l} = -GM\frac{(X+l\rho)}{r^3} + G(M+m_e)\frac{X}{R^3} \quad (2.40)$$

or

$$\left(\ddot{\rho} + \frac{GM\rho}{r^3}\right)l + 2\dot{\rho}\dot{l} + \rho\ddot{l} = -GX\left(\frac{M}{r^3} - \frac{M+m_e}{R^3}\right). \quad (2.41)$$

There will be two other equations in Y and Z. These equations may be solved to give $(\ddot{\rho} + (GM\rho/r^3))$, $2\dot{\rho}$ and ρ. Except for r, all other quantities are known or can be derived from the observed quantities; r can also be expressed in terms of ρ. From the triangle whose sides are R, r and ρ (Fig. 2.4) the following relation may be obtained:

$$r^2 = R^2 + \rho^2 - 2\rho R\cos\theta \quad (2.42)$$

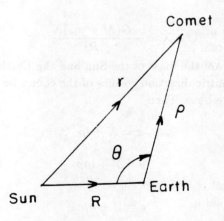

Fig. 2.4. Shows the geometry of the triangle with distances marked.

where θ is the angle between R and ρ. The projection of R in the direction of ρ gives
$$R\cos\theta = -(lX + mY + nZ). \tag{2.43}$$
Therefore
$$r^2 = R^2 + \rho^2 + a_1\rho \tag{2.44}$$
where
$$a_1 = 2(lX + mY + nZ).$$
Equations (2.41) and (2.44) can be solved for r and ρ. Knowing the values of r and ρ, the comet's heliocentric coordinates (x,y,z) and the velocity components $(\dot{x},\dot{y},\dot{z})$ can then be obtained from the relations
$$\begin{aligned} x &= X + l\rho \\ \dot{x} &= \dot{X} + \dot{l}\rho + l\dot{\rho} \end{aligned} \tag{2.45}$$
and with similar equations for $y, \dot{y}, z$ and $\dot{z}$. Knowing the position and velocity of the comet, the orbital elements and hence the orbit can be determined from the method already discussed. The method outlined above is generally referred to as *Laplace's method*.

The actual computation involves a knowledge of the value l, m, n, X, Y, Z and their derivatives. The values of X, Y, Z are given in the Ephemeris. From this data the values of the first derivative can be found out. The calculation of geocentric direction cosines, their first and second derivatives can be deduced from three observations of the comet which are not too far off from each other. Let the dates of observation be t_1, t_2 and t_3. The right ascension and declination for these three times are known. Here we will just show an approximate method of getting the first and second derivatives. In actual practice one can use various refined methods.

The average value of the first derivative of l for the time between t_1 and t_2 is given by
$$\dot{l}_{12} = \frac{l_2 - l_1}{t_2 - t_1}. \tag{2.46}$$
Similarly
$$\dot{l}_{23} = \frac{l_3 - l_2}{t_3 - t_2}. \tag{2.47}$$
If the time interval $(t_2 - t_1) \approx (t_3 - t_2)$, then the value of $\dot{l}$ at time t_2 is approximately equal to
$$\dot{l}_2 = \frac{1}{2}\left[\dot{l}_{12} + \dot{l}_{23}\right]. \tag{2.48}$$

Similarly

$$\ddot{l}_2 = \frac{\dot{l}_{23} - \dot{l}_{12}}{\frac{1}{2}(t_3 - t_1)}. \tag{2.49}$$

Similar relations for the first and second derivatives and m and n can be obtained.

The elements obtained from three sets of observations define the initial orbit of the comet. For getting a better orbit it is necessary to have many more observations. The initial orbit can be improved further as more and more observations become available. In fact equations can be set up for the difference between the predicted and the observed positions. These can then be solved to get the corrections for the preliminary orbit elements.

In the discussion so far, it is assumed that the comet is only under the influence of the Sun's gravitational field. But in actual practice the orbit gets perturbed due to the planets as comets enter the solar system. The dominant effect arises mainly from the planets Jupiter and Saturn because of their large masses. When the comet is far off, the perturbation produced due to stars, has also to be considered. The calculations which include many of these perturbations have been carried out numerically. Through these efforts the dynamical evolution of comets has been studied.

Problems

1. The components of velocity of a body are (0,1,3) corresponding to the position (4,2,1). Calculate a and e. What is the nature of the orbit? Assume for simplicity $\mu = 1$.
2. The ecliptic heliocentric coordinates of position and velocity of a comet are (4,2,3) and (2,2,1) respectively on March 16, 1959. Find the elements of the orbit of the comet. Here again assume $\mu = 1$.
3. The time period of the Earth around the Sun is 1 year and its orbital velocity is 30 km/sec. Compute the distance from the Earth to the Sun.
4. Calculate the lifetime of comets which have aphelion distances of 5 au and 5×10^4 au, are almost in parabolic orbits and can survive 1000 perihelion passages.
5. The comet moving in an elliptical orbit has an eccentricity of 0.985. Compare its velocity at perihelion and aphelion.
6. Show that in elliptic motion about a focus under attraction μ/r^2, the

radial velocity is given by the equation

$$r^2 \dot{r}^2 = \frac{\mu}{a}\{a(1+e) - r\}\{r - a(1-e)\}$$

7. Calculate roughly the distance from the Sun beyond which Comet Halley spends about half of its total time period of 76 yrs.
8. Estimate the average values of r, dr/dt, dv/dt and the kinetic energy in an elliptical orbit taking time as an independent variable.
9. A satellite is orbiting in a circle at an altitude of 600 km. Knowing the radius and surface gravity of the Earth, calculate its orbital velocity and period of revolution. If it is brought to an altitude of 400 km, what is the time period?

References

The solutions of the equations of Sec. 2.2.3 are specially discussed in
1. Moulton, F.R. 1970. *An Introduction to Celestial Mechanics*, New York: Dover Publications Inc.
2. Roy. A.E. 1965, *The Foundations of Astrodynamics,* New York: The Macmillan Company.

CHAPTER 3
PHYSICAL ASPECTS

The gaseous material of a comet is immersed in the radiation field of the Sun. The study of the interaction between the two requires knowledge of some of the basic laws of radiation as well as of the spectroscopy of atoms and molecules. Here we will briefly review some of the relations which are used in subsequent chapters for interpreting the cometary observations.

3.1. Black Body Radiation

The energy radiated at different wavelengths by a black body (which absorbs all of the incident radiation) at temperature T is given by Planck's law

$$B_\lambda d\lambda = \frac{2hc^2}{\lambda^5} \frac{1}{e^{hc/\lambda kT} - 1} d\lambda \qquad (3.1)$$

where h, k and c denote the Planck constant, Boltzmann constant and the velocity of light respectively. Figure 3.1 shows a plot of the Eq. (3.1) for several temperatures which shows clearly the shift of the maximum of the curve to shorter wavelengths with an increase in temperature. The wavelength corresponding to the peak of the curve $\lambda_{\max}$ can be represented by the equation

$$\lambda_{\max} T = 0.2897 \qquad (3.2)$$

known as *Wien's displacement law*. Figure 3.1 also shows that the total amount of the radiation emitted shifts gradually from the ultraviolet to the visible and to the infrared spectral regions as the temperature goes from a higher value to a lower value.

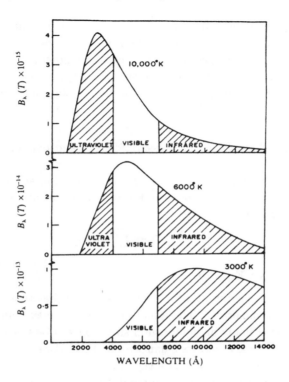

Fig. 3.1. The Planck function for various temperatures. The regions of ultraviolet, visible and infrared are clearly marked.

The energy density u_ν of the radiation field is related to B_ν through the relation

$$u_\nu = \frac{4\pi}{c} B_v(T) \tag{3.3}$$

The total energy density is given by

$$u = \int u_\nu dv = aT^4 \tag{3.4}$$

where a is the radiation constant. The relation (3.4) known as the *Stephan-Boltzmann law*, shows that the energy density of the black body radiation depends upon the fourth power of temperature.

3.2. Perfect Gas Law

The particles of gas in a container which are constantly in motion collide with each other as well as with the walls of the container, giving rise to a resultant force and hence the pressure. The pressure is a function of the density of the gas and the temperature. For a perfect gas, in which the interatomic or intermolecular forces can be ignored, there is a simple mathematical relation between the pressure, density and temperature of the gas, called the *equation of state*. The equation of state for a perfect gas is given by

$$p = nkT \tag{3.5}$$

where p, n and T represent the pressure, density and temperature, respectively.

For a gas containing a number of non-interacting species of various types each exerting its own pressure, the total gas pressure is just the sum of the various components, i.e.,

$$P = \sum P_i = \sum n_i kT. \tag{3.6}$$

3.3. Dissociative Equilibrium

If the gaseous medium is at a low temperature, the atoms can combine together to form molecules. These molecules in turn may dissociate giving back the atoms. Finally an equilibrium situation will be reached when the direct and the reverse processes balance each other. For such an equilibrium situation, it is possible to calculate the number of molecules formed out of the individual atoms for a given temperature.

Consider two atoms x and y combining together to form the diatomic molecules xy, i.e.,

$$x + y \rightleftarrows xy.$$

For the equilibrium situation one has the relation

$$\frac{p(x)p(y)}{p(xy)} = K(xy) \tag{3.7}$$

where $p(x)$, $p(y)$ and $p(xy)$ are partial pressures of x, y and xy respectively. $K(xy)$ is called the *equilibrium constant* or the *dissociation constant* of the reaction. The equilibrium constant depends upon the temperature and on various parameters of the molecule. An expression for the equilibrium constant can be obtained in an explicit form as

$$\log_{10} K_{xy}(T) = \log_{10} \frac{p_x p_y}{p_{xy}} = -\frac{5040.4D}{T} + \frac{5}{2} \log_{10} T$$
$$+ \frac{3}{2} \log_{10} M + \log_{10} \frac{Q_x Q_y}{Q_{xy}} + 4.41405. \tag{3.8}$$

Here M is the reduced mass of the molecule equal to $(m_{xy}/(m_x+m_y))$ where m_x, m_y and m_{xy} are the masses of x, y and xy respectively. The Q's are the *partition functions* and D is the energy required to dissociate the molecule called the *dissociation energy*. For many molecules of astrophysical interest the equilibrium constant can be calculated from Eq. (3.8) as all the relevant spectroscopic parameters are known. Since the equilibrium constant is a function of the temperature, one usually fits the calculated data with a polynomial expression of a suitable form.

3.4. Doppler Shift

The frequency of the emitted radiation depends upon the relative velocity of the source and the observer. The effect is produced only by the component of velocity in the direction towards or away from the observer called the *radial velocity*. The shift of the lines produced as a result of the above motion is termed as *Doppler shift* or *Doppler effect*. The expected shift for a source moving with the velocity v is given by

$$\frac{\Delta\lambda}{\lambda_0} = \frac{(\lambda - \lambda_0)}{\lambda_0} = \frac{v}{c} \qquad (3.9)$$

where λ and λ_0 are the observed and the laboratory wavelengths respectively. The relative velocity of the source is denoted as positive if it is moving away from the observer and negative if it is moving towards the observer. In the former case, the shift is towards longer wavelengths while in the latter case it is towards shorter wavelengths. From the measurement of the shift of the lines, using Eq. (3.9), it is possible to determine the velocity of the source. Doppler shift has been used extensively in astronomy to derive the relative velocities of the various astronomical objects.

3.5. Spectroscopy

3.5.1. *Atomic spectroscopy*

Bohr's formalism of the absorption and emission processes in atoms is that the lines arise out of transitions between well-defined electron energy levels having definite quantum number. The wavelength of the emitted radiation arising out of the two energy levels E_1 and E_2 is represented by

$$\lambda = \frac{hc}{E_2 - E_1}. \qquad (3.10)$$

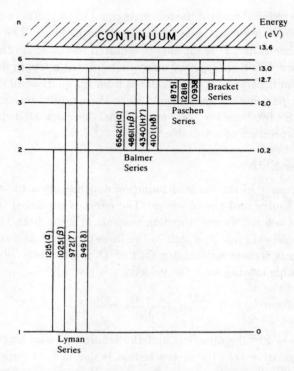

Fig. 3.2. Energy level diagram of the hydrogen atom showing the various series.

When the transition takes place from a lower level to a higher level the energy is lost from the incident radiation and it gives rise to an *absorption* line. If the reverse process takes place it releases the energy and is known as the *emission* line.

The energy level diagram for the hydrogen atom, which has a single orbital electron around a proton, is shown in Fig. 3.2. The energy levels are defined by the quantum number n which can take values $n = 1, 2, \ldots, \infty$. The *Lyman* series arises out of the transition from $n = 2, 3, \ldots$ to $n = 1$. The Lyman α line corresponding to transition $n = 2$ to $n = 1$ has a wavelength of 1216Å and is in the ultraviolet region. This line is very strong in comets. Most of the lines of the *Balmer* series lie in the visible spectral region. Similarly there are *Paschen, Brackett* and other series, whose lines lie mostly in the infrared region.

As n increases, the energy levels come closer and closer together and finally they coalesce. The transitions arising out of these highest levels give rise to a continuum. The excitation potential of a line is the energy required to excite the line. The excitation potential for the Lyman α line is 10.15 eV. The *ionisation potential* is the energy required to remove the electron *completely* from the atom. For the hydrogen atom this energy is 13.54 eV. For atoms with more electrons, the spectra becomes complicated. However, the basic model on which they can be explained remains the same except that one has to consider various types of transitions.

Since the electrons in an atom have orbital and spin angular momenta around the nucleus, they are therefore characterised by three quantum numbers, n, l and j. They denote *total* quantum number, the *orbital angular momentum* quantum number and the *total angular momentum* quantum number. The quantum number j for the case of l and s coupling is given by

$$j = l + s$$

where s is the *spin* quantum number representing the spin of the electron and can take the values $+\frac{1}{2}$ and $-\frac{1}{2}$. From many electron systems, the vector sum of the above quantities has to be taken and is represented by the capital letters L, S and J. Therefore

$$L = \sum l_i, \quad S = \sum s_i \quad \text{and } J = L + S.$$

The J levels are in general degenerate with $(2J+1)$ levels. They split up into $(2J+1)$ levels in the presence of an external magnetic field. L can take values 0, 1, 2 up to $(n-1)$ and they are represented as S, P, D, F, ... terms. The value with $S = 0$, $\frac{1}{2}$, 1, ... denotes the *multiplicity* of the levels and refers to lines as *singlets, doublets, triplets*, etc. The level is generally written as

$$n^{(2S+1)}L_J$$

where n is the total quantum number, $(2S+1)$ gives the multiplicity, J the total angular momentum and L is the term symbol.

In general, the transitions between various levels have to satisfy certain *selection* rules. For electric dipole transitions the selection rules are the following:

$\Delta J = 0, \pm 1$ with $J = 0 \nrightarrow J = 0$

$\Delta L = 0, \pm 1$

$\Delta S = 0$.

In situations where dipole transitions are forbidden it is possible to observe magnetic or quadrupole transitions. These are called *forbidden* lines. The transition probabilities of these lines are much smaller than those of *allowed* transitions. Many of the forbidden lines have been observed in various astrophysical situations because of low density present in them. Many of these lines cannot usually be observed under normal laboratory conditions. The forbidden lines are generally denoted with a square bracket. For example the forbidden line $\lambda = 6300$ Å of oxygen arising out of ^1D level is denoted as $O[^1D]$. If the atom is neutral, it is designated by putting I in front of the chemical symbol, like OI, NI, ... The symbols II, III ... represent atoms in singly ionized, doubly ionized states like NII, NIII, ...

3.5.2. *Molecular spectroscopy*

Many of the diatomic and complex molecules are abundant in objects like the Sun, the cool stars and comets. If the two atoms in the diatomic molecule are of the same type, it is called a *homonuclear* molecule. Examples of this type are H_2, N_2, O_2, etc. If the two atoms are of different types, like CN, CH, OH, etc., It is called a *heteronuclear* molecule. The spectrum of even the simplest diatomic molecule shows complicated behaviour comprising different bands and each band itself is made up of many lines. This is due to the fact that the two atoms in the molecule can vibrate individually along the common axis as well as rotate along the axis perpendicular to the common axis. The total energy of the molecule is the sum total of the kinetic and potential energies of the electrons and the nuclei. This is generally represented in terms of the potential energy curves which give the variation of the energy as a function of the internuclear separation (Fig. 3.4). The energy required to separate a stable molecule into its components is called the *dissociation energy* of the molecule. The various vibrational energies of the molecule are denoted by the *vibrational* quantum number v and can take values 0, 1, 2 ... Each vibrational level is further split up into various rotational levels denoted by the rotational quantum number J. Therefore each electronic state has many vibrational levels, each of which in turn has several *rotational* levels as shown schematically in Fig. 3.3. A transition can take place between vibrational and rotational levels of the two electronic states. The total energy E, of the molecule can be represented as a sum of the electronic (E_{el}), vibrational (E_{vib}) and rotational (E_{rot}) energies namely

$$E_{\text{total}} = E_{el} + E_{vib} + E_{rot}. \tag{3.11}$$

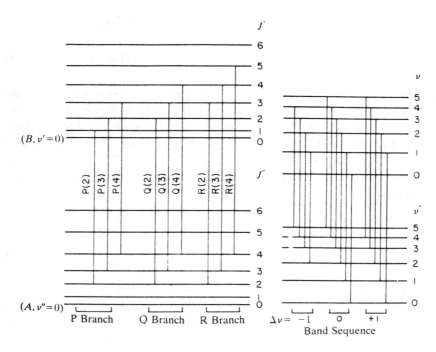

Fig. 3.3. Schematic representation of the vibrational and rotational levels of two electronic states A and B of a molecule. The left one shows the transitions involving P, Q and R branches. The right one shows the transitions which define the band sequence (see text).

In terms of actual energies

$$E_{\text{el}} > E_{\text{vib}} > E_{\text{rot}}.$$

The electronic transitions give lines which fall in the visible and UV regions while those of rotational transitions lie in the infrared and far-infrared regions. The vibrational energy of the molecule can be represented as

$$G(v)hc \equiv E_{\text{vib}} = hc\omega_e \left(v + \frac{1}{2}\right) - hc\omega_e x_e \left(v + \frac{1}{2}\right)^2 \\ + hc\omega_e y_e \left(v + \frac{1}{2}\right)^3 + \cdots \quad (3.12)$$

where ω_e, $\omega_e x_e$ and $\omega_e y_e$ are the spectroscopic constants of the molecule.

The expression for the rotational energy is of the form

$$F(J)hc = E_{\text{rot}} = \frac{h^2}{8\pi^2 I} J(J+1) = hcBJ(J+1) \qquad (3.13)$$

where $B = [h/(8\pi^2 cI)]$, J is the rotational quantum number and it can take values 0, 1, 2, ... I is the moment of inertia of the molecule. B is known as the *rotational constant* of the molecule. All the quantum numbers of the lower electronic level (T''') are denoted as double prime $('')$, like v'' and J'', while that of the upper level (T') as prime $(')$ like v' and J'. Therefore, the wave number (cm^{-1}) of the transition between the two electronic states is given by

$$\nu = \frac{1}{hc}\{[E_{\text{el}}(T') - E_{\text{el}}(T'')] + [E_{\text{vib}}(v') - E_{\text{vib}}(v'')]$$
$$+ [E_{\text{rot}}(J') - E_{\text{rot}}(J'')]\}. \qquad (3.14)$$

The classification of the electronic states of a molecule is based on a scheme similar to that employed for atoms. L, i.e., $(L_1 + L_2)$ and S, i.e., $(S_1 + S_2)$ are the total angular momenta of the two atoms in the molecule. The projections along the axis of the molecule are denoted as Λ and Σ. Λ can take values $0, 1, 2, \ldots, L$. The designations used for there states are

for $\Lambda = 0 \quad 1 \quad 2 \quad 3 \quad \cdots$
$\qquad\qquad \Sigma \quad \Pi \quad \Delta \quad \Phi \quad \cdots$ (for molecules)
similar to $\quad$ S $\quad$ P $\quad$ D $\quad$ F $\quad \cdots$ in the atomic case.

The total angular momentum Ω is given by

$$\Omega = \Lambda + \Sigma$$

which is similar to the case of quantum number J in the atomic case. The selection rules for the electronic transitions are

$$\Delta\Lambda = 0, \pm 1$$
$$\Delta\Sigma = 0$$

and

$$\Delta\Omega = 0, \pm 1.$$

The transition between any two electronic states, is determined by the rotation-vibration structure of the two states involved. There are no rigorous selection rules for the vibrational quantum number and so the transition can take place between any two vibrational levels of the two electronic states. A pure vibrational transition between the two electronic states, called a *band* is denoted as (v', v''), i.e., the quantum number of the upper level is written first. Thus for example (2, 0) means a transition from the upper vibrational level $v' = 2$ to the lower vibrational level $v'' = 0$. Pure vibrational transitions in a given electronic state are allowed for heteronuclear molecules but not for homonuclear molecules like C_2, N_2, etc.

Each of the vibrational bands is further split up into a large number of rotational lines. The selection rule for the rotational quantum number J is given by $\Delta J = J' - J'' = 0, \pm 1$. However, if $\Lambda = 0$, only $\Delta J = \pm 1$ is allowed. Therefore, the rotational transitions give rise to three series of lines called P, Q and R branches corresponding to $\Delta J = -1, 0, +1$ (Fig. 3.3).

The rotational lines of a given vibrational band cannot in general be resolved in low resolution spectra and therefore it results in a blended feature. It can be resolved with a higher spectral resolution. For some molecules the bands arising for the same change in the vibrational quantum number in going from the upper to the lower electronic state have wavelengths very close to each other. Therefore these bands cannot be resolved and give rise to blended features known as *band-sequences* (Fig. 3.3). For example, $\Delta v = 0$ of the Swan band sequence for the C_2 molecule occurs around $\lambda = 5165$ Å (Fig. 4.1). To separate the bands from the band-sequences, it is necessary to go in for even higher resolutions. To resolve rotational structure, still higher resolutions are required (Fig. 5.10).

As in the atomic case, the molecular term can be written as

$$Z^{(2S+1)}\Lambda_\Omega$$

where Z represents the designation of the electronic state, $(2S + 1)$ the multiplicity where S is the electron spin, Ω the total angular momentum and Λ is the kind of term like Σ, Π, etc. As a typical case, the designation of some of the band systems are given below for illustrative purposes.

$(B^2\Sigma - X^2\Sigma)$ of the CN system and $(A^2\Pi - X^2\Sigma)$ of the CO^+ system and so on.

3.6. Isotope Effect

The diatomic molecules which have the same atomic number but have slightly different masses can give rise to a separation of the lines. This is known as *isotopic shift*. The difference in the emitted frequencies between the two isotopic molecules arises through the difference in the reduced masses of the two molecules. The spectroscopic constants of the isotopic molecules (i) and the ordinary molecule are related by the following relations.

$$\omega_e(i) = \rho \omega_e$$
$$\omega_e x_e(i) = \rho^2 \omega_e x_e$$

and

$$\omega_e y_e(i) = \rho^3 \omega_e y_e \qquad (3.15)$$

where

$$\rho = \left[\frac{\mu}{\mu(i)}\right]^{1/2}.$$

Here μ denotes the reduced mass of the molecule. The amount of the shift depends upon the ratio of the masses and on the value of the vibrational quantum number v. In the case when only the first term in the energy term is included, the isotopic shift is given by

$$\Delta \nu_{\text{shift}} = (1 - \rho) \left[\left(v' + \frac{1}{2}\right) \omega'_e - \left(v'' + \frac{1}{2}\right) \omega''_e \right] \qquad (3.16)$$

Therefore, the isotopic shift is a function of the value of $(1 - \rho)$. The larger the deviation from unity, the greater will be the shift of the lines. The sign of $(1 - \rho)$ indicates the type of shift expected. The negative sign implies a shift towards shorter wavelengths. For example the (1, 0) band of the blue degraded Swan $\Delta v = +1$ sequence occurs around 4737 Å while that of ^{12}C ^{13}C occurs around 4745 Å.

The isotopic mass difference also has an effect on the rotational constant B which in turn changes the rotational levels as

$$F^i(J) = \rho^2 B J(J+1). \qquad (3.17)$$

Therefore, the isotopic effect could also be seen in the rotational spectra of the molecules.

3.7. Franck-Condon Factors

The electronic transitions in a molecule give rise to several types of intensity patterns depending on the type of the molecule. For example, for some molecules (0, 0) transitions is the strongest, while for others the strongest line may be for a different value of (v', v''). The observed variations in the intensity distributions can be understood in terms of the *Franck-Condon principle*. The basic idea is the following. The electron jump takes place from one electronic state to another preferentially at the turning points of any vibrational level. This is due to the fact that the time of passage between these two turning points is much shorter than the time spent at the turning points. Therefore the relative position and velocity after the transition is the same as before the transition. Hence, if a transition takes place from a given value of v' of the upper state, the quantum number of the jumped lower state depends on the location and shape of the two potential curves. Figure. 3.4 shows the expected results for two cases. In one case, the minimum of the potential curves is nearly one above the other, while in the second case, one is shifted for higher values of r. Based on the Franck-Condon principle the (0, 0) transition should be the strongest for case 1. For case 2, the strongest lines will be arising from the higher values of v'. For emission lines, there will be two values of v'', corresponding to points B and D in Fig. 3.4 for which the intensity

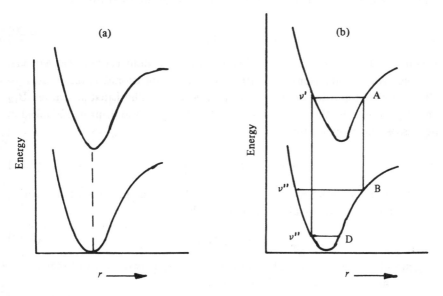

Fig. 3.4. Understanding of the expected intensity distribution in emission according to the Franck-Condon principle (see text).

will be maximum. The locus of the strongest bands is a parabola called the *Condon parabola*. The Condon parabola can be calculated theoretically provided the potential curves are known. The Franck-Condon factors give a measure of the relative band intensities for an electronic transition.

3.8. Intensity of Emitted Lines

The intensity of a spectral line in emission, defined in ergs/sec, is given by

$$I_{21} = N_2 A_{21} h\nu_{21} . \tag{3.18}$$

Here N_2 is the number of the species in state 2 and ν_{21} is the frequency of the emitted radiation from state 2 to state 1. A_{21} is the *Einstein coefficient* which gives the probability for a spontaneous transition from state 2 to state 1, even without any external influence. The Einstein coefficient may be expressed in terms of a parameter called the *strength* S_{21} of a line as

$$g_2 A_{21} = \frac{64\pi^4 \nu_{21}^3}{3hc^3} S_{21} \tag{3.19}$$

where g_2 is the *statistical weight* or the *degeneracy* of the upper level. Other quantities have their usual meanings. The mean life of state 2 is

$$\tau_2 = \frac{1}{A_{21}} . \tag{3.20}$$

It is instructive to see the order of value of τ for various types of transitions. For electronic transitions $\tau \sim 10^{-8}$ sec. For vibrational transitions $\tau \sim 10^{-3}$ sec and for rotational transitions $\tau \sim 1$ sec. The Einstein values B_{12} and B_{21} represent the probability for the absorption and induced emission processes and is given by

$$g_2 B_{21} = g_1 B_{12} = \frac{8\pi^3}{3h^2} S_{21} . \tag{3.21}$$

The *total line strength* S_{21} of the molecular line is the product of electronic, vibrational and rotational components. Therefore

$$S_{21} = S_{\text{el}} S_{\text{vib}} S_{\text{rot}} . \tag{3.22}$$

In general the probability of a transition between two states of eigenfunctions ψ' and ψ'' is given by the equation

$$R = \int \psi' \mu \psi'' dr \tag{3.23}$$

where μ is the dipole moment. R^2 is proportional to the transition probability. Usually S_{el} and S_{vib} are combined together as

$$(S_{el} S_{vib}) \equiv S_{el}^{vib} = \left| \int \psi_{v'} R_e \psi_{v''} dr \right|^2. \tag{3.24}$$

Here R_e is called the *electronic transition moment*. $\psi_{v'}$ and $\psi_{v''}$ are the eigenfunctions for the vibrational states v' and v''. R_e itself is defined by the expression

$$R_e = \int \psi'_e \mu_e \psi''_e d\tau_e \tag{3.25}$$

where ψ_e's are the electronic wavefunctions and μ_e refers to the electric dipole moment for the electrons. In general the electron wavefunction ψ_e also depends to some extent on the internuclear distance r. Hence R_e should also depend on r. However, since the variation of R_e with r is slow, this variation is often neglected and R_e is replaced by an average value of $\bar{R}_e$. Therefore Eq. (3.24) becomes

$$S_{el}^{vib} = \bar{R}_e^2(\bar{r}_{v'v''}) \left| \int \psi_{v'} \psi_{v''} dr \right|^2 \tag{3.26}$$

where $\bar{r}_{v'v''}$ is called the *r-centroid* and it is a characteristic internuclear separation which can be associated with a given band (v', v'') and is given by

$$\bar{r}_{vv''} = \frac{\int \psi_{v'} r \psi_{v''} dr}{\int \psi_{v'} \psi_{v''} dr}. \tag{3.27}$$

The integral over the products of the vibrational wavefunctions of the two states of Eq. (3.26) is known as the *overlap integral* and is generally called the *Franck-Condon factors* of the (v', v'') band. Therefore, Eq. (3.26) can be written as

$$S_{el}^{vib} = \bar{R}_e^2(\bar{r}_{v'v''}) q_{v'v''} \tag{3.28}$$

where

$$q_{v'v''} = \left| \int \psi_{v'} \psi_{v''} dr \right|^2.$$

Hence Eq. (3.22) becomes

$$S_{21} = \bar{R}_e^2(\bar{r}_{v'v''}) q_{v'v''} S_{rot}. \tag{3.29}$$

Hence the electronic transition moment $\bar{R}_e^2$ refers to r-centroid $\bar{r}_{v'v''}$ and S_{rot} is usually known as *Höln-London factors*. Equation (3.29) shows that the total strength of a molecular line is essentially given by the product of the three-strength factors, namely the electronic transition moment, Franck-Condon factors and the Höln-London factors.

The values of the electronic transition moment for any band basically have to come from the laboratory measurements of the intensity of the lines, as theoretical calculations are very difficult. Enormous amount of work has been carried out in various laboratories over the world to extract this basic data from the line intensity measurements of various bands. But the data is still very meagre. So for most of the cases one ends up using either the mean value or an approximate value for the electronic transition moment.

The Franck-Condon factors can be calculated from a knowledge of the wavefunction of the vibrational levels, which comes out of the solution of the Schrödinger equation. For this, the potential function $U(r)$ has to be expressed in a convenient and mathematical form. There are various representations of the potential curve. The well-known function is that of Morse and is generally called *Morse function*. This has been extensively used in the literature as it is quite simple and convenient. However it does not represent the potential curves for all the cases exactly. So other expressions for the potential have been suggested. The one that is becoming commonly used in recent years is known as the *RKR potential* referring to the authors Rydberg, Klein and Rees, who proposed it. In this method, the potential curve is constructed point by point from the laboratory measured values of the vibrational and rotational levels. The potential curves obtained in this way are much superior in many cases compared to Morse potential representation. The advantage of this method is the fact that it uses experimentally determined values of the quantities. However, the disadvantage is that the experimental values are often not of good quality and, in addition, this method is quite cumbersome. Many people have written computer programs to evaluate $q_{v'v''}$ and $\bar{r}$-centroids for given values of input parameters. For many molecules of astrophysical interest, the values of $q_{v'v''}$ have been published in the literature.

On the other hand, the rotational strength factors have to be calculated from the theory. They depend upon the structure of the molecule, type of coupling, type of transition involved, etc. Until recently there existed a lot of confusion in the definition and normalization of Höln-London factors. This has been clarified by several workers. The expression for the Höln-London factors has been evaluated for various cases of interest and is available in the literature. Computer programs have also been written to compute these factors.

It is also possible to get an estimate of the Einstein A value or the oscillator strengths f, directly from the laboratory measurements of the intensity of lines. Various techniques and methods have been employed to measure these quantities. The measurements are however hard to make and so there are not many measurements available at the present time. Even if they are available, the values by different methods or by different observers often disagree. Therefore, one has to make use of laboratory measured values along with the calculated quantities in any particular situation.

3.9. Boltzmann Distribution

The calculation of the intensities of lines involves a knowledge of the relative population of atoms or molecules in different excited states. In a thermal equilibrium in which every process is balanced by its inverse, the population distribution among various levels is described according to *Boltzmann formula*. The population distribution between two discrete states 1 and 2 with an energy separation $\epsilon_{12} = E_2 - E_1$ is given by the expression

$$\frac{n_2}{n_1} = \frac{g_2}{g_1} e^{-\epsilon_{12}/kT} \tag{3.30}$$

where g_1 and g_2 are the *statistical weights* of the two levels and for an atom it is equal to

$$g(J) = 2J + 1$$

where J is the total angular momentum quantum number. The expression can be used to calculate the population distribution in different levels of a given temperature. Conversely, with a knowledge of the population in two or more levels, the excitation temperature can be determined.

3.10. Λ-Doubling

The assumption that the interaction between the angular momentum due to rotation of the molecule as a whole and the electron orbital angular momentum can be neglected is valid only for the case of $\Lambda = 0$. This corresponds to the case of Σ states. However, for the electronic states $\Pi, \Delta, \ldots$, for which $\Lambda \neq 0$, this interaction splits the rotational energy levels. This is called Λ-*doubling*. The splitting of the levels is generally very small and is of the order of 1 cm^{-1} or less. Therefore the transitions between these levels will give rise to lines which lie in the radio frequency region. For transitions of the type $(^1\Sigma - ^1\Sigma)$ there is no Λ doubling. But for others like $(^2\Sigma - ^2\Pi)$,

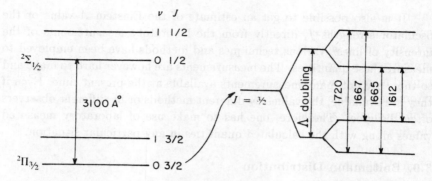

Fig. 3.5. Energy level diagram of the OH molecule and the associated Λ-doubling.

Table 3.1. Radio lines of OH molecule.

Frequency (MHz)	Relative intensities of lines	Transition probability (sec^{-1})
1612.23	1	4.50 (−12)
1665.40	5	2.47 (−11)
1667.36	9	2.66 (−11)
1720.53	1	3.24 (−12)

the effect of Λ doubling can be seen. Λ doubling is the characteristic feature of molecules like OH, CH, etc., which have large rotational constants. The OH radical has been studied more extensively than any other radical. The energy level diagram of the hydroxyl radical, with only a few vibrational levels is shown in Fig. 3.5. The strong (0, 0) band which is generally seen in emission in comets arises from the transitions as shown in Fig. 3.5. The lowest rotational level, corresponding to $v'' = 0$ and $J'' = 3/2$ is further expanded in Fig. 3.5 to show the Λ-splitting. The two Λ-doubling states are further split by the hyperfine interaction with the hydrogen nucleus. Therefore in all there are a total of four lines (Table 3.1). All these lines have been seen from many astronomical sources.

3.11. Solar Radiation

The subject of the study of the total amount of energy as well as the spectral distribution in the solar spectrum is an old one. It is of great interest in various fields. It has therefore been studied extensively in the entire range of the electromagnetic spectrum. Yet the information at the present time is not complete. The detailed and extensive measurements are carried out in the case of the Sun because of its proximity, it being a typical main sequence star and also because of its effects on the Earth's environment. It also turns out that most of the phenomena that one observes from a comet is associated directly or indirectly with the radiation field of the Sun.

The surface temperature of the Sun is about 6000°K. According to Planck's distribution law [Eq. (3.1)] most of the energy is concentrated in the visible region of the spectrum. Therefore the amount of energy measured from the Earth corrected for the atmospheric transmission gives a good estimate of the total energy of radiation. Of course, one should try to allow for the UV and IR radiation which is cut off by the Earth's atmosphere. The intensity of the solar radiation as seen from the Earth's atmosphere at the Earth's mean distance from the Sun is called the *solar constant*. To get the solar constant value, extensive ground based measurements have been carried out. The average value for the solar constant is found to be ~ 1.95 cal/cm^2/min.

The spectral energy distribution in the visible region is quite smooth and can be represented well by a black body of temperature 6000°K. Superposed on it are the absorption lines of the gases of the solar atmosphere. For measuring the solar emission in the UV region, it is necessary to go above the Earth's atmosphere. Such observations have been carried out using rockets and satellites. As one goes towards the UV spectral region, the number of emission lines arising out of different ionization levels of the various elements increases. In fact below around 1500 Å, the major portion of the energy is in emission lines rather than in the continuum (Fig. 3.6). As one goes towards shorter and shorter wavelengths, one is essentially seeing the higher and higher temperature regions, namely the chromosphere and the corona of the Sun, from where the emission lines arise. Therefore the shape of the continuum radiation in the UV and far UV regions depend to certain extent on the resolution of the instrument. The higher the resolution of the instrument the more the number of weak lines can be resolved. The intense lines in the UV region are the Lyman α and Lyman β occurring at 1215.7 Å and 1025.7 Å, respectively. They have fluxes of about 3×10^{11} and 2×10^9 photons/cm^2/sec at the top of the atmosphere as compared to the continuum level in the same region of about 10^8 photons/cm^2/sec. Lyman α and Lyman β emission lines are quite important as they can dissociate, ionize

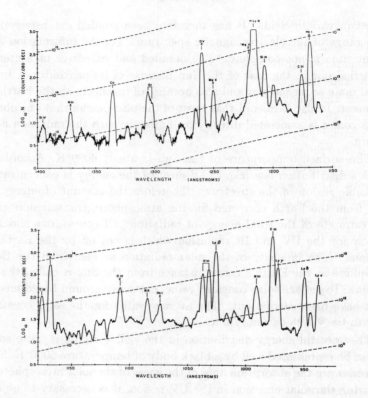

Fig. 3.6. A representative spectral distribution of the quiet Sun between 1150 to 1400 Å (Dupree, A. K. and Reeves, E. M. 1971. *Ap. J.* **165** 599).

or excite the species. The observational data in the UV region obtained by various observers or groups do not always agree with each other due to the difficulties associated with the measurement. So this uncertainty should be kept in mind in using the published data. In addition, the emission could vary with the solar conditions as for example, whether the Sun is quiet or flares are present and so on. In recent years efforts are being made to synthesize the solar spectra for the UV spectral region. The agreement with the high resolution observations is very encouraging. One of the major difficulties faced in this investigation is the lack of availability of laboratory spectroscopic data pertaining to various elements and molecules.

It will become clear later on that the emission and absorption lines, strong as well as weak present in the solar spectrum play a crucial role in the formation of spectral line in a cometary atmosphere.

3.12. Solar Wind

The solar plasma from the corona that flows out continuously into the interplanetary medium is called the *solar wind*. The possible existence of such plasma flows came from the pioneering investigations of Biermann based on the study of plasma tail of comets. The theoretical work of Parker in the 1950's based on the hydrodynamic expansion of the solar corona predicted the expected nature of the solar wind. The theory also showed that the magnetic field originating in the photosphere and dragged into the interplanetary medium takes the form of an Archimedean spiral structure. The observations carried out with satellites later on confirmed Parker's theory of the solar wind. The actual interplanetary magnetic field measurements have established the Archimedean spiral structure for the magnetic field. The mean values obtained from observations for some of the physical parameters of interest for the solar wind near the Earth are the following: number density of electrons or protons, $N_e \sim N_p \approx 5/\text{cm}^3$; velocity ≈ 450 km/sec; electron temperature $\sim 1.5 \times 10^5$ °K; magnetic field $\approx 2.5\gamma$ where $\gamma = 10^{-5}$ gauss. However, it is well known that the Sun is not quiet all the time and a lot of activities of violent nature do take place on the surface of the Sun, like flares, bursts, etc. These have a great effect on many of the geomagnetic activities. In fact there is an almost one-to-one relation between the disturbance occurring on the Sun and the effect observed on the terrestrial atmosphere. The existence of such a correlation has been known since early times. These effects arise basically due to the fact that a disturbance on the surface of the Sun releases suddenly a large amount of plasma which is coupled through interplanetary magnetic field. Therefore, the average physical properties of the solar wind change with the solar conditions. During disturbed periods the solar wind velocity and the density at 1 au could vary by a factor of two or so. The magnitude as well as the direction of the magnetic field also change. The extensive measurements made with satellites have given an enormous amount of information with regard to the variation of physical quantities with heliocentric distance for quiet as well as for disturbed solar conditions. These measurements have also shown the presence of magnetic sector boundaries in the magnetic field which have opposite polarity. These regions are separated by null surfaces where the magnetic field is zero. Since the plasma tail of a comet and the solar wind are coupled to each other through interplanetary magnetic fields (Chap 10), the interpretation of the observed cometary features requires a knowledge of the exact conditions of the solar wind at the time of observation.

Problems

1. Derive Eq. (3.1) in frequency units.
2. Calculate the wavelength of vibrational bands of the Swan system ($d^3\Pi$ - $a^3\Pi$) for the C_2 molecule and $(B - X)$ transitions of the H_2 molecule for $v'' = 0, 1, 2$ and $v' = 0, 1$. This will help in understanding the meaning of band sequence. What instrumental resolution is required to resolve vibrational and rotational structure of the Swan bands.
3. Derive the expression for the isotope shift for the vibrational and rotational transitions including second order terms.
4. Calculate the isotope shift between $^{12}C\ ^{12}C$ and $^{12}C\ ^{13}C$ for the (1, 0) band of the Swan system.
5. Calculate the frequency and wavelength of the line arising out of the transition between $n = 109$ to 108 in hydrogen and carbon atoms. Discuss its significance.
6. Is there a Doppler effect when the observer or the source moves at right angles to the line joining them? How can the Doppler effect be determined when the motion has a component at right angles to the line?
7. Make an estimate of the temperature of the Sun from the fact that it can be seen.
8. Explain why the spectrum of the Sun at wavelengths less than 1500A consists entirely of emission lines.
9. What evidence do we have, direct and indirect, of the existence and properties of solar wind?
10. What is the probable mechanism responsible for the generation of the solar wind?
11. Sun's radiation striking the Earth has an intensity of 1400 watts/meter^2. Assuming the Earth is a flat disk at right angles to the Sun's rays and that the incident energy is completely absorbed, calculate the radiation force acting on the earth. Compare it with the force due to the Sun's gravitational attraction.

12. Calculate the Einstein A-value for the rotational transitions 2 to 1 in the ground electronic state and compare it with the value for the electronic transitions between A and X levels for the CO molecule.

References

1. Herzberg, G. 1950. *Molecular Spectra and Molecular Structure I. Spectra of Diatomic Molecules*. New York: Van Nostrand Reinhold Company.

CHAPTER 4

SPECTRA

The main goals of the study of spectra are to identify the species responsible for the observed lines and to obtain information about the physical conditions present in the source. It is no exaggeration to say that our present understanding about various aspects of cometary phenomena has come directly or indirectly from the study of their spectra.

4.1. Main Characteristics

The identification of the spectra of comets is one of the active areas of study. This is due to the combination of many factors such as better instruments, availability of space vehicles, computation techniques, etc. The identification of the lines in the spectra of a comet is quite a complex and difficult task. The usual procedure is to look for the coincidences between the laboratory wavelength of lines and the observed wavelenghts in the spectra of a comet which have been corrected for the velocity effect (Chap.3). With this method many of the well-known atoms and molecules which are seen in other astronomical objects have been identified in the spectra of comets. Many more are still not identified. However, with better and better resolution as well as going into new spectral regions many more new species have been identified.

The extensive spectral observations carried out on Comet Halley which have been compiled in the Atlas of Cometary Spectra, show the vast extent of the lines which could not be identified. Therefore much work remains to be done in the area of identification of lines. It is appropriate at this point to mention the fact that the problem of identification of lines is intimately connected with the availability of spectroscopic data pertaining to various atoms and molecules. Laboratory data is not available for each and every transition for all the atoms and molecules. Therefore, there could always be lines in the astronomical spectra which cannot be identified. It is interesting to note that several transitions were observed first in the cometary spectra before being studied in the laboratory. The well-known case is the bands of the ion CO^+, generally called the Comet-Tail System. Other examples are the bands of C_3 and H_2O^+.

The observations carried out in the visual spectral region of around 3000 to 8000Å have been the main source of information for the study of spectra of cometary atmosphere for the last several decades. These studies had limitations due to the fact that photographic techniques and also most of the spectroscopic observations could be made only on bright comets at small heliocentric distances. Early observations were also limited to spectra taken at low spectral resolution. In recent years many of these limitations have been overcome. Based on the spectra in the visual region, it is possible to arrive at some general pattern regarding the main characteristic features of the spectra of comets.

At large heliocentric distances, the comet appears as a point source. This is due to the reflection of the solar radiation by the nucleus. For heliocentric distances, $r \gtrsim 3au$, the spectrum is mainly the continuum radiation arising mostly due to the scattered solar radiation by the dust particles present in the cometary atmosphere. The emission lines of the various molecules appear roughly in a sequence as the comet approaches the Sun. The molecular bands first to appear are those of CN at $r \sim 3$ au followed by the emission from C_3 and CH. Thereafter, the emission from C_2, OH, NH and NH_2 appear in the spectrum. They are often strong enough to reveal their structure. The coma ions generally appears last. The relative intensity of emission bands and continuum varies from comet to comet. In sun-grazing comets, say for $r \sim 0.1$ au, numerous metallic emission lines crowd the spectrum.

The spectrum of a plasma tail shows mostly the presence of ionized species and generally at $r \leq 2au$. The notable exceptions include the Comet Humason (1962 VIII) where CO^+ was seen even at $r \geq 6$ au. Among the observed ions in the plasma tail of a comet, the emission from CO^+ dominate.

As a typical case, Fig. 4.1 shows the spectra of Comet Encke in the visual region taken at a resolution (full width at half maximum intensity) of

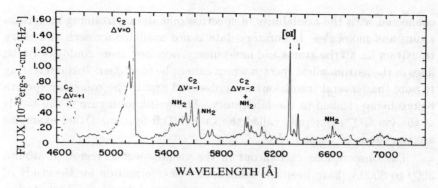

Fig. 4.1. Typical scanner spectrum of Comet Encke using the 3-m Lick Observatory Telescope in the visual region. (*Courtesy of Spinrad.*)

$\Delta \lambda = 7\text{Å}$. The Swan band sequences corresponding to $\Delta v = -1, 0$ and $+1$ of the C_2 molecule ($d^3\Pi - a^3\Pi$), whose wavelengths lie around 5635, 5165 and 4737 Å respectively, are the strongest in the spectra. The vibrational structure could be resolved with better resolution spectra. The resolution of the rotational structure requires still higher resolution. In fact it has been possible to resolve completely the rotational structure of $\Delta v = +1$ band sequence in the spectra of Comet West (1976 VI) taken with a resolution of $\Delta \lambda = 0.3\text{Å}$(Fig. 5.10). The spectra of Comet Halley taken at a spectral resolution of 0.07Å beautifully shows the rotational structure of the (0,0) Swan band of the C_2 molecule(Fig. 5.14). Since the Swan bands of the C_2 molecule dominate the spectrum in the visual region, to a first approximation, it also determines the 'visual diameter' of the head of the comet. It may also be noted that the wavelength of $\Delta v = +1$ band sequence of the isotopic molecule $^{12}C\ ^{13}C$ (4745Å), is shifted by about 7Å with respect to that of $^{12}C\ ^{12}C$ (4737 Å) and it has been seen in the spectra of many comets. Unfortunately this isotopic feature is blended strongly with the emission from the NH_2 molecule (($\tilde{A}_2A_1 - \tilde{X}_2B_1$) system). Other bands of NH_2 can also be seen (Chap. 5).

The emission due to C_3 molecule has a broad feature extending roughly from 3950 to 4140 Å, with a strong peak around 4050Å. The identification of C_3 feature in comets was difficult as the laboratory analysis was not available. Various transitions of the CN molecule, both the red ($\lambda \sim 7800\text{Å}$-$1\mu m$) and the violet ($\lambda \sim 3600$-$4200$ Å) systems corresponding to transitions ($A^2\Pi - X^2\Sigma^+$) and ($B^2\Sigma^+ - X^2\Sigma^+$) have been identified. The rotational structure of CN(0,0) band is well-resolved (Fig. 5.1). For CH molecule both the systems ($A^2\Delta - X^2\Pi$) and ($B^2\Sigma - X^2\Pi$) which lie in the ultraviolet ($\lambda \sim 3900\text{-}4300$Å) have been seen. The (0,0) and (1,1) bands corresponding to ($A^2\Sigma^+ - X^2\Pi$) of OH at 3090 Å and 3140Å, which is close to the atmospheric cut-off were detected from ground-based observations. The rotational structure was partially resolved. Numerous lines belonging to ($A^3\Pi - X^3\Sigma$) system of OH$^+$ ($\lambda \sim 3500$Å) have been identified. The lines of H_2O^+ ($\lambda \sim 5500 - 7500$Å) were identified for the first time in Comet Kohoutek. The bands of ($A^2\Pi - X^2\Sigma$) of CO$^+$ around $\lambda \sim (3400\text{-}6300$Å) has been seen. The sodium D lines at 5890 and 5896Å also show up for $r \leq 1.4$ au. The good quality spectra that exists for Comet Halley has shown a large number of unidentified lines.

The use of rockets and satellites has made it possible to extend the observations into the ultraviolet region. This is the region of the spectrum say from 1000 to 4000Å where the abundant atomic and molecular species have their resonance transitions. In fact the spectra of comets taken in the ultraviolet region has clearly demonstrated the richness of molecular emissions in this spectral region. The spectral resolution available at the present time is still not good. Comet West in 1976 provided a good opportunity to secure high quality spectra in the UV region as the Comet was quite bright. For illustrations, the rocket spectra of this comet covering the wavelength region from 1600 to 4000Å is shown in Fig. 4.2. Many molecules like CS, CN$^+$ and others were identified for the first time based on the spectra of Fig. 4.2 and others. The (0,0) transition, (A-X) of OH at 3090 Å which was weak when observed from the ground is found to be very strong when observed outside the Earth's atmosphere. The rotational structure of OH bands has also been resolved completely. Shortward of 1800Å, the spectrum is dominated by multiplets of CI at 1561 Å, 1657Å, OI at 1304Å and CII at 1335Å. Many weak CI features are also present. The lines of SI multiplet at 1807, 1820 and 1826Å were detected for the first time in the Comet West. The strong emission ($A^1\Pi - X^1\Sigma^+$) bands of CO commonly referred to as the Fourth Positive system were also identified for the first time in Comet West. Figure 4.3 shows the spectrum covering this spectral range. The presence of CO$_2^+$ ion in comets came from the detection of the bands of ($\tilde{A}^2\Pi$- $\tilde{X}^2\Pi$) system in the region 3300-3900Å. The ($\tilde{B}^2\Sigma - \tilde{X}^2\Pi$) double band system whose wavelength occurs around 2890Å has also been identified. In the region of 2000 to 2600Å the First Negative bands ($B^2\Sigma^+ - X^2\Sigma^+$) of CO$^+$ are generally very strong. The

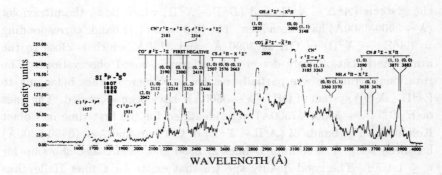

Fig. 4.2. Rocket spectrum of Comet West in the 1600 to 4000 Å spectral region. The richness of molecular emissions can be seen. *(Courtesy of Andy Smith.)*

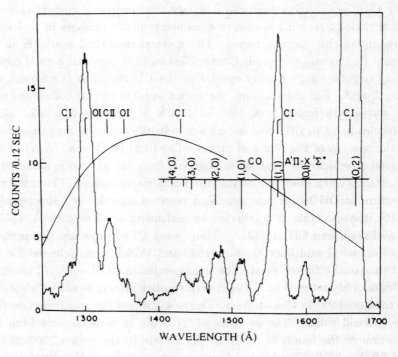

Fig. 4.3. Low resolution spectrum of Comet West taken from an Aerobee rocket in the wavelength region 1300 to 1700 Å which shows bands of CO molecule (Feldman, P.D. and Brune, W.H. 1976, *Ap. J*, **209**, L45).

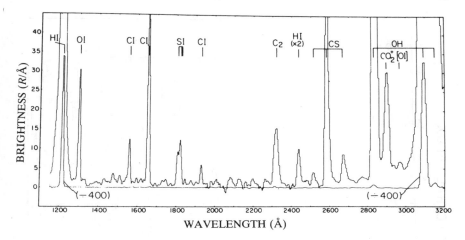

Fig. 4.4. The ultraviolet spectrum of Comet Bradfield 1979 X obtained from the International Ultraviolet Explorer (Weaver, H.A., Feldman, P.D., Festou, M.C. and A'Hearn, M. F. 1981, *Ap. J*, **251** 809).

Mulliken System of the C_2 molecule corresponding to $\Delta v = 0$ has a wavelength around 2300Å. This feature is generally blended with the strong bands of CO^+ lines. Hence, this feature could show up only if CO^+ lines are weak or absent as it happened in the case of Comets Seargent (1978 XV) and Bradfield (1979 X). In the spectra of these comets, the wavelengths at which bands of CO^+ occur were clear of any lines except for a feature at $\lambda \sim 2300$Å. This feature is attributed to the Mulliken system of $\Delta v = 0$. The relevant spectra is shown in Fig. 4.4. Several strong emission bands of $(B^3\Sigma_u - X^3\Sigma_g)$ of the S_2 molecule have been identified in the wavelength region of 2800-3100Å based on the beautiful spectra of the Comet IRAS-Araki-Alcock (1983VII) as shown in Fig. 4.5.

The observations made with the orbiting astronomical observatory (OAO-2) satellite in 1970 on Comet Bennett (1970 II) and on Comet Tago-Sato-Kosaka (1969 IX) in the light of the hydrogen Lyman α line at 1216 Å led to the discovery of a hydrogen halo around the visible coma. This important observation also led to the realization that the mass loss rates from comets are much higher than previous estimates which were based on observations in the visual spectral region. After OAO-2, several satellites and rockets have been used extensively for the study of comets in the ultraviolet region. In particular, mention must be made of the International Ultraviolet Explorer (IUE) satellite which has been operating since 1978.

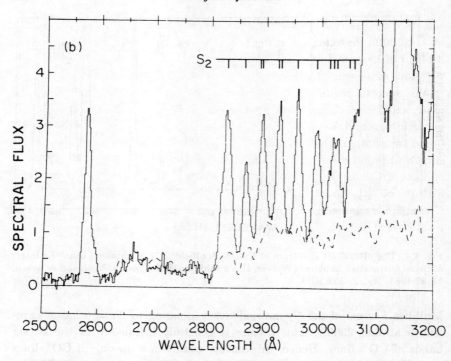

Fig. 4.5. The spectrum of Comet IRAS-Araki-Alcock in the wavelength region 2500 to 3200 A obtained with the International Ultraviolet Explorer Spacecraft. The image No. LWR 15908 was taken on May 11, 1983 with the nucleus centered, when the comet was 0.032au from the Earth. The new features corresponding to (B-X) system of S_2 are identified (A'Hearn, M.F., Feldman, P.D. and Schleicher, D.G. 1983, *Ap. J. Letters* **274**, L99).

The instruments on board this satellite cover the spectral region from 1150 to 3400Å. This much coverage is obtained from two spectrographs, one concerning the spectral range from 1150 to 1950Å and the other from 1900 to 3400 Å. It could be used on comets as faint as 10th magnitude. Hence IUE satellite has been used extensively for making observations on many comets, in the ultraviolet region and covering a wide range of heliocentric distances. So far around 40 comets have been observed with the IUE satellite. They all seem to show similar UV spectra. The Cameron bands arising from the forbidden triplet-singlet transitions ($a^3\Pi - X^1\Sigma^+$) have been observed near 2200Å in Comet Hartley 2 (1991 XV) with the Hubble Space Telescope. This telescope is well suited for the observation of weak lines due to its high sensitivity. Observations of Comets Austin (1990V) and Levy (1990 XX) at $\lambda < 1200$Å have indicated the presence of a feature at 1025.7Å which is a blend of Lyman β line of HI and OI line.

The extension of observations into the infrared region is important as this spectral region is characterized by vibration-rotation transitions of many molecules. Some of the bands of the CN red system($A^2\Pi - X^2\Sigma$) occurring around $\lambda \sim 8000$Å were first seen in the Comet Mrkos (1957 V) based on photographic spectra. This has later been observed from many comets. Several other bands of CN can also be seen. Figure 4.6 shows the spectra of Comet Bradfield (1979X) covering up to about 1 μ. Many weak features of NH_2 and the Phillips bands of C_2 are present. The spectra of Comet Halley taken with resolution of about 0.45 Å has resolved fully the rotational structure of the (2,0) Phillips band.

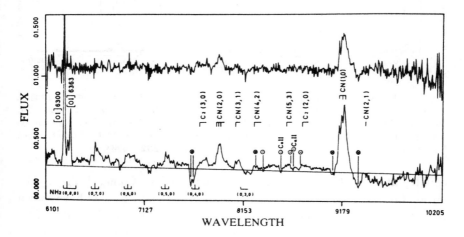

Fig. 4.6. The near infrared spectrum of Comet Bradfield (Danks, A.C. and Dennefeld, M. 1981, A. J 86, 314).

At wavelengths larger than 1μm, lines of several molecules have been detected. In particular the first and direct detection of H_2O in Comet Halley came from the high resolution observation of ν_3 vibrational band at 2.7μm with the Kuiper Airborne Observatory (Fig. 4.7). A broad emission feature around 3.4 μm observed from ground and in situ measurements of Comet Halley and subsequently in other comets, is attributed to C-H stretch vibrations of organics (Fig. 4.8). The ν_3 band of CO_2 at 4.25 μm was detected in Comet Halley from the Vega Spacecraft. The ν_3 band

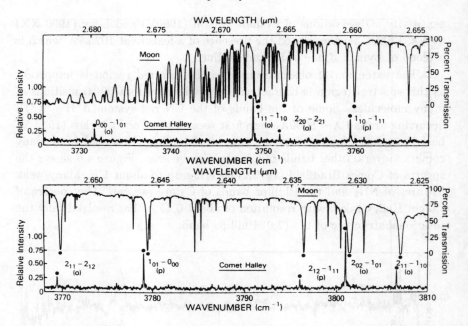

Fig. 4.7. High resolution infrared spectroscopic observations of the ν_3 band (2.7 μm) of H_2O in Comet Halley taken from NASA's Kuiper Airborne Observatory on December 24, 1985. The lines originating from Ortho (O) and Para (P) H_2O are resolved completely. The spectra of the Moon is also shown for comparison (Mumma, M.J., Weaver, H.A., Larson, H.P., Davis, D.S. and Williams, M. 1986. *Science* **232**, 1523).

of CH_4 at $\lambda \sim 3.3 \mu m$ was marginally detected in Comet Halley. The infrared observations in the region 1 to 30 μm of many comets have shown the presence of two strong broad emission features around 10 and 20 μm. These features are generally attributed to silicate materials (Chap. 9).

Most of the polyatomic and complex molecules have rotational transitions whose wavelengths lie in the millimeter and centimeter region, i.e. radio frequency regions. Studies based on interstellar matter showed that interstellar clouds are dominated by molecules of simple, complex as well as of biological importance. It was immediately felt that many of these molecules may also be present in the cometary material. If found, it will have a great implication with regard to the origin of comets. They could also possibly be the parent molecules (Chap.6) of cometary radicals. Therefore the radio searches for many of these molecules have been carried out on many comets. Only a few of the molecules like HCN, HCO, H_2CO etc were identified. The molecule methanol (CH_3OH) has been identified recently

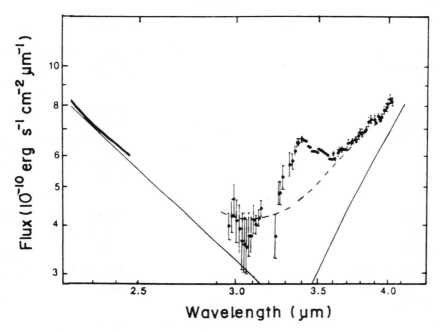

Fig. 4.8. The infrared spectrum of Comet Halley taken on March 31, 1986 (r=1.17 au) shows the presence of a broad emission feature around 3.4μm. The dashed line refers to the continuum for a black body of 350K and the scattered sunlight (Wickramasinghe, D.T. and Allen, D.A. 1986. *Nature* **323**, 44).

through several rotational transitions at millimeter wavelengths in Comets Austin (1990 V) and Levy (1990 XX). A list of the molecules that have been looked for in Comet Halley in given in Table 4.1. An extensive search for various kinds of molecules which was carried out on Comet Halley, not only to confirm earlier identifications but also to look for possible new features of molecules, met with limited success.

With regard to the spectra of diatomic molecules, the Λ- doubling of OH around 18 cm was first detected in Comet Kohoutek in 1974. Since then these lines have been monitored regularly in almost all the comets. The molecule CO is quite abundant in comets. However, the rotational transition of CO at 2.60 mm corresponding to a $J = 1 \rightarrow 0$ transition could not be seen in Comets Kohoutek and Bradfield and antenna temperature for the two comets $\leq 0.5^\circ K$. However the higher rotational transitions of CO, such as $J = 2 \rightarrow 1$ (1.3 mm) and $J = 3 \rightarrow 2$ (0.87 mm) although

Table 4.1. Spectral lines searched for in Comet Halley.

Transition	Frequency (MHz)	Telescope	Reference
Cl $^3P_1$-$^3P_0$	492162	KAO	Keene et al.*
$C_3H_2 1_{10}$-1_{01}	18343	NRAO 43 m	Mathews el al.*
CH_3CN 3-2	36392	Haystack 37 m	Webber and Haschick*
CH_3CN 5-4	91959	IRAM 30 m	Bockelée-Morvan el at. (1987)
		FCRAO 14 m	Swade et al. (1987)
CH_3CN 6-5	110360	OSO 20 m	Winnberg et al. (1987)
CH_3CN v_8=1 6.5	110709	FCRAO 14 m	Swade el at. (1987)
CH_3OH 2_0-3_4 E	12178	NRAO 43 m	Bairla*
CH_3OH 15_3-14_4 A+	88590	OSO 20 m	Winnberg et al. (1987)
CH $^2\Pi_{1/2}$ J=1/2 F=0.1	3264	NRAO 43 m	Tumer*
CH $^2\Pi_{1/2}$ J=1/2 F=1.1	3335	NRAO 43 m	Tumer*
CH $^2\Pi_{1/2}$ J=1/2 F=2.1	3349	NRAO 43 m	Tumer*
CO^+ N=2.1 J=5/2-3/2	236063	NRAO 12 m	Baum and Hoban (1986)
CO 2-1	230538	IRAM 30 m	Bockel'ee-Morvan et al. (1987)
Hl 21 cm	1420	RATAN 600	Bystrova et al.*
H_2CO 1_{10}-1_{11}	4830	NRAO VLA	Snyder et al (1989)
		MPI 100 m	Bird et al. (1987)
H_2CO 2_{11}-2_{12}	14488	NRAO 43 m	Bairla*
H_2O 3_{13}-2_{20}	183310	KAO	Gulkis et al.*
H_2O 4_{14}-3_{32}	380197	KAO	Gulkis et al. (1989)
		KAO	Keene et al.*
H_2O 6_{16}-5_{23}	22235	MPI 100 m	Bird et al. (1987)
		IRO 14 m	Scalise et al. (1987)
		NRAO 43 m	Bairla*
		LPI 22 m	Berulis et al. (1987)
H_3O^+ 1_1-2_1	307192	NRAO 12 m	Wooden et al. (1986)
		MWO 5 m	Wooden et al. (1986)

Table 4.1. (*Continued*)

Transition	Frequency	Telescope	Reference
HC_3N 4-3	36796	Haystalk 37 m	Webber and Haschick*
		LPI 22 m	Berulis et al. (1987)
HC_3N 5-4	45000	NRO 45 m	Kaifu et al.*
HC_3N 10-9	90971	IRAM 30 m	Bockel'ee-Morvan et al. (1987)
HC_3N 11-10	100076	IRAM 30 m	Bockel'ee-Morvan et al. (1987)
HC_3N 12-11	109174	FCRAO 14 m	Swade et al. (1987)
HCN 1-0	88638	FCRAO 14 m	Schloerb et al. (1986)
		IRAM 30 m	Despois et al. (1986)
		OSO 20 m	Winnberg et al. (1987)
		HCRAO Interferometer	Wright et al.*
		NRO 45 m	Kaifu et al.*
HCO^+ 1-0	89189	FCRAO 14 m	Swade et al. (1987)
HNC 1-0	90664	FCRAO 14 m	Swade et al. (1987)
$NH_3 1_1\text{-}1_1$	23694	MPI 100 m	Bird et al. (1987)
$NH_3 3_3\text{-}3_3$	23870	NRAO 43 m	Bairla*
		NASA DSN (Tidbinbilla) 64 m	Gulkis et al.*
OCS 8-7	97301	IRAM 30 m	Bockel'ee-Morvan et al. (1987)
$OH\,^2\Pi_{1/2} J=1/2\ F=1\text{-}0$	4765	MPI 100 m	Bird et al. (1987)
$OH\,^2\Pi_{3/2} J=7/2\ F=4\text{-}4$	13441	NRAO 43 m	Bairla*
SiO v=1 1-0	43122	Haystack 37 m	Webber and Haschick*

*unpublished data.
(Crovisier, J. and Schloerb, F. P. 1991. In *Comets in post-Halley Era*, eds. R. L. Newburn, Jr. et al. Kluwer Academic Publishers, p. 149.)

weak and difficult to detect could be stronger compared to the $J = 1 \rightarrow 0$ transition.

The above discussion shows that the search for molecules in comets has given limited results. This does not mean that they are not present in comets, since the detection of a line depends upon various factors like column density of the species, instrumental limitations, excitation mechanisms etc. The timing of the observation is also relevant for the search of a molecule. But, most of the molecules that were not detected in Comet Halley, although looked for, are in overall agreement with the expected low abundances of the molecules and the excitation mechanism of these lines.

In addition to line radiation, continuum observations in the radio frequency region have been looked for in several comets. The continuum emission at 3.71 cm was detected for the first time in Comet Kohoutek. It has been looked for in several other comets. Here again, the success is only partial as it has been seen only in a few comets like West, IRAS-Araki-Alcock, Halley, Brorsen-Metcalf (1989o) and Okazaki-Levy-Rudenko(1989r).

The *in situ* mass spectrometer studies of Comet Halley has given lot of new information about the species present in the coma. They have given rise to the identification of a large number of new species, which is difficult to detect through traditional spectroscopic means.

Table 4.2. Some of the observed species in comets.

neutrals :	H, C, O, S, Na, K, Ca, V, Cr, Mn, Fe, Co, Ni, Cu
	C_2, CH, CN, CO, CS, OH, NH, S_2
	C_3, NH_2, NH_3, H_2O; HCN, HCO, H_2CO, CH_3CN, CH_3OH, H_2S
Ions :	C^+, CO^+, CH^+, CN^+, N_2^+, H_2O^+, CO_2^+, OH^+
From mass spectra	H_2O, $CO/N_2/C_2H_4$, CO_2, H_2CO, H^+, C^+, CH^+
	CH_2^+/N^+, CH_3^+/NH^+, $O^+/CH_4^+/NH_2^+$,
	$OH^+/NH+_3/CH_5^+$, H_2O^+/NH_4^+, H_3O^+, H^3S^+,
	$C_3H_3^+$, C_3H^+
Dust :	Continuum
	10 and 20 μm features (silicate),
	3.4 μm (C-H stretch feature)

(Adapted from Huebner, W. F., Boice, D. C., Schmidt, H. U. and Wegmann, R. 1991, in *Comets in the post-Halley Era*, eds. R. N. Newburn, Jr. et al. Kluwer Academic Publishers, p. 907.)

A compilation of the observed atomic and molecular species in comets is given in Table 4.2. It should, however, be pointed out that the cometary coma must contain many more molecules than those given in Table 4.2. The following interesting points may be noted from Table 4.2.

1. The molecules detected are composed of the most abundant elements in the universe, namely H,C, O and N.
2. Most of the species detected are organic, indicating the importance of carbon, similar to the case of interstellar molecules.
3. The presence of NH and NH_2 implies that NH_3 should be present. This is also expected from the point of view of chemistry. After many unsuccessful searches, it was identified in the Comet IRAS-Araki-Alcock and more recently in Comet Halley.
4. Methane (CH_4) is tentatively identified.
5. The presence of CO_2 in comets was inferred indirectly from the presence of CO_2^+, as it does not absorb photons in the visible or in the near ultraviolet down to less than 2000A. But the direct detection of CO_2 come from the infrared observations of Comet Halley.
6. One important element missing from the list is nitrogen. This is due to the fact that the resonance transition of nitrogen, which is at 1200 Å, is very close to the strong hydrogen Lyman α line. Hence, it is very hard to detect this line. But nitrogen can be inferred from the strong CN emission lines.
7. Most of the observed species are radicals or ions which are physically stable, but chemically highly reactive. This means that they cannot exist as such in the nucleus. So the general belief is that these radicals are the by-products of the break-up of some complex molecules. This has led to what is generally termed as 'parent-daughter' hypothesis for the observed species. Several possible parent molecules have been detected in Comet Halley.
8. The predominance of radicals implies low densities in the coma which in turn means that the collisions may not be very important. However, very close to the nucleus they have to be considered.
9. All the products of water photodissociation have been observed.
10. Grains mostly made up of C,H,N and O called 'CHON' particles are present in comets. These grains contain highly complex molecules.
11. The molecule S_2 is not a common molecule present in Comets.

4.2. Forbidden Transitions

So far we have been discussing the spectra arising out of the allowed dipole transitions which have a mean lifetime of $\sim 10^{-8}$ sec.

Many forbidden lines arising out of several atoms have been seen from various types of astronomical objects, including comets. The mean lifetime for magnetic dipole and electric quadrupole transitions are $\sim 10^{-3}$ and 1 sec respectively. The well-known auroral red lines of 6300.23Å and 6363.87Å, the green line of 5577.35Å of neutral oxygen atom have been seen in many comets. The red doublet of [OI] was identified for the first time in Comet Mrkos. In the beginning there was some confusion as to whether these lines were due to the Earth's atmosphere or were intrinsic to the Comet. This was finally resolved based on high dispersion spectra. The red lines observed in the spectra of a comet should be Doppler shifted in a manner similar to other cometary molecular emissions, if they are intrinsic to the comet, while that of atmospheric origin should not be shifted. Based on such arguments it was conclusively shown that [OI] lines are of cometary origin. The red lines originate from the ^1D upper level while the green lines arise from the ^1S level as can be seen from the energy level diagram of neutral oxygen atom (Fig. 4.9). The intensity of the red doublet is quite strong as can be seen from Fig. 4.1. The green line is weak compared to the red line and is affected more by the atmospheric component. So an accurate sky-subtraction becomes harder. In addition, the green line is badly blended with the cometary C_2 (1,2) Swan band. Even so, the green line has been detected. The red doublet lines are also blended slightly with the emission from NH_2. The spectra of Comet Halley taken at very high resolution of 0.17Å has resolved the various NH_2 lines and of the forbidden oxygen lines. Since red doublet lines are strong in comets, they have been observed regularly and studied extensively in comets. The other line at $\lambda = 2972$Å has also been detected in Comet Bradfield.

All the conventional methods used to obtain the spectra of comets cannot give very high resolution. But it is well-known that high resolution spectra contain enormous information with regard to the velocity structure, the mechanism of excitation, etc. For securing such high resolution observations, Fabry-Perot instruments can be used. This can be used profitably as comets have velocities of a few km/sec relative to the Earth, which make the cometary lines to be Doppler shifted with respect to the lines originating from the Earth's atmosphere. Such observations have been carried out starting with Comet Kohoutek in 1974 for the H_α line of 6565 Å and 6300 Å

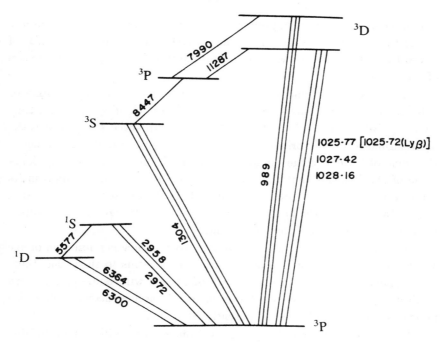

Fig. 4.9. Energy level diagram of oxygen atom showing various transitions.

line of [OI]. The Fourier Transform Spectrometer used for the detection of H_2O in Comet Halley had a spectral resolution of $\lambda/\delta\lambda = 10^5$.

4.3. Line-to-Continuum Ratio

Based on the spectra in the visual spectral region, one can classify the observed spectra of comets into two categories: (1) strong continuum and (2) strong molecular line emissions. Although at present a large number of observations exist on comets taken since early times they are however not homogeneous. This is due to the fact that the spectra have been taken with different instruments as well as at different heliocentric distances. Also all the available spectra are not of a very good quality. This is particularly so for earlier spectral observations. Therefore, it is hard to make any real meaningful ratio of continuum to line emission in comets. Nonetheless an attempt was made to use the available observations in order to see any trend in the line to continuum ratios. The study was based on about 85 comets covering the period between 1865 to 1975. All the observation were divided into two broad groups, corresponding to heliocentric distances of less than or greater than 1 au. Qualitatively the correlation which is of interest refers to whether the line to continuum ratio changes with the age of the comet.

Age of a comet is generally measured by the reciprocal of the semimajor axis, i.e. a^{-1}. New comets coming from the Oort cloud for the first time have $a > 10^4$ au or $(1/a) < 100 \times 10^{-6} au^{-1}$. With successive passages the orbit shrinks gradually primarily due to planetary perturbations and hence the value of $(1/a)$ becomes larger and larger. Therefore statistically, large value of $(1/a)$ means that the comet has gone through many times in its orbit. Consequently, the increasing value of $(1/a)$ corresponds to increasing time with respect to the first approach to the solar system. In order words, $(1/a)$ gives a measure of the comet's age.

The analysis of the ratio of continuum to emission as a function of age showed very similar distribution patterns. This seems to show that there is no qualitative change in the spectra of comets with age. This in turn implies that the composition of the nucleus is homogeneous, which also indicates the nucleus to be undifferentiated.

It may be noted that the spectra of comets is entering a new era due to the availability of sensitive instruments and detectors which can provide high spatial, spectral and temporal observations. It is also possible to get 2-dimensional images. In addition, the Earth's atmosphere is no more a hindrance for making observations in various spectral regions due to the use of rockets and satellites, Kuiper Airborne Observatory and Space Telescope.

As can be seen from the general discussion of cometary spectra presented so far, comets are very rich in emissions arising out of various kinds of molecules. The next logical step is to extract the physical conditions of the gaseous material present in the coma from a study of these lines, which will be discussed in the following chapter.

Problems

1. Describe the spectra of a Comet at far of distance from the Sun and very close to the Sun. Why is there a difference?
2. Cometary spectra show only the emission lines. Explain.
3. Suppose one were to observe an absorption line in the spectra and the optical depth required is 0.1. Calculate the column density of sodium atoms for the D1 line.

4. What are the various criteria that may be used for a firm identification of an unknown line in a cometary spectrum.
5. What is meant by forbidden lines? Describe the conditions under which forbidden lines occur. Mention some specific examples and discuss what we can learn from them.

References

A good account of the spectra of comets can be found in the following:
1. Arpigny, C. 1965, *Ann. Rev. Astr. Ap.* **3**, 351.
2. Swings, P. and Haser. L., 1956, *Atlas of Representative Cometary Spectra* Liege: The Institute of Astrophysique.
3. Arpigny, C., Rahe. J., Donn. B., Dossin, B. and Wyckoff, S. 1995 *Atlas of Cometary Spectra* (in press)

A discussion of the observations in the Ultraviolet (4,5), Visible (6), Infrared (7,8) and Radio (9) region can be found in
4. Feldman, P.D. 1982. In *Comets*, ed. L.L. Wilkening, Tucson: University of Arizona Press. p. 461.
5. Feldman, P.D. 1989. In *World wide investigations, Results and Interpretation*, ed. Ellis Horwoodlla, Chichester
6. A'Hearn, M.F. 1982. In *Comets*, ed. L.L. Wilkening, Tucson: University of Arizona Press. p. 433.
7. Hanner, M.S. and Tokunaga, A.T. 1991. In *Comets in the Post Halley Era*. eds. R.C. Newburn, M. Neugebauer and J. Rahe. Kluwer Academic Publishers p. 93.
8. Weaver, H.A., Mumma, J. and Larson. H.P. 1991. In *Comets in the Post Halley Era*. eds. R.C. Newburn, M. Neugebauer and J. Rahe, Kluwer Academic Publishers p. 93.
9. Crovisier, J. and Schloerb, F.P. 1991. In *Comets in the Post Halley Era*. eds. R.C. Newburn, M. Neugebauer and J. Rahe. Kluwer Academic Publishers, p. 149.

The first observational results for Comet Kohoutek using Fabry-Perot instrument are given in
10. Huppler, D., Reynolds, R.J., Roesler, F.L., Scherb, F. and Tranger, J. 1975. *Ap.J.* **202**, 276.

The material of Sec. 4.3 is contained in
11. Donn, B. 1977, In *Comets, Asteroids and Meteorities: Interrelations, Evolution and Origins*, ed. A. H. Delsemme. Toledo: University of Toledo Press p. 15.

CHAPTER 5

SPECTRA OF COMA

In the last chapter, we discussed the general characteristics of the spectra of comets. As was pointed out, the analysis of these spectral lines can give information about the temperature, pressure and the physical processes responsible for producing the lines, in addition to the abundances of the species. Here we would like to discuss in some detail a few of these aspects, with special reference to the cometary spectra.

5.1. Fluorescence Process

In order to analyse cometary spectra it is necessary to know the mechanism responsible for the excitation of cometary emissions. The absorption of solar radiation in their resonance transitions which then trickle down to give the radiation is called the *resonance fluorescence process*. Whether the mechanism of excitation is due to the collision process with atoms, molecules or ions, or due to the resonance fluorescence process is governed by the relative time scales for the two processes. The collision time scale is given by

$$\tau_{coll} \approx \frac{1}{n\sigma v} \tag{5.1}$$

where n, σ and v denote the number density, the collision crosssection and the velocity of the species respectively. Using for the collision cross-section with neutral atoms or ions, a typical value $\sigma \sim 10^{-16}$ - 10^{-17} cm^2 and $v \approx 1$ km/sec, $n \approx 10^5/cm^3$, the characteristic collision time scale turns out to be

$$\tau_{coll} \approx 10^7 sec. \tag{5.2}$$

This time scale is much larger than the typical time scale for absorption of solar radiation in the visible region of say C_2 and CN at a distance of 1 au, which is about 10 to 100 sec. Therefore for typical cometary conditions the absorption of solar radiation is the main excitation process. Another observation which also supports the above conclusion is that the line emissions seen in a cometary spectra arise mostly from the ground electronic state of the molecule and this involves a lot of energy. It can therefore be brought about only by the absorption of the solar radiation. However, the striking success for the process came from the work of Swings. He noticed that the relative intensity of rotational lines of the CN (0,0) band seemed to have peculiar minimum and maximum intensities at certain locations. The pattern also changed with the Sun-Comet distance, as can be seen from Fig. 5.1, which shows a marked difference between the two spectra taken about 10 days apart. Comparing the position of the observed lines with the solar spectrum, it was found that the position of minimum intensity corresponded to the regions of less flux in the solar radiation field as compared to that of maximum intensity positions. This in turn was related to the presence or absence of absorption lines called *Fraunhofer lines*. In other words, there existed a definite correlation between the absorbed solar radiation by the molecule and the corresponding emission intensity. Since the comet has a variable velocity around its orbit, the frequency of the radiation that the molecule in the comet absorbs depends upon this velocity due to the Doppler shift effect. Therefore the observed intensity pattern should change depending upon whether the Fraunhofer lines come in the way of absorption or not. These are also consistent with the observed intensity patterns. This effect is generally called the 'Swings effect'.

Let us further examine the validity of the resonance fluorescence excitation process from a detailed analysis of the rotational and vibrational spectra of molecules. The calculation of intensities of lines or a synthetic spectrum requires the knowledge of the population distribution in different energy levels of the molecule. This is a complicated problem as it is necessary to simultaneously consider the different electronic, vibrational and rotational levels of the molecule in the level population calculation. The nature of the energy levels depends upon the type and structure of the molecule. Just as an example to show the complexity of the energy level diagram of the molecules, Fig. 5.2 shows the various energy levels and the transitions that have to be considered for the OH molecule. Therefore it is rather difficult to give a general set of equations for the calculation of the

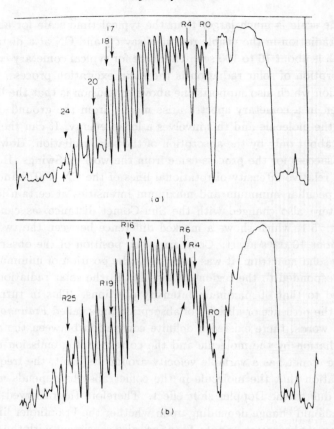

Fig. 5.1. The microdensitometer tracings in the spectral region of CN violet band of Comet Ikeya 1963 1.(a) and (b) refer to the spectra taken on the 3rd and 13th of March 1963 respectively. The variations in the band structure between the two spectra can be seen. (Taken from Whipple, F. L. 1978, in *Cosmic Dust*. ed. J. A. M. McDonnell, New York; John Wiley and Sons, p. 1.)

population distribution of molecules. Instead, one has to consider each molecule on a more or less individual basis depending on its energy level structure. However, there are some simple diatomic molecules for which the energy levels are such that it is a good approximation to consider vibrational and rotational transitions separately. Therefore it is appropriate to outline the method for the cases of vibrational and rotational levels treated separately. The formalism can easily be adapted to take into account the complexities of the molecule of interest. Earlier studies indicated that the

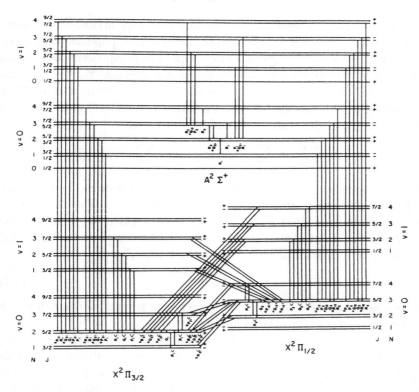

Fig. 5.2. The energy level diagram of OH molecule showing electronic, vibrational and rotational structure and the various transitions between them (Schleicher, D. G. and A'Hearn, M. F., 1988, *Ap. J.* **331**, 1058).

C_2 molecule appears to deviate from the general behaviour of all other molecules. Therefore, the case of C_2 molecule will be considered separately.

5.1.1. *Rotational structure*

Consider the case of rotational structure for a given vibrational band of a diatomic molecule. The intensity of a line depends, among other things, on the fraction of the species in different rotational levels. The Boltzmann's law for the distribution of population in different rotational levels for an assumed temperature is not a good approximation due to the fact that the presence of Fraunhofer lines in the solar spectrum tends to make the population distribution in various rotational levels irregular. In addition, the collisions which tend to produce the Boltzmann distribution, are rather infrequent due to low densities present in a cometary atmosphere. Therefore, it becomes necessary to determine the resulting population distribution from the solution of the statistical equilibrium equations which take into account the absorption and the emission processes. For the pure radiative case, the number of molecules in an upper rotational state N_j is determined by the balance between the number of molecules entering and leaving this particular state. This can be written as

$$\frac{dN_j}{dt} = \sum_{P,Q,R} N_i B_{ij} \rho_{ij} - N_j \sum_{P,Q,R} A_{ji} \tag{5.3}$$

where N_i is the number of molecules in the lower rotational state i, such that $i < j$. The summation is over P, Q and R branches. The $A's$ and $B's$ are the Einstein coefficients and ρ is the radiation density corresponding to the wavelength of the transition. ρ has to be corrected for the radial velocity effect (Chap.3) of the comet through the relation

$$\lambda_{exc} = \lambda_{lab} \left(1 - \frac{v}{c}\right) \tag{5.4}$$

where v is the radial velocity of the comet with respect to the Sun which can be calculated as the orbit is known.

The detailed energy distribution of the solar radiation field with the Fraunhofer lines present has also to be used in the calculation of ρ. This requires a very high dispersion scan of the solar spectrum in the region of interest. Sometimes for simplicity and to a first approximation, 'blocking coefficients' (γ_λ) are used, which give a measure of the degree of absorption in a given range of wavelengths. Therefore multiplying the mean radiation field with the blocking coefficients, takes into account the effect of Fraunhofer lines in an approximate manner. However, the detailed energy distribution should be used as far as possible.

Since the physical conditions do not change drastically within the time of observation, one can use $(dN_j/dt) = 0$ which implies a steady state condition. Therefore, Eq. (5.3) reduces to

$$\sum_{P,Q,R} N_i B_{ij} \rho_{ij} = N_j \sum_{P,Q,R} A_{ji}. \tag{5.5}$$

A similar expression can be written for the transitions in the lower rotational state as

$$\sum_{P,Q,R} N_j A_{ji} + N_{i+1} A_{i+1}^{rot} = N_i \left[\sum_{P,Q,R} B_{ij} \rho_{ij} + A_i^{rot}\right] \tag{5.5a}$$

where A^{rot} represents the transition rate for pure rotational transitions in the lower electronic state, which should be included for a heteronuclear molecule. The total number of equations involved depends upon the number of rotational levels in the upper and lower states. A simultaneous solution of these equations with the added condition that the sum total of populations in all the levels is equal to unity, i.e.

$$\sum_{t=1}^{i+j} N_t = 1 \tag{5.6}$$

gives the relative populations in various rotational levels. For illustrative purposes, Fig. 5.3 shows the relative populations of the rotational levels of (0,0) band of $B^2\Sigma^+ - X^2\Sigma^+$ transition of the CN molecule for various heliocentric distances. It shows clearly that with decrease in distance lot more rotational levels are populated.

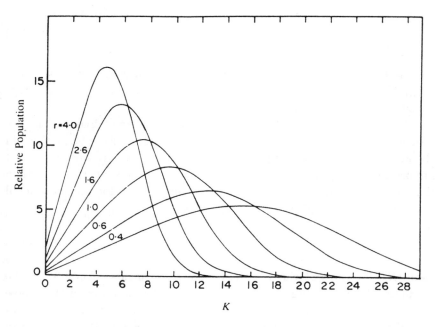

Fig. 5.3. The figure gives the relative population distribution of rotational levels for the ground state of CN in comets for different heliocentric distances. (Arpigny, C. 1964, *Ann. d'Ap.* **27**, 393).

5.1.2. *Vibrational structure*

Let us now consider the vibrational bands of diatomic molecules. The relative populations in different vibrational levels have to be obtained from the solution of the statistical equilibrium equations written for each vibrational level. Since many electronic transitions can arise from the same ground state, it is necessary to take into account all these electronic transitions in addition to the vibrational transitions. The statistical equilibrium of any vibrational level is determined as before, by a balance between the transitions into and out of that particular level. The equations for the lower and upper electronic states i and j respectively for a homonuclear molecule can be written as follows. For the vibrational levels of the lower electronic state

$$N_i \sum_{k=m}^{p} B_{ik}\rho_{ik} = \sum_{k=m}^{p} N_k(B_{ki}\rho_{ki} + A_{ki}), \ i = 1, 2, \cdots (m-1) \quad (5.7)$$

and for the vibrational levels of the upper electronic state

$$N_j \sum_{l=1}^{m-1} (B_{jl}\rho_{jl} + A_{jl}) = \sum_{l=1}^{m-1} N_l B_{lj}\rho_{lj}, \ j = m, p. \quad (5.8)$$

Here N is the fraction of molecules in different vibrational levels, such that $\sum_{t=1}^{p} N_t = 1$. The $A's$ and $B's$ are the Einstein coefficients, ρ is the energy density of the solar radiation.

For heteronuclear molecules, the vibrational transitions within the ground state of the molecule are permitted. This term has to be included in the above equations. For upper electronic states, these transitions are not important as the probability for electronic transitions is much greater than that of vibrational transitions. The Einstein A value for the vibrational transitions $v \rightarrow v'$ is given by the expression

$$A_{vv'}^{vib} = \frac{64\pi^4 \nu^3}{3hc^3}|R_{vv'}|^2. \quad (5.9)$$

Here $|R_{vv'}|^2$ represent the vibrational matrix elements. These could be calculated from quantum mechanics provided the variation of the dipole moment function is known. However for most of the molecules the variation of the dipole moment function is not known. For such cases, one can approximate $|R_{vv'}|^2$ as

$$|R_{vv'}|^2 = \mu_1^2|R'_{vv'}|^2 \quad (5.10)$$

where μ_1 is the coefficient of the linear term in the dipole moment expansion and

$$R'_{v+p,v} = \left[\frac{r_e^2 B}{\omega_e} x_e^{p-1} \frac{(v+p)!}{p^2 v!}\right]^{1/2} \quad (5.11)$$

where $p = v' - v$ and the other parameters are the usual spectroscopic constants of the molecule. This expression is a good approximation if $(v + v')x_e \ll 1$. This condition is generally satisfied for most of the molecules of astrophysical interest.

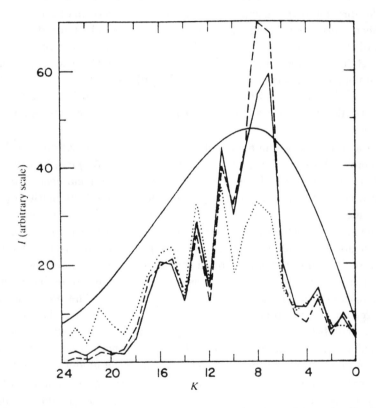

Fig. 5.4. Comparison of the expected and the observed (solid curve) intensity distribution in the R-branch of the violet (0,0) band of CN system for Comet Mrkos. The continuous curve refers to results based on Boltzmann distribution for 500°K. The dotted and dashed curves refer to results based on statistical equilibrium calculations without and with Fraunhofer lines taken into account, (Arpigny, C. 1964, *loc. cit*).

5.1.3. Comparison with observations

The intensities or the synthetic profile of the bands can easily be calculated from the knowledge of the population distribution in different rotational or vibrational levels. The calculated profiles can then be convolved with the Gaussian profile of instrumental resolution corresponding to the observations. The results of such calculations for the rotational profile of the (0,0) band of $B^2\Sigma^+ - X^2\Sigma^+$ transition of the CN molecule for various cases is compared with the observed profile for Comet Mrkos in Fig. 5.4. They clearly indicate that the observed profile agrees well with the expected profile when the effects of Fraunhofer lines are taken into account.

A more detailed and sophisticated synthetic spectrum of the same band has been calculated which takes into account several hundred rotational levels of the CN molecule distributed between the three electronic states X, A and B. The calculated profile agrees very well with the observed profile of several comets. Such comparisons for the (0,0) band or bands arising from $A^1\Pi - X^1\Sigma^+$ transitions for several other molecules such as CO, CS and OH have also been carried out for various comets. They also show good agreement.

In a similar way, the expected intensities of various vibrational bands of the molecules CO, CO^+, CS, CN, CN^+ etc (Table 5.1) can be calculated based on the resonance fluorescence process. As a typical case, Table 5.2 shows a comparison of the expected and the observed intensities of lines for the molecules CO^+ and CS. The results for several other molecules also show a similar agreement. The extension of cometary spectra into the ultraviolet region has made it possible to look for some of the transitions which lie in the ultraviolet region, although the model based expected fluorescence intensities may be weak. In some cases there could be some disagreement between the calculated and the observed intensities. The reason for such a discrepancy could arise partly due to the fact that there is a disagreement in the observations itself. In addition, there could be uncertainties in the oscillator strengths, variation of electronic transition moment with the r-centroid and other parameter which enter into the model calculation.

For those molecules for which the oscillator strengths are not available, it may be possible to reverse the problem and get an estimate for the oscillator strength of the transitions from a comparison of the expected and the observed intensities of lines.

From the discussion so far, it is clear that the resonance fluorescence process is very successful in explaining the observed band spectra of various

Table 5.1. Transitions included in the population calculation of some molecules.

Molecule	Transitions	$\lambda(0,0)$ Å
CN	$A^2\Pi - X^2\Sigma^+$	10925
	$B^2\Sigma^+ - X^2\Sigma^+$	3883
CN$^+$	$C^1\Sigma - a^1\Sigma$	3185
	$f^1\Sigma^+ - a^1\Sigma^+$	2180
CO	$A^1\Pi - X^1\Sigma^+$	1544
	$B^1\Sigma^+ - X^1\Sigma^+$	1150
CO$^+$	$A^2\Pi - X^2\Sigma^+$	4910
	$B^2\Sigma^+ - X^2\Sigma^+$	2189
	$B^2\Sigma^+ - A^2\Pi$	3973
CS	$A^1\Pi - X^1\Sigma^+$	2576
CH	$A^2\Delta - X^2\Pi$	4314
	$B^2\Sigma^- - X^2\Pi$	3889
NH	$A^3\Pi - X^3\Sigma$	3360
OH	$A^2\Sigma^+ - X^2\Pi$	3090

Table 5.2. (A-X) bands of CO$^+$ and CS.

Band (CO$^+$)	Computed*	Observed[a]	Band (CS)	Computed+	Observed
(0,0)	0.11	0.12	(0,0)	1.0	1.0[b] 1.0[c]
(1,0)	0.63	0.60	(1,0)	0.03	0.05
(2,0)	1.00	1.00	(0,1)	0.13	0.12 0.11
(3,0)	1.20	1.16	(1,1)	0.07	0.08
(4,0)	0.67	0.64	(0,2)	0.01	0.01
(5,0)	0.28	0.40	(1,2)	0.03	0.012

[a] Comet Humason (1961e) (Arpigny, C. 1964. Ann. d'Ap. **27**, 406).
[b] Comet West (Smith, A.M., Stecher, T.P. and Casswell, L. 1980. Ap. J. **242**, 402).
[c] Comet Bradfield (Feldman, P.D. et al. 1980, Nature **286**, 132).
* (Krishna Swamy, K.S. 1979, Ap.J. **227**, 1082; Magnani, L. and A'Hearn, M.F. 1986, Ap. J. **302**, 477).
+ (Krishna Swamy, K.S. 1981. Astr. Ap. **97**, 110; Sanzovo, G.C., Singh, P.D. and Huebner, W.F. 1993. AJ. **106**, 1237).

molecules in comets. So far, only the effect of the radial component of the velocity of the comet in its orbit leading to Doppler shift of lines has been considered. This is the well known Swings-effect. In addition, it is necessary to consider the velocity of the gas in the coma. As the gas comes out of the nucleus, it expands outwards in all directions. Therefore, the resultant Doppler shift of the solar radiation depends upon these two components. The resultant velocity will be different at different locations in the comet and hence the relative intensities of the same feature should vary with position in the coma. Such an effect was first noticed by Greenstein in the Comet Mrkos. This is generally called Greenstein effect. This can be seen clearly from Fig. 5.5 where the relative intensities of CN(0,0) lines at three different locations in the comet are given. To take care of the effect of internal motions in the solution of statistical equilibrium equations, it is necessary to know the distribution of velocities of the species as a function of the distance from the nucleus. This has to be derived on the basis of a simple physical model. One such model could be that the gaseous material is flowing out of the nucleus with a uniform velocity. The model could be made more complicated by including the resistance force due to the solar wind and so on. In addition to the orderly motion discussed so far, there may also be motions present purely of random character. It is even harder to consider this type of velocity in the model calculation. Therefore the subject of the various kinds of gaseous motions in comets is quite complicated and it is hard to disentangle one from the other. In principle, if one has a very high dispersion spectrum, it is possible to construct a detailed model including the effect of all the velocities, which can reproduce the observations well.

There are also other complications that have to be taken into account. In the above treatment it is assumed that the coma is optically thin. For reasonable value for the absorption cross section, $\sigma \approx 10^{-17} - 10^{-18} \ cm^2$ for molecules, the optical depth effect ($\tau = N\sigma \gtrsim 1$) has to be considered for the molecules with a large column density N, which imply large production rates of the molecules. Since for most of the cometary species the optical depth $\tau \ll 1$, the optically thin assumption is a very good approximation for most of the lines. But for some resonance lines such as Lyman α line of hydrogen at 1216A and the rotational lines of the H_2O molecule, the optical depth could be significant and therefore could affect the molecular excitation through radiative trapping. Another complication arises for molecules with short lifetimes. It is possible that such molecules may decay even before the fluorescence equilibrium is reached, which require several

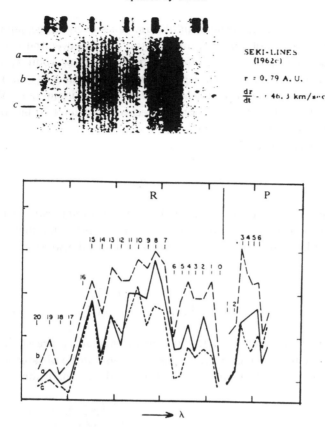

Fig. 5.5. Top section shows the rotational lines of (0.0) band of the violet CN system in Comet Seki-Lines. The lower portion represents the tracings of the spectra in the regions a, b and c. (Arpigny, C. 1965. *Mem. Acad. Roy. Belg. Cl 8°*, Vol. 35, No. 5).

cycles of excitation and de-excitation. These have to be treated based on time dependent calculations.

In addition to the radiative process collisional excitation and de-excitation could also be important. Since the density to a first approximation varies with distance from the nucleus as r^{-2}, the collisions can be an important excitation process close to the nucleus, $\sim$ a few thousand kilometers for molecules with production rates $\sim 10^{29}/sec$. In the inner coma, the molecules will be more in thermodynamic equilibrium rather than fluorescence equilibirum. As one moves outwards there will be a region where both thermodynamic and fluorescence equilibria will be operating. Finally at large distances, fluorescence equilibrium takes over. Since the observed intensities or profile of emission lines are an average over the line of sight in the coma, it is necessary to integrate over the whole volume, including both collision and radiative processes. For collision effects, one generally considers the collision of the molecule with H_2O as it is the most abundant molecule in the coma. However, other species might have to be considered if they turn out to be important.

The collision de-excitation rate is given by

$$C_{coll} = n_{H_2O} \sigma_{coll} \overline{v} \qquad (5.12)$$

where σ_{coll} is the collision cross section for rotational transitions of the molecules produced by collisions with H_2O, n_{H_2O} is the number density of H_2O molecules and $\overline{v}$ is the mean relative velocity between H_2O and the molecule and is given by

$$\overline{v} = \frac{8kT}{\pi} \left[\frac{1}{m_{H_2O}} + \frac{1}{m_{mol}} \right]. \qquad (5.13)$$

Here T is the temperature of the gas, m_{H_2O} and m_{mol} are the masses of the two molecules. The excitation cross sections can be obtained from the de-excitation crosssections from the detailed balance condition

$$\sigma(l \to u) = \sigma(u \to l) \frac{g_u}{g_l} \exp\left[-\frac{E_u - E_l}{kT_{kin}}\right] \qquad (5.14)$$

where u and l refer to upper and lower levels and T_{kin} the temperature that characterises the colliding particles.

The modelling of the collision process in the coma is rather uncertain due to uncertainties in the collision cross sections between the molecular species and the H_2O molecule and the temperature of the coma gas. However the best available data is generally made use of in such calculations. The variation of the H_2O density with distance from the nucleus is also required and is generally calculated based on Haser's model (Sec. 6.1.2). The time evolution of the profiles, until it attains the equilibrium state, can be investigated based on the time dependent population distribution. As an example, the results for the case of A-X(0,0) band of CS is shown in Fig. 5.6 for different initial population distributions. Such calculations are important when comparing with observations made closer to the nucleus.

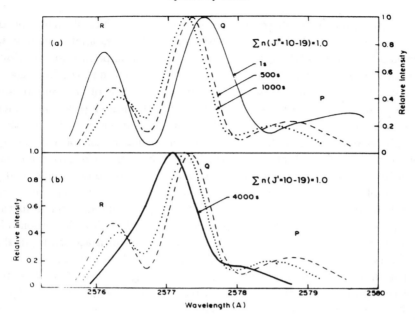

Fig. 5.6. Time evolution of the synthetic profile of P, Q and R branches of the (0,0) band of CS molecule for the case when the initial population ($t = 0$) is distributed over the rotational levels in the lower level $J''=10$ to 19 (a) or upper levels $J'=10$ to 19 (b) at 4000 km from the nucleus. The time for each curve is marked. The curves are for a spectral resolution of 0.8A corresponding to Halley's data (r=1.1 au, Q (H_2O)=$3.0\times10^{29}s^{-1}$, T=300K; Krishna Swamy, K. S. and Tarafdar, S. P. 1993. *Astr. Ap.* **271**, 326).

5.1.4. Case of C_2 molecule

One exception to the general behaviour of the resonance fluorescence excitation process appeared to be the case of the C_2 molecule. This was noticed as early as 1953 and showed that the calculated relative band strengths did not agree with the observed relative strengths of the Swan system. There was a discrepancy of nearly a factor of two between the expected and the observed intensities. Early investigators had expressed this discrepancy in terms of the vibrational temperature of the C_2 molecule. Therefore it is also called 'the problem of low vibrational temperature of C_2'. The observed difference was also found to be dependent upon the heliocentric distance of the comet.

The excitation mechanism which populates the upper energy states also determine the vibrational level population in the lowest electronic state.

The initial population distribution produced at the time of formation of the molecule is masked by the excitation and decays, finally leading to the equilibrium population distribution. The resulting population distribution in the bottom triplet state should therefore be close to the Boltzmann distribution corresponding to the colour temperature of the Sun in the Swan band region. However the observed distribution gives a lower value $\sim 3500K$. This discrepancy was rather disturbing in view of the fact that the C_2 Swan bands are the strongest in the visual region of a cometary spectrum and this had questioned the basic fluorescence excitation mechanism itself. The problem of low vibrational temperature of C_2 has finally been resolved based on a new physical effect which seems to operate very effectively in the case of the C_2 molecule as can be understood from the energy level diagram of the C_2 molecule shown in Fig. 5.7. The molecule C_2 has singlet and triplet states. The Swan bands which are strong in the cometary spectra arise from the triplet state, which lies about 714 cm^{-1} higher than the ground singlet state. This had given rise to another interesting problem which also eluded a reasonable explanation. i.e. why is it that Swan bands are strong in a cometary spectrum even though they do not arise from the ground state of the molecule, contrary to the normal situation. It was therefore not clear whether this reflected the preferential formation of the triplet state at the time of formation of the C_2 molecule or the fact that only the weak lines of the singlet series are expected in the spectra. In Fig. 5.7 the vibrational levels of the electronic states $a^3\Pi_u$ and $b^3\Sigma_g^-$ of the triplet states are also shown. It shows that the vibrational levels of the ground triplet electronic state, for $v'' > 4$, lies at a higher energy level compared to those of the b $^3\Sigma_g^-$ state. Therefore the transitions from higher vibrational levels of the a $^3\Pi_u$ electronic state can decay to lower lying vibrational levels through the vibrational levels of the b $^3\Sigma_g^-$ electronic state. The result of such transitions is to depopulate the higher vibrational levels of the a $^3\Pi_u$ state giving rise to more concentration in the lower vibrational levels which implies a non-Boltzmann type of population distribution (Fig. 5.8). This effect is in a sense equivalent in principle to vibrational transitions in the ground state of the molecule. Therefore even though C_2 is a homonuclear molecule wherein transitions between the lowest lying vibrational levels are forbidden, there is a physical mechanism by which the population from higher levels cascade radiatively to lower vibrational levels in the ground electronic state of the molecule. This increases the concentration in the lower levels thereby simulating a lower excitation

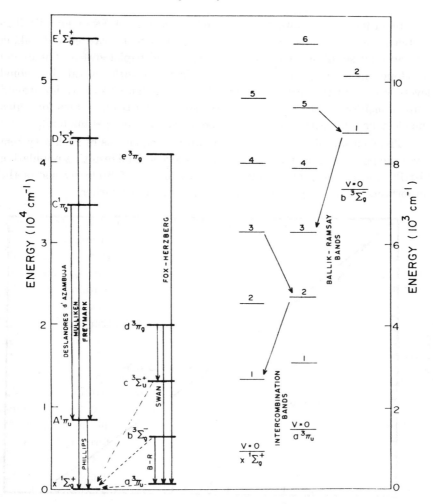

Fig. 5.7. Energy level diagram for the C_2 molecule. The electronic structure is shown on the left-hand side. The vibrational levels of three lowest electronic states are shown on the right-hand side. B-R indicates the Ballik-Ramsay bands.

temperature. Since the rate of spontaneous downward transitions would be fixed, this process would become relatively more important in the lower radiation field at greater heliocentric distances, so that the vibrational temperature would be expected to drop with increasing heliocentric distance.

In order to calculate the expected intensities of Swan and Phillips systems, the modelling of the C_2 molecule has to be carried out taking into account simultaneously both the singlet and triplet states. The model developed includes a total of ten electronic states with several vibrational levels in each of the electronic state. The forbidden singlet-triplet ground state transitions is the link between the singlet and triplet states for which the electronic transition moment is taken as a variable parameter.

The solution of the statistical equilibrium equations of such a system gives the population distribution in various levels. The resulting population distribution in the ground state as shown in Fig. 5.8 clearly shows the importance of the depopulation mechanism discussed earlier.

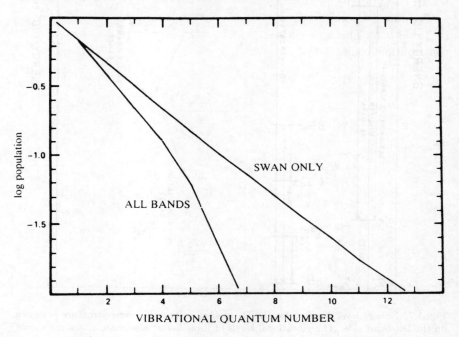

Fig. 5.8. The expected vibrational population distribution in the $a^3\Pi_u$ state is shown for the cases when Fox-Herzberg, Ballik-Ramsay and Swan bands (referred as All bands) or only Swan bands are included in the calculations. (Krishna Swamy, K. S. and O'Dell, C. R. 1977. *Ap. J.*, **216**, 158). The deviation occurs due to new depopulation mechanism discussed in the text.

The expected band sequence flux ratios can be calculated from a knowledge of the steady state relative populations in different vibrational levels. The total energy emitted in a band sequence for a given value of $\Delta v = (v' - v'')$ can be written as

$$F(\Delta v) \approx \sum_{V''} n_{V''} \sum_{\Delta v} B_{V''v'} \rho_{V''v'} P_{v'v''} v_{v',v''}. \tag{5.15}$$

Here $n_{V''}$ is the level population and $P_{v'v''}$ is the probability that when an absorption takes place from the state v'' to v' state, it will come back to state v''. The above expression can be written as

$$F(\Delta v) \approx \sum_{V''} n_{V''} PR(V'', \Delta \nu)$$

where $PR(V'', \Delta \nu)$ denotes the production rate. The production rates can be calculated from a knowledge of the molecular parameters and are given in Table 5.3. This table shows that for the band sequence $\Delta v = 0$ the contribution comes mostly from the lowest states, while for $\Delta v = +1$,

Table 5.3. Production rates for Swan band sequences.

v''	$\frac{PR(v'', \Delta v)}{PR(v''=0, \Delta v=0)}$		
	$\Delta v = -1$	$\Delta v = 0$	$\Delta v = +1$
0	0.2745	1.000	0.0813
1	0.3167	0.6333	0.3484
2	0.3255	0.4153	0.5335
3	0.3109	0.2450	0.6509
4	0.2813	0.1351	0.7168
5	0.2438	0.0700	0.7449
6	0.2030	0.0368	0.7454
7	0.1643	0.0249	0.7328
8	0.1299	0.0265	0.7182
9	0.1004	0.0371	0.7108
10	0.0747	0.0523	0.7151
11	0.0531	0.0669	0.7180
12	0.0383	0.0673	0.6402
13	0.0277	0.0564	0.0517

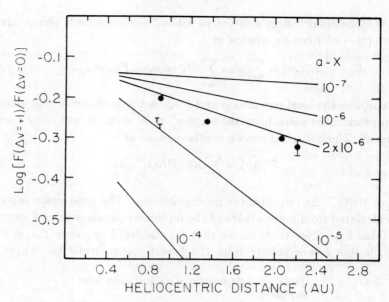

Fig. 5.9. A comparison of the expected and the observed Swan band flux ratios plotted as a function of the heliocentric distance. The continuous lines are the model based dependence shown for several values of transition moment for the lowest singlet triplet transitions. The filled circles show the observed dependence for Comet Halley (O'Dell, C. R., Robinson, R. R., Krishna Swamy, K. S. McCarthy, P. J. and Spinrad, H. 1988, Ap. J. **334**, 476).

higher states also contribute appreciably. Figure 5.9 shows a comparison of the expected $\Delta v = +1$ Swan band sequence flux ratios for several values of the electronic transition moment, $|Re|^2$ for the ground state singlet-triplet transitions, with observations of Comet Halley. The expected variation from the model for transition moment $\sim 2.5 \times 10^{-6}$ gives a good fit to the observed variation. The calculated intensities can also explain the observed flux ratios for the Mulliken system of the C_2 molecule. The results of these calculations also explain the observed result that the intensities of the Swan bands are stronger compared to those of the Phillips bands. These investigations also show that the forbidden singlet-triplet transitions are quite important. The wavelength of these transitions lies in the infrared spectral region. Unfortunately these transitions have not yet been analysed in the laboratory.

The molecule CO has a structure very similar to that of the C_2 molecule. However, the strong bands of CO seen in a cometary spectra arise out of singlet states in contrast to the case of the C_2 molecule. It is therefore of interest to know the physical reason for the inversion in intensities of the singlet and triplet series for the case of the C_2 molecule. This arises basically due to the difference in the energy level structures of the two molecules as can be seen below.

Consider the simple case of two levels with population n_1 and n_2 corresponding to the lowest singlet and triplet states of the molecule with energy separation ΔE. The resulting population in the upper and lower levels, n_2 and n_1 depend essentially on the rate of the absorption and emission processes and this can be written as

$$\frac{n_2}{n_1} \propto \left(\frac{B\rho}{A}\right). \qquad (5.17)$$

Here A and B are the Einstein coefficients and ρ is the energy density of the solar radiation. The ratio of the change in population for two values of ΔE is given by

$$\left[\frac{Triplet}{Singlet}\right]_{\Delta E_1} / \left[\frac{Triplet}{Singlet}\right]_{\Delta E_2} = \left(\frac{\lambda_1}{\lambda_2}\right)^5 \frac{F_\odot(\lambda_1)}{F_\odot(\lambda_2)}. \qquad (5.18)$$

Here $F_\odot(\lambda)$ represents the solar flux. The above equation shows a strong dependence on the wavelength. Therefore the relative intensities of singlet-singlet to triplet-triplet transitions of a molecule depend strongly on the energy separation between the lowest singlet and triplet states which are the 714cm^{-1} for C_2 and 49000 cm^{-1} for CO.

Let us now consider the rotational structure of C_2 molecule. The Swan bands of the C_2 system arise out of $3\pi_0$, $3\pi_1$ and $3\pi_2$ states. The relative population in different rotational levels of each of these substates for a given vibrational transition can be obtained from the solution of statistical equilibrium equations of lower and upper states. For vibrational population distribution, the values based on statistical equilibrium equations can be used. The synthetic profile resulting out of the superposition of many lines arising out of the various bands and corrected for the instrumental effect is compared with the observed profile in Fig. 5.10.

5.1.5. *Molecules other than diatomic*

The analysis of polyatomic molecules is more complicated than that of diatomic molecules. Therefore, they are not studied in any great detail. The

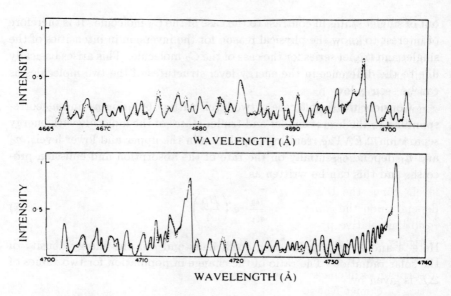

Fig. 5.10. Comparison of the calculated (continuous) and observed (dashed) spectra of $\Delta v = +1$ sequence of the C_2 Swan system for Comet West. (Observations are from Lambert, D. L. and Danks, A. C. 1983, *Ap. J.* **268** 428; spectral resolution (FWHM)=0.3Å).

work done on molecules like H_2O, NH_2, CO_2^+, NH_3, etc., seems to indicate that the resonance fluorescence process is the major excitation mechanism for these molecules as well. Most of the lines of polyatomic molecules lie in the infrared region. In fact, the volume integrated molecular emissions from the coma including the effect of collisions have been performed for several molecules like CO, CO_2, C_2H_2, OCS, HC_3N, C_2N_2 and so on to look for these features in the spectra of comets.

5.1.6. *OH radio lines*

The hydroxyl radical gives rise to lines in the radio region due to Λ-splitting of the levels (Table 3.1). The line at 1667 MHz which occurs at a wavelength of 18 cm was first detected in Comet Kohoutek in 1974. The most studied lines in comets are the 1665 and 1667 MHz lines. The intensity of the lines are found to be variable. Curiously, one finds that these lines have been seen in the emission as well as in the absorption which depend on the heliocentric distance of the comet. These observations can be explained by the fluorescence excitation process. Since the transitions between the Λ doublet levels are highly forbidden, in the absence of any other excitation mechanism, the levels should be populated according to their statistical weights at the coma temperature. The deviation from this equilibrium value is generally referred to as inversion or anti-inversion. The Cosmic microwave background radiation of 2.7K and the galactic sources can induce transitions. The line may be seen in absorption if the population is anti-inverted or in emission if the population is inverted due to stimulated emission. The Λ doublet levels are actually populated through the process of UV solar absorption lines from the ground state $^2\Pi_{3/2}$ ($v'' = 0, J = 3/2$) to the state $^2\Sigma^+$, v=0 and v=1 levels and then cascade back to the ground level. Since the UV absorption lines or the UV fluorescence efficiencies depend upon the comet heliocentric velocity (Swings effect, Fig.5.11), the resulting relative population distributions in the four levels also depends critically on the heliocentric velocity of the comet and hence on the heliocentric distance. This can result in the lines of OH to be produced either in absorption or in emission. Such an effect can clearly be seen from the observation of Comet Kohoutek which is shown in Fig. 5.12. The upper section

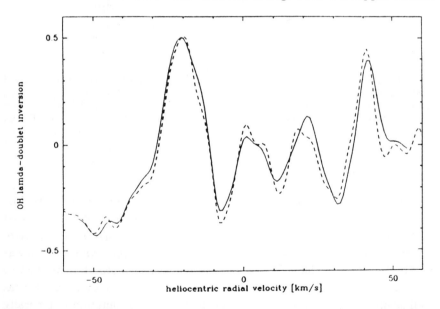

Fig. 5.11. The g-factor or the Fluorescence efficiency factor for the (0,0) band of OH is shown as a function of the heliocentric radial velocity. The variation in the value can clearly be seen arising due to Swings effect (Schleicher, D. G. and A'Hearn, M. F., 1988. *Ap.J.* **331**, 1058).

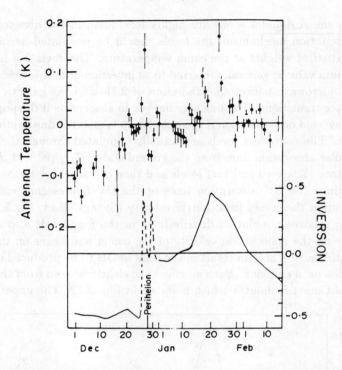

Fig. 5.12. The radio OH data for Comet Kohoutek. The top portion denotes the observed peak antenna temperature of the 1667 MHz line as a function of time. The lower portion denotes population inversion predicted by the ultraviolet pumping model as a function of time (Adapted from Biraud, F. Bourgois, G., Crovisier, J. Fillif, R., Gerard, E. and Kazes, I., 1974 *Astr. Ap.* **34**, 163).

of the curve shows the observed peak antenna temperature of 1667 MHz line as a function of time. The lower section shows the expected inversion based on the ultraviolet pumping model and is given by $[(n_u - n_l)/(n_u + n_l)]$. Here n_u and n_l represent the projected densities of the upper and lower levels of the Λ-doublet. There is a one-to-one relation between the two curves. This mechanism has been verified for a large number of radial velocities as well as observations based on several comets. This explanation is generally accepted and is used in interpreting the 18 cm observations. Therefore, the ultraviolet pumping of the OH radio lines is the dominant excitation mechanism. However, the collisions can also be quite effective as the Λ

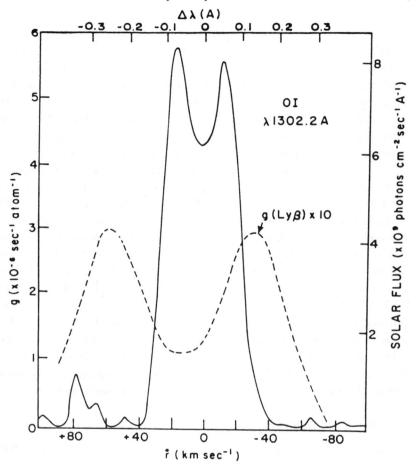

Fig. 5.13. High resolution profile 1302.2A line of OI in the Sun (Solid curve). The dashed curve represents the g-factor for fluorescence excited by solar Ly β as a function of heliocentric velocity (Feldman, P. D., Opel. C. B., Meier, R. R. and Nicolas, K. R. 1976. *The Study of Comets*, eds. B. Donn et al. NASA SP-393, Washington, D.C. p. 773).

doublet transitions are forbidden and the radiation field is weak. It is found that collisions with ions and electrons can be more important than even neutrals. This could have a severe effect in influencing the inversion in the inner coma of comets with high production rates. This is recognized as an important process in the calculation of radio emission from comets.

5.1.7. Oxygen lines

Allowed transitions. The triplet lines of oxygen at $\lambda=1304$ Å(1302.17 Å) are very strong in the UV region of the cometary spectra (Fig. 4.4). Of particular interest are the Comets Kohoutek and West in which these lines have been seen and these comets had heliocentric velocities in the range of about 44 to 55 km/sec. To see whether the excitation process of these lines is through resonance fluorescence process, a high dispersion spectrum of the solar line at 1302.2Å is shown in Fig. 5.13. The line width of ±40 km/sec corresponds to a Doppler shift of about ±0.17Å. For $\dot{r} > 40$ km/sec, the cometary absorption wavelength will be completely shifted from the solar line. Therefore, for comets with heliocentric velocity $\dot{r} \lesssim 30$ km/sec, the emission is produced by the resonance scattering of the solar lines. For higher velocities, either the line should be very weak which is not the case or there should be some other excitation mechanism. From the energy level diagram of the oxygen atom (Fig. 4.9) it can be seen that the line 1025.77 is very close to the solar Lyman β line of hydrogen of 1025.72. Therefore excitation is produced mainly through the solar Lyman β induced fluorescence. The intensity of OI 1304Å multiplet produced by resonance scattering of the solar lines is much stronger than those produced by the Lyman β fluorescence. Therefore the intensity of the emission line 1304Å depends critically on the radial velocity (Swings effect), the width of the solar emission line and the width of cometary absorption line.

5.1.8. Forbidden transitions

The identification of the red oxygen doublet lines at 6300 and 6364 Å in comets immediately raises the question of the excitation mechanism of these lines as they are electric dipole forbidden transitions. First of all, it is interesting to investigate whether the fluorescence process can or cannot explain these lines. The direct method to test this hypothesis is to compare the fluorescence efficiency rate for the red oxygen lines with those of ultraviolet resonance lines which is given in Table 5.4. The values show that the excitation to $^1S_0$ level (2972 Å) is negligible (Fig. 4.9). Also the expected intensities of the line 6300Å can be calculated based on the observed intensities of the 1304 line and the g-factors of Table 5.4. The expected intensities of red lines are very much smaller than those of the observed values. Therefore, the resonance fluorescence process cannot be responsible for the excitation of forbidden oxygen lines. The collisional

Table 5.4. Fluorescence efficiency factors at 1 au for oxygen.

Line (Å)	g-factor* (photons/sec/atom)
989	1.6 (−8)
1027	0.4 to 1.5 (−6)
1304	0.6 to 1.5 (−5)^a
	1.1 to 3.9 (−7)^b
2972	3.1 (−15)
5577	6.21(−14)
6300	4.2 (−10)
6364	1.3 (−10)

* See Chap. 6
Number in the bracket refers to power of 10 (a) for solar absorption; (b) Lyman β induced fluorescence; (Festou, M.C. and Feldman, P.D. 1981, *Astr. Ap.* **103**, 154).

excitation is also inadequate for explaining the observed emission.

Therefore it is generally assumed that the photo dissociation of some molecules leave oxygen atoms in the excited ^1D state which then decay to ^3P ground state giving rise to 6300A emission. ^1D state can be generated by the processes involving the absorption of a photon by the molecules H_2O, OH, CO_2 and CO (Sec. 6.2.6). However, the contributions from the photodissociation reactions of CO_2 and CO are much smaller than that of H_2O.

5.1.9. *Molecular band polarization*

The resonance fluorescence excitation process which is responsible for the observed emission lines in comets leads to two other important effects. The first being that even though the incident radiation is a natural one, the fluorescence lines should be linearly polarized. Secondly the total intensity of emission is not isotropic but depends upon the phase angle.

The linear polarization of emission bands in comets was first observed in Comet Cunningham 1941 I from the study of C_2 and CN bands. This was followed by observations of other comets. More recently, the polarization of C_2 (5140), C_3 (4060), CN (3871) and OH (3090) bands have been measured in Comets Halley and Hartley - Good 1985 XVII. The polarization of the bands of CO^+ (4260) and H_2O^+ (7000) has also been measured in Comet Halley. The measured linear polarization in percentage for Comet Halley at a phase angle $\sim 50°$ is around 3.9, 4.0, 3.8 and 1.8 for the molecules CN, C_2, C_3 and OH respectively.

There exists several studies pertaining to the polarization of fluorescence lines, both from experimental and theoretical points of view. The theoretical studies are based on the application of the principle of Zeeman splitting in a magnetic field. Therefore, the important parameter in determining the polarization in a molecular band $u \to l$ (u is the upper level and l is the lower level) is the Larmor frequency ν_L of the molecule. In the case of a strong magnetic field, $2\pi\nu_L > A_{u \to l}$ where $A_{u \to l}$ is the Einstein coefficient for the transition, the molecule will precess strongly in the upper state before de-excitation takes place and therefore it essentially looses memory of its unidirectional excitation process. The resultant effect being that the fluorescence emission line will be isotropic and unpolarized. For the other case of a weak magnetic field, $2\pi\nu_L < g_{l \to u}$, where $g_{l \to u}$ is the excitation rate of the band, the fluorescence cycle will lead to alignment of the molecule with respect to the direction of the excitation field. The typical values for a cometary molecular band is, $A_{u \to l} \sim 10^6/sec$ and $g_{l \to u} \sim 10^3$ to $10^{-1}/sec$. The general value of the magnetic field as measured by spacecrafts in Comet Halley is of the order of a few 10 nT. This corresponds to $\nu_L \sim 10^2/sec$. These values show that

$$g_{l \to u} < 2\pi\nu_L < A_{u \to l}.$$

The resultant effect being that the molecules will precess in the lower state between the fluorescence cycles, but not in the upper state. This therefore preserves the polarization for the case of anisotropic excitation.

The excitation by unidirectional natural radiation leading to linear polarization of an emission line has maximum polarization for a phase angle of 90° (i.e. observed perpendicular to the exciting radiation). The expected variation of polarization with phase angle θ is given by the expression

$$P(\theta) = \frac{P_{max} \sin^2 \theta}{1 + P_{max} \cos^2 \theta} \tag{5.19}$$

and the total intensity should vary as

$$I_{total} = \frac{P_{max}\cos^2\theta + 1}{\frac{1}{3}P_{max} + 1}. \quad (5.20)$$

Here P_{max} represents the maximum polarization corresponding to $\theta = 90°$. The value of P_{max} depends upon different rotational quantum number J for the P, Q and R branches of the band. Theoretical calculations are available only for a limited number of transitions at the present time.

The value of P_{max} for various molecular bands can be determined from the observed polarization and phase angle corresponding to the time of observation. Since at the present time the cometary polarimetry does not resolve molecular bands, the observed polarization is an average over individual polarization of several lines. The derived value of P_{max} from observations for the C_2 and CN molecules is in agreement with the expected theoretical value of P_{max} of 7.7%. The observed polarization variation also established the validity of the theoretical dependence given by Eq. (5.19). The P_{max} values derived from observations for the C_3 molecule $\sim 6\%$ which is similar to that of C_2 and CN, but smaller than the expected value of 19%. This indicates that a more accurate theoretical calculation is needed. The derived P_{max} value for H_2O^+ and CO^+ is quite high $\sim 20\%$. At the present time theoretical calculations have been performed for only a few molecules. Therefore theoretical investigations of polarization for molecules of cometary importance is of great interest. It is also interesting to note that fluorescence emission is anisotropic in character and therefore might have some effect on the observed line intensities.

5.2. Excitation Temperature

The population distribution under thermal equilibrium is given by the Boltzmann law. Even though it does not represent a real situation, it can be used to a first approximation for estimating the excitation temperature. This excitation temperature is found to be not much different from the temperature obtained based on a detailed analysis of the level population in many cases.

5.2.1. *Rotational temperature*

The temperature determined from the application of the Boltzmann law to rotational lines is generally called the *rotational temperature*.

For rotational levels, the exponential factor $e^{-E/kT}$ is given by $e^{-BJ(J+1)hc/kT}$ where B is the rotational constant of the molecule (Chap. 3). In addition, for each level there are $(2J+1)$ states. The number of molecules in any rotational level J in terms of the total number N can be expressed as

$$N_J = \frac{N}{Q_{rot}}(2J+1)e^{-BJ(J+1)hc/kT}. \tag{5.21}$$

Here Q_{rot} is the rotational partition function and is given by

$$Q_{rot} = \sum (2J+1)e^{-BJ(J+1)hc/kT}. \tag{5.22}$$

The rotational population N_J, goes through a maximum and then decreases with an increase in J. For a given temperature, the maximum value of J for which the population is maximum can be evaluated from the above equation and is given by

$$J_{max} = \sqrt{\frac{kT}{2Bhc}} - \frac{1}{2}. \tag{5.23}$$

The intensity of a rotational line in the emission can be written as (Chap. 3)

$$I_{em} = C\nu^4 S_J e^{-BJ(J+1)hc/kT}. \tag{5.24}$$

Here C is a constant and S_J is the Höln-London factor. T is the rotational temperature. A plot of log (I_{em}/S_J) as a function of $J(J+1)$ results in a straight line. From the slope of the line the rotational temperature can be evaluated.

The above method has been applied to the case of cometary spectra to obtain the rotational temperature of various molecules. One can get a better estimate of the rotational temperature if the rotational levels are populated to high quantum numbers. In addition one requires the spectra of high dispersion for the lines to be resolved. Studies have been carried out for molecules like CH, CN, OH, C_2, etc. Except for the C_2 molecule, the rotational temperatures for all other molecules have values in the range of about 200 to 400°K. The rotational temperature for the C_2 molecule based on the profile analysis of well-resolved lines of Swan bands gives the temperature values in the range of about 3000 to 4000°K (Fig. 5.10). The vast difference in the rotational temperature between the C_2 and other molecules arises due to the fact that for heteronuclear molecules like CH, CN, etc., the transitions in the ground state of the molecule are possible;

these radiate away the energy. This makes the temperature low. But for the homonuclear molecule C_2, such types of transitions are forbidden and hence give higher temperatures.

The (0,0) band spectra of Swan system for Comet Halley taken at a very high resolution of 0.06-0.08A (FWHM) have completely resolved all the rotational lines of the R branch. The observed intensities of these rotational lines indicated the existence of two vastly different well defined Boltzmann temperature population distributions corresponding to those of lower and higher rotational levels. In particular, the low rotational levels with $J' \leq 15$ gave a rotational temperature, $T_{rot} \sim 600 - 700K$, while the higher levels gave a rotational temperature, $T_{rot} \sim 3200K$. The synthetic spectra based on time-dependent rotational population distribution with resonance fluorescence excitation process show that a reasonably good fit over the whole wavelength region of the observed spectra, 5162-5132A, can be obtained for a time interval of around 2×10^3 secs as shown in Fig. 5.14. This time scale is consistent with the time $\sim 2.7 \times 10^3$ secs taken by the C_2 molecules with a velocity $\sim 1 km/sec$, to cross the equivalent radius of the projected area on to the sky. This indicates that the observations have picked up only molecules with the time interval $\sim 2 \times 10^3$ secs from the time of formation indicating that the level populations does not appear to have reached the steady state values.

5.2.2. *Vibrational temperature*

A mean vibrational excitation temperature may be defined from the Boltzmann law representing the population distribution in various vibrational levels. An expression for vibrational temperature can be written in an analogous manner to that of rotational temperature as

$$N_v = \frac{N}{Q_{vib}} e^{-G(v)hc/kT} \tag{5.25}$$

where $Q_{vib} = \Sigma e^{-G(v)hc/kT}$ is called the vibrational partition function. $G(v)$ represents the energy of the vibrational level v. The emission intensity of a vibrational band can be written as (Chap. 3).

$$I_{em} = C q_{v'v''} e^{-G(v)hc/kT} \tag{5.26}$$

where $q_{v'v''}$ is the Franck-Condon factor for the band (v', v'') and C is a constant. A plot of log $(I_{em}/q_{v'v''})$ versus $G(v)$ gives a straight line from which the vibrational temperature can be determined.

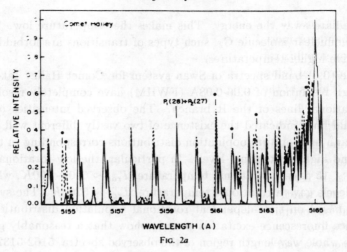

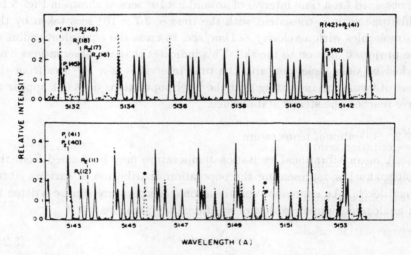

Fig. 5.14. Comparison of the calculated time dependent spectra (Solid line) with the observed spectra (dashed line; Lambert et al. 1990. Ap. J, **353**, 640) for the (0,0) Swan band in the wavelength region 5166 to 5131A of Comet Halley. The calculated curves are for a time interval of 2×10^3 secs and FWHM=0.06A. The short line with filled circles show the location of the NH_2 features. Several of the P, Q and R lines are also marked (Krishna Swamy, K. S. 1991. Ap. J, **373**, 266).

Several attempts have been made to get an estimate of the vibrational temperature of molecules. However, one difficulty with this method is that it is hard to get many bands where the contamination from other bands or from other lines is negligible. So one usually uses the ratio of intensities of two bands or more bands if available. For the case of C_2 molecule, the band sequence flux ratios corresponding to $\Delta v = 0$ and $+1$ is generally used. The vibrational temperature obtained for C_2 molecule is around 3000 to 5000°K. This is of the same order as the rotational temperature.

5.3. Abundances of Heavy Elements

The emission lines of heavy elements are seen only for the heliocentric distance $\simeq 0.15$ au. Comets passing through such small perihelion distances are very rare. One such comet was the Sun grazing Comet Ikeya-Seki in 1965 which had a perihelion distance $\sim 0.005 au$. This comet provided the opportunity for observing emission lines of various elements like $Na, K, Ca, Ca^+, Fe, Ni, Mn$ and so on. This made it possible to make an abundance analysis of these elements.

In a steady state, the relative populations in the upper levels are governed by the Boltzmann's formula reduced by the dilution factor W, which is a measure of the deviation of the energy density of radiation from the equilibrium value i.e. $u_\lambda \propto W \lambda^{-5} 10^{-\theta \chi}$. Therefore the intensity of an emission line is given by

$$I = \left(\frac{8\pi^2 e^2 h}{m}\right) \frac{Ngf}{u\lambda^3} 10^{-\theta \chi} \qquad (5.26a)$$

where χ is the excitation potential of the line and $\theta = (5040/T)$. T is the excitation temperature corresponding to the colour temperature of the exciting solar radiation. The term represented by the power of 10 corresponds to the exponential factor in the Boltzmann distribution function. N is the total number of atoms/cm^2 in the line of sight and u is the partition function of the line. The other quantities have their usual meanings. The above equation can be written as

$$\log\left(\frac{\lambda^3 I}{gf}\right) = -\theta \chi + constant. \qquad (5.27)$$

Therefore, a plot of $\log (\lambda^3 I/gf)$ against the excitation potential χ for all the observed lines will result in a straight line whose slope gives the value of θ. Figure 5.15 shows one such typical plot for FeI lines. The straight line drawn through the points is for $\theta = 1.0$. Knowing the excitation temperature, the total number of atoms can be calculated easily from

$$\begin{aligned}\log N = {}& constant + \log(\lambda^3 I) - \log(gf) \\ & + \log u + \theta \chi\end{aligned} \qquad (5.28)$$

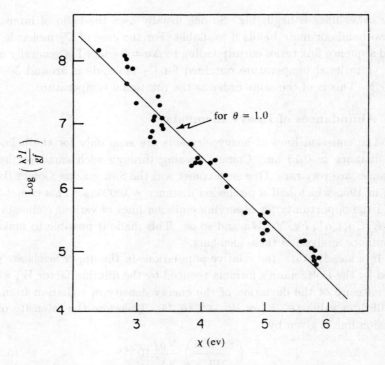

Fig. 5.15. A plot of log $(\lambda^3 I/gf)$ as a function of the excitation potential of the line. The straight line drawn through the points is for $\theta_{ex} = (5040/T_{ex}) = 1.0$ (Adapted from Arpigny, C. 1977, *Proceedings of the Robert A. Welch Foundation Conferences on Chemical Research XXI*, Cosmochemistry. Houston, p. 9; see also Preston, G. W. 1966, *Ap.J.* **147**, 718.)

The procedure can be repeated for all the observed lines. The value of $\theta = 1.0$ also fitted the other lines. Therefore the number of atoms of various kinds can be determined. The calculated relative abundances of various elements relative to iron is similar to that of the solar value. This is in accordance with the abundances derived from in situ measurements of Comet Halley made by spacecrafts. The instruments on board Vega and Giotto spacecrafts were able to measure directly the abundances of elements present in the dust grains in the coma. With a reasonable value for the gas to dust ratio and using the measured abundance ratios of light elements in comets, the elemental abundances in Comet Halley have been derived. Except for hydrogen, all other abundances are consistent with the solar values.

5.4. Isotopic Abundances

The study of the isotopic abundances in comets has attracted considerable attention as it has significant amount of information in it as to the conditions which prevailed at the time of formation of these objects. Fortunately, the most abundant elements, namely, H,C,N and O do have many stable isotopes. Therefore a comparison of the isotopic ratios of these elements in different kinds of objects will reveal the history of the whole evolutionary process. The detection of several complex molecules in comets has given the general feeling that the cometary material and the interstellar material could be very similar in nature. The isotopic ratios could help in clarifying this problem as well.

For a diatomic molecule, the vibrational and rotational isotopic shifts are proportional to $\left(1 - \sqrt{\frac{\mu}{\mu_i}}\right)$ and $\left(1 - \frac{\mu}{\mu_i}\right)$ respectively, where μ and μ_i are the reduced masses of the normal and the isotopic molecule. This shows that the expected isotopic shifts from molecules are quite small. Therefore, in order to study the isotopic molecule, it will be a great advantage if the abundance of the isotopic molecule is large, the isotopic line is well separated from the line of the normal isotope and also lie in the less crowded region of the spectrum. This introduces severe constraints on the selection of the molecules. In addition, the spectra should be of higher resolution to separate the isotopic line from the normal line. These restrictions automatically limit the study to bright comets. Due to the difficulties involved in the spectral and radio astronomical observations the studies have been limited to only a few isotopic species. In principle, the isotopic ratios can be determined better than the absolute abundances as they are not model dependent.

The only isotopic ratio that has been determined for several comets is the $^{12}C/^{13}C$ ratio based mainly on the analysis of the Swan band intensities of the C_2 molecule. It is well suited for the study as the (1,0) band of the Swan system which occurs at 4737Å is well separated from the $^{12}C/^{13}C$ band which occurs around 4745A.

A typical medium resolution scan in the region of interest for the Comet Kohoutek is shown in Fig. 5.16. From a comparison of the observed intensity ratios of these two bands, it is possible to get an estimate for the isotopic ratio of $^{12}C/^{13}C$. This method has been applied to many comets. Unfortunately the isotopic line at 4745 Å is strongly blended with the emission lines of NH_2 molecule. The high resolution scan around 4745 Å

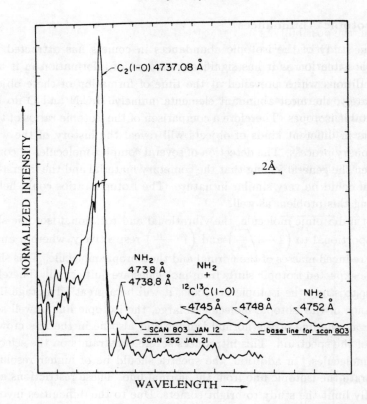

Fig. 5.16. Photoelectric scans of Comet Kohoutek in the region of 4737Å corresponding to (1,0) band of the Swan System. Resolution is 0.4 Å. The $^{12}C/^{13}C$ feature is around 4745Å which is blended with NH_2, (Danks, A. C., Lambert, D. L. and Arpigny, C. 1974, *Ap. J.* **194**, 745.

line shows that the blending comes from 4 lines of NH_2. Two of the lines are well separated from the isotopic feature. From a fitting of the observed line profile, the relative contribution of $^{12}C^{13}C$ to the whole emission feature can be determined. This information is used along with the observed intensity ratio of 4737 and 4745 Å obtained from the low resolution spectra to get the isotope ratio of $^{12}C/^{13}C$. The (1,0) Swan band is the best target for the determination of $^{12}C/^{13}C$ ratio as the lines of $^{12}C^{13}C$ and $^{12}C_2$ are well separated. However to obtain a good estimate of $^{12}C/^{13}C$ ratio, it is necessary to have spectra of improved resolution, better signal-to-noise and apply proper correction for blending of NH_2 lines. To avoid the blending problem, the possibility of using the (0,0) band of the Swan system has been considered. With a high resolution spectra of Comet West taken at 0.15Å it was possible to resolve completely the rotational structure of the (0,0) band. However, the isotopic features are rather weak, which makes the derived isotopic ratio uncertain. The determination of the carbon isotopic ratio from the study of 3883Å(0,0) band of $B^2\Sigma - X^2\Sigma$ system of CN is also rather difficult as one requires high resolution spectra. In addition, the lines of $^{12}C^{15}N$, $^{13}C^{14}N$, weak lines of (1,1) band and P-branch lines of the CN system also lie in this wavelength region (Fig. 5.17). The results

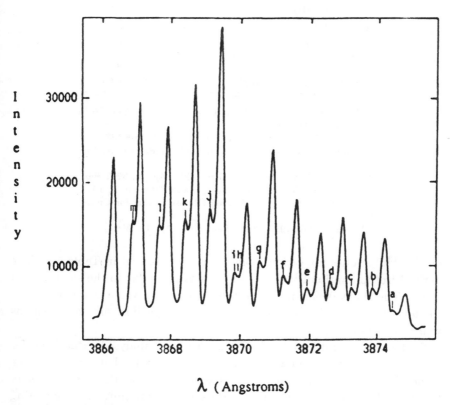

Fig. 5.17. The high resolution spectrum of Comet Halley in the wavelength region from 3866 to 3874A. The R-branch lines of ^{12}CN and ^{13}CN overlap in this region. The features marked a to g and i belong to the (0,0) band of the ^{13}CN molecule while those marked with h and j to m are P-branch lines from the (1,1) band of the ^{12}CN molecule (Jaworski, W. A. and Tatum, J. B. 1991, Ap. J., **377**, 306).

Table 5.5. Observed isotopic ratio of carbon.

Comet	Molecule	$^{12}C/^{13}C$
Ikeya (1963 I)	C_2	70 ± 15
Tago-Sato-Kosaka (1969 IX)	C_2	100 ± 20
Kohoutek (1973 XII)	C_2	115^{+30}_{-20}
		135^{+65}_{-45}
Kobayashi-Berger-Milon (1975 IX)	C_2	110^{+20}_{-30}
West (1976 VI)[a]	C_2	60 ± 15
Hallay (1986 III)[b]	CN	89 ± 17
Halley (1986 III)[c]	CN	90
Levy (1990 C)[c]	CN	90
Austin (1989 C)[c]	CN	90
Solar value		90
Interstellar Matter (Mean)[d]		~ 60

(Adapted from Vanysek, V. and Rahe, J. 1978. *The Moon and Planets,* **18**, 441).

[a] Lambert, D.L. and Danks, A.C. 1983, *Ap. J.*, **268** 428: The low value for Comet West could possibly be due to blending problems.

[b] Jaworski, W.A. and Tatum, J.B., 1991, *Ap. J.*, **377**, 306.

[c] Wyckoff, S. Kleine, M. Wehinger, P and Peterson, B. 1993. Bull. Am. Astr. Soc. **25**, 1065 (Revised value for Comet Halley).

[d] Wannier, P.G. 1980. *Ann. Rev. Astr. Ap.* **18**, 399.

of $^{12}C/^{13}C$ studies for various comets are summarised in Table 5.5. The derived isotopic ratio of $^{12}C/^{13}C$ for comets is similar to the solar system value. The slight departures from the solar system value could be due to the uncertainties in the observations as the isotopic lines are very weak, as well as in modelling.

The isotopic ratio of $^{12}C/^{13}C$ has also been analysed extensively for various objects in the solar system and in the interstellar medium. The values for the interstellar matter is based largely on the observation in the radio frequency region. There are some uncertainities in converting the observed intensities to the actual abundances and the values vary from about 10 to 100. Nevertheless, the mean range of values seems to be around 30 to 50. This is smaller compared to the cometary value of around 90. The equilibrium value of $^{12}C/^{13}C$ based on the hydrogen burning in stars

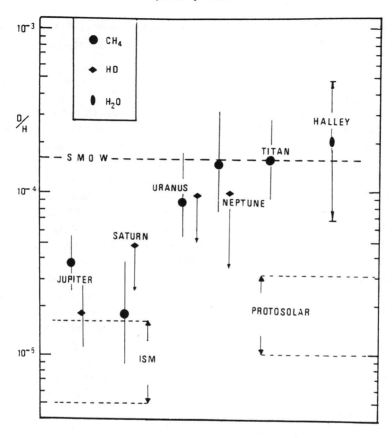

Fig. 5.18. The deuterium to hydrogen ratio in Comet Halley is compared with the values observed in planets, diffuse interstellar matter (ISM), solar nebula (protosolar) and with the value for the standard mean ocean water (SMOW). The D/H ratio in Comet Halley was derived from in situ measurements of relative abundances of HDO (Vanysek, V. 1991. In *Comets in Post-Halley Era*, Eds. R.L. Newburn, Jr. et al. Kluwer Academic Publishers, Vol. 2, p. 879).

(i.e. CNO cycle) is around 4. The general reduction of the ratio of $^{12}C/^{13}C$ in interstellar matter compared to the solar system value is more likely to be produced as a result of the addition of ^{13}C. This could be brought about by the processing of ^{13}C by the CNO cycle in stars. On the basis of the present picture, the processed material of the stars find their way finally into the interstellar medium and thus enrich the medium with ^{13}C isotopes. The observed isotopic ratio of $^{12}C/^{13}C$ between the solar system and the interstellar value appears to be consistent to a first approximation with the scenario of chemical evolution which has taken place for the last 4.6×10^9 years.

The deuterium-to-hydrogen ratio is of great interest as deuterium is most likely to be synthesized in the early universe. Therefore attempts were made to derive this ratio from the analysis of OH lines from Comet Halley. Only an upper limit of OD/OH $< 4 \times 10^{-4}$ was derived. The neutral mass spectrometer on board the Giotto spacecraft could make measurements of the composition of neutral and ion species in Comet Halley with high mass resolution. Therefore it has made it possible to get the isotopic ratio of several elements. The D/H ratio derived from the study of HDO/H_2O is in the range 5×10^{-5} to 5×10^{-4}. This is inconsistent with the values of D/H found in the giant atmospheres of Jupiter and Saturn and of the Primordial solar nebula. Therefore, there appears to be two distinct values of D/H that exist in the solar system (Fig. 5.18). The cometary material is enriched in deuterium by a factor of around 10 as compared to the D/H ratio assumed in the protoplanetary nebula. It is likely that the enrichment of deuterium might have arisen from the ion-molecular reactions which are fast at lower temperatures and in a dense medium. The other measured isotopic ratios of $^{16}O/^{18}O \sim 450$ and $^{32}S/^{34}S \sim 22$ from in situ mass spectrometry are consistent with the solar values of 500 and 23 respectively. The ratio of ortho to para H_2 has also been determined from Comet Halley observations. The derived ratio of 2.2 to 2.3 differs from the equilibrium value of 3.0 (Sec. 12.3).

Lastly, mention may be made of the isotopic ratios of light elements like Li, Be, B, etc., as these elements will be destroyed easily in stars, even if they were present in the original material. The detection in comets imply that they ought to have been produced more recently by different physical processes. However, the detection of light element isotopes in comets is a difficult problem.

The analysis of the spectra presented so far shows that the resonance fluorescence process is the excitation mechanism of the observed emission lines in comets. The abundance of various elements obtained from the study of the spectra of Comet Ikeya-Seki as well as *in situ* measurement of Comet Halley show values very similar to those of solar abundances. In addition, the isotopic ratio of $^{12}C/^{13}C$ in comets has a value of about 90, which is the same as the solar system value.

Other isotopic ratios such as $^{16}O/^{18}O$ and $^{32}S/^{34}S$ are found to be similar to the solar values except the D/H ratio. It appears that the abundance of deuterium is enhanced compared to the solar values by a factor of 10 or so. The likely mechanism of such an enhancement could be through the ion-molecular reactions.

Problems

1. Consider the (B-X) transitions of the molecular hydrogen. Write down the equation between the upper and lower populations, $N(v')$ and $N(v'')$, assuming only the absorption and spontaneous emission terms are important. If the population in the lower vibrational levels is determined by the Boltzmann distribution at temperature $T = 2000°K$, calculate the population in the upper vibrational levels and the resulting emission.
2. Consider the ground electronic state $(X^1\Sigma^+)$, the first singlet (A $^1 II$) and the triplet (a $^3 II$) electronic states of the CO molecule. From the statistical equilibrium equations, obtain an expression for the intensity ratio of $(a^3 II - X^1\Sigma^+)$ to $(A^1 II - X^1\Sigma^+)$ transitions in terms of the molecular parameters of the molecule and its numerical value. Assume $\rho B_{21} < A_{21}$ and $\rho B_{31} < A_{31}$. This problem will give an idea of the expected intensities of forbidden transitions compared to those of allowed transitions.
3. Calculate the wavelength of a solar photon of $\lambda = 5000 Å$ at the comet which is at $r = 0.52 au$. The comet is moving in an elliptical orbit with $e = 0.8$ and has a period of 5 years.
4. The relative intensities of the rotational lines of molecular spectra in a comet will vary irregularly with rotational quantum number and the intensity of an individual line will vary irregularly with the distance of the comet from the Sun. Explain why these effects occur and describe how in principle, you could compute the line intensities.
5. In terms of the spectral features observed in the coma of a comet, is it possible to infer the abundances of molecules in the primordial nebula? Explain.

References

The effect of Fraunhofer lines on cometary spectra was first pointed out in
1. Swings, P. 1941, *Lick Obs. Bull.* **19**, 131.

The application of Resonance Fluorescence Process to molecules can be found in
2. Arpigny, C. 1965, *Mem. Acad. Roy. Belg*, Cl. 8°, **35**, 5.

For some later work, the following papers may be referred.
3. Krishna Swamy, K.S. 1981, *Astr. Ap.* **97**, 110.
4. Weaver, H.A. and Mumma, M.J. 1984, *Ap.J.* **276**, 782.
5. Kleine, M., Wyckoff, S., Wehinger, P.A. and Peterson, B.A. 1994, *Ap. J.* **436**, 885.
6. Magnani, L. and A'Hearn. M.F. 1986, *Ap.J.* **302**, 477.
7. Schleicher D.G. and A'Hearn. M.F. 1988. *Ap.J.* **331**, 1058.

For the calculation of vibrational transition probabilities, the following book may be referred to:
8. Penner, S.S. 1959, *Molecular Spectroscopy and Gas Emissivities*, Reading: Addison-Wesley Publishing Company.

The early work on C_2 molecule is discussed in
9. Stockhausen, R.E. and Osterbrock, D.E. 1965, *Ap.J.***141**, 287.

For later work, the following papers may be referred:
10. Krishna Swamy K.S. and O'Dell. C.R. 1987, *Ap. J.* **317**, 543.
11. O'Dell, C.R., Robinson, R.R., Krishna Swamy, K.S., Spinrad, H. and McCarthy, P.J. 1988, *Ap.J.* **334**, 476.
12. Gredel, R., van Dishoeck, E.F. and Black, J.H. 1989, *Ap. J.* **338**, 1047.

The following papers deal with polarization of molecular bands.
13. Ohman, Y. 1941, *Stockholm Obs.* Ann. **13**, No. 11, p.1.
14. Le Borgne, J.F. and Crovisier, J. 1987, *Proc. Symp. on Diversity and Similarity of Comets*. Brussels, Belgium.
15. Sen, A.K., Joshi, U.C. and Deshpande, M.R. 1989, *Astr. Ap.* **217**, 307.

CHAPTER 6

GAS PRODUCTION RATES IN COMA

The gaseous molecules vaporized from the nucleus are subjected to a variety of physical processes which could dissociate them step by step. Hence, the species observed in the spectra of comets could comprise contributions arising out of the dissociated products, as well as those released directly by the nucleus or even by dust particles. Therefore, the physical quantity of interest is the production rate of the species and its variation with the heliocentric distance. The production rate of a molecule can be determined either from the observation of the total luminosity or from the surface brightness of a given spectral line, band or a band-sequence.

If the vaporization of H_2O is mainly due to the absorption of the solar energy, then the production rate of H_2O should also have a r^{-2} dependence with the heliocentric distance. Therefore, if the observations show a deviation from the r^{-2} dependence, it is of great significance and has important information with regard to the mechanism of evaporation of the material from the nucleus. It is also of interest to see whether the production rate of the molecules differs from comet to comet and whether it has any correlation with the morphology, dust content, age, etc. of the comet. From the results, one may be able to understand better the evaporation of material from the nucleus as well as the source of the observed radicals and molecules in comets. Some of these aspects will be the subject of discussion in this chapter.

6.1. Theoretical Models

6.1.1. *From the total luminosity*

The total production rate of a particular species can be determined from the measurement of the total flux of the radiation from the entire coma of a given line or band. If the mean lifetime of the molecule is τ_i then the coma will contain N_i molecules of this type at any given time so that

$$N_i = Q_i \tau_i, \qquad (6.1)$$

where Q_i is the production rate of the species. Since the excitation of the lines is due to the resonance fluorescence process, for an optically thin case, the luminosity of any line or a band is proportional to the total number of the species in the coma and to the fluorescence efficiency factor called the '$g - factor$'. The g-factor actually represents the probability of scattering of a solar photon per unit time per molecule. The luminosity at wavelength λ_i is therefore given by

$$L_i = g_i N_i = g_i Q_i \tau_i. \qquad (6.2)$$

Since the observations include the whole extent of the coma, the flux received on the Earth F_i is given by

$$F_i = \frac{L_i}{4\pi \Delta^2}. \qquad (6.3)$$

Therefore, the production rate is given by

$$Q_i = \frac{4\pi \Delta^2 F_i}{g_i \tau_i} \qquad (6.4)$$

where Δ is the geocentric distance. The production rate can therefore be calculated provided g_i, τ_i and F_i are known. In general, the measurements covering only a given instrumental field of view is available and not over the whole coma. For such cases the total number of species in the field of view, N_i can be calculated from the relation

$$N_i = \frac{4\pi \Delta^2 F_i}{g_i}$$

and hence the column density can be calculated.

The g-factor represents the number of photons per second scattered by a single atom or molecule exposed to the unattenuated sunlight. If the solar photon flux is πF_ν per unit frequency in the neighbourhood of the line, the g-factor for a strict resonance scattering is given by

$$g_\nu = F_\nu \int \alpha_\nu \, d\nu \qquad (6.5)$$

where α_ν is the absorption cross-section. Therefore,

$$g_\nu = \pi F_\nu \left(\frac{\pi e^2}{mc}\right) f \qquad (6.6)$$

or

$$g_\lambda = \left(\frac{\pi e^2}{mc^2}\right) \lambda^2 f_\lambda (\pi F_\lambda) \; photons/sec/molecule \qquad (6.7)$$

where f_λ is the absorption oscillator strength and πF_λ the solar flux per unit wavelength. For molecules, the relative transition probabilities for downward transitions must be taken into account. Therefore, Eq. (6.7) can be written as

$$g_\lambda = \left(\frac{\pi e^2}{mc^2}\right) \lambda^2 (\pi F_\lambda) f_\lambda \tilde{\omega} \qquad (6.8)$$

where

$$\tilde{\omega} = \frac{A_{v'v''}}{\sum_{v''} A_{v'v''}}.$$

Here the A's are the Einstein coefficients. The g-factor depends upon the oscillator strength and the solar flux. As pointed out in Chap. 3, the solar

Table 6.1a. g-factor for ultraviolet line of some species at 1 au.

Species	Wavelength(Å)	g-factor (a) (photon/sec/atom) or molecule
OI	1302-1304	$1.15 \pm 0.58 \times 10^{-6}$
CO(3,0)	1447	$2.02 \pm 0.50 \times 10^{-7}$
CO(2,0)	1478	$4.25 \pm 1.06 \times 10^{-7}$
CO(1,0)	1510	$3.60 \pm 0.90 \times 10^{-7}$
CI	1561	$7.28 \pm 1.82 \times 10^{-6}$
CO(0,1)	1597	$2.18 \pm 0.55 \times 10^{-7}$
CI	1657	$4.42 \pm 1.11 \times 10^{-5}$
SI	1807-1826	$4.40 \pm 1.10 \times 10^{-5}$
CS(0,0)	2576	9.5×10^{-4} (b)

[a] For $\dot{r} = 31.46 \; km/sec$ for the atomic species: Sahnow, D. J., Feldman, P. D., McCandliss, S. R. and Martinez, M. E. 1993, *Icarus*, **101**, 71.
[b] Sanzovo, G. C., Singh, P. D. and Huebner, W. F. 1993, *AJ.* **106**, 1237.

radiation in the ultraviolet region is dominated by line emissions. Therefore, the Doppler shift due to the comet's heliocentric radial motion can produce

Table 6.1b. g-factor for microwave lines of several species at 1 au.

Species	Transition	Band origin (cm^{-1})	Band g-factor (photon/sec/mol)
H_2O	ν_2	1595	2.4(−4)
	ν_3	3756	2.8(−4)
CO_2	ν_2	667	9.0(−5)
	ν_3	2349	2.9(−3)
CO	(1,0)	2143	2.6(−4)
NH_3	ν_1	3337	2.9(−5)
	ν_2	950	3.3(−4)
	ν_3	3444	1.9(−5)
	ν_4	1627	1.2(−4)
CS_2		1532	2.4(−3)
OCS	ν_3	373	3.0(−3)
HCN	ν_3	3311	3.5(−4)
CH_4	ν_3	3019	3.4(−4)
	ν_4	1306	1.1(−4)
H_2CO	ν_1	2782	3.9(−4)
	ν_2	1746	2.6(−4)
	ν_5	2843	4.6(−4)

Crovisier, J. 1992. In *Infrared Astronomy with ISO*. Eds. Th. Encrenaz and M. F. Kessler (Nova Science Publishers), p 221.

large changes in the g-factor as the comet moves in its orbit (see Fig. 5.11). This has to be taken into account in the analysis of the observations. But the average g-factor has been calculated for many molecules and is available in the literature. Since the g-factor involves the solar radiation field which varies with the heliocentric distance, the values for several molecules of interest are given in Table 6.1 for a heliocentric distance of 1 au.

The calculation of the *lifetime* of the species involves a knowledge of the photodissociation and photoionization rates, J_d and J_i respectively. These can be calculated from the relation

$$J_{i,d} = \int_{\lambda_c}^{0} \sigma(\lambda)\pi F_\odot(\lambda)d\lambda \ sec/molecule \tag{6.9}$$

where $\sigma(\lambda)$ is the photodestruction cross-section and $F_\odot(\lambda)$ is the incident solar flux. λ_c represents the threshold for the dissociation or ionization process. For most of the species, the photodissociation lifetime is relevant since the dissociation energies are smaller than those of ionization potentials. However, for the case of the CO molecule, the two are comparable as the dissociation potential $\sim 11eV$ and the ionization potential $\sim 14eV$. The lifetime of the molecules is just the inverse of Eq. (6.9), i.e.,

$$\tau_i = J^{-1} \; sec. \tag{6.10}$$

It should be noted that the dependence of the solar flux with heliocentric distance (Eq. 6.9) also enters in the lifetime of the molecules. Usually, the lifetime is referred to $1 au$. Since both g_i and τ_i depend directly on the solar flux which varies as r^{-2}, the product $g_i \tau_i$ [Eq. (6.4)] becomes independent of r. Because of the presence of large uncertainties in the value of cross-sections as well as the solar fluxes in the extreme ultraviolet region, the calculated J values and in turn the lifetime of the species are also uncertain. For example, the solar flux itself could vary by factors of

Table 6.2. Lifetimes of several molecules at 1 au heliocentric distance.

Species	Lifetime (sec)	Species	Lifetime (sec)
H_2	1.2(7)	NH_3	5.9(3)
CH	1.1(2)	OH	4.0(5)
			4.3(5)*
C_2	1.0(7)	HNCO	6.6(4)
NO	4.5(5)	CH_4	1.4(5)
H_2O	8.3(4)	HC_3N	2.6(4)
	9.6(4)*		
CN	3.1(5)	CH_3OH	1.0(5)
		C_2H_4	2.1(4)
CS_2	3.5(2)	C_2H_6	1.25(5)
NH_2	4.8(5)	CO	3.2(6)

Huebner, W. F., Keady, J. J. and Lyon, S. P. 1992. *Astro. Space Sci.* **195**, 1.
* Budzien, S. A., Festou, M. C. and Feldman, P. D. 1994, *Icarus*, **107**, 164.

2 to 4 during the solar cycle. Therefore, the accurate calculation of the lifetime of a species is one of the main problems at present. The lifetime of a few molecules is given in Table 6.2. The uncertainty in the lifetime of the molecules directly affects the calculated production rates.

6.1.2. From surface brightness distribution

It is possible to get much more information than just the production rates, if the variation of brightness of various emissions with radial distance

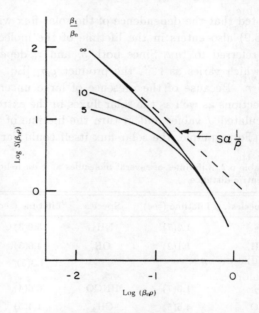

Fig. 6.1. Theoretical surface brightness distributions based on Haser's model for different ratios of the scale lenghts of daughter and the parent molecule. (Adapted from O'Dell, C. R. and Osterbrock, D. E. 1962. *Ap. J.* **136**, 559.)

from the centre of the comet could be observed. It is a common practice nowadays to make such observations using a diaphragm whose field of view is much smaller than the projected size of the coma. A typical observed brightness profile has a flat region near the centre and falls rather steeply in the outer region (Fig. 6.1). If the average brightness of a molecule is denoted as B_i at the projected distance ρ as seen in the sky then

$$B_i(\rho) = \overline{N}_i(\rho) g_i, \tag{6.11}$$

where $\overline{N}_i(\rho)$ is the average column density of the molecule along the line of sight. To get the total number of a species one has to integrate over the model dependent variation of density as a function of the projected distance ρ from the nucleus. This is then to be convolved with the instrumental field of view.

It is generally believed that the observed species are the dissociated or the decayed products of the original complex molecules which were vaporized from the nucleus (Chap. 4). Because of the finite lifetimes of these dissociated products, they in turn break up further as they move out in the coma. Therefore, the coma can be divided roughly into three zones: a productive zone, an expansion zone and a destructive zone.

Let us start with the simplest model in which molecules are released by the nucleus with a production rate

$$Q = 4\pi R_0^2 E \tag{6.12}$$

where R_0 is the radius of the nucleus and E is the evaporation rate ($/cm^2/sec$). It then expands isotropically with a velocity v. For such a case, the density falls off as R^{-2} where R is the distance from the centre of the coma and it is given by

$$n(R) = \frac{Q}{4\pi v} \frac{1}{R^2}. \tag{6.13}$$

At a projected distance ρ as seen in the sky, the column density is obtained by integrating along the line of sight and is given by

$$N(\rho) = \frac{Q}{4v} \frac{1}{\rho}. \tag{6.14}$$

The column density, therefore, varies as $(1/\rho)$. The above formalism is very simple and gives only rough estimates.

As the molecules move outwards, they are generally dissociated by the solar radiation. They have a mean lifetime τ. This corresponds to a distance $R_d = v\tau$, where v is the average velocity of the molecule. The density distribution is now given by

$$\begin{aligned} n(R) &= \frac{Q}{4\pi v R^2} e^{-(R/R_d)}, \\ &= \frac{C}{R^2} e^{(-R/R_d)} \end{aligned} \tag{6.15}$$

where
$$C = \frac{Q}{4\pi v}.$$

A slightly more complicated model is based on the parent-daughter hypothesis in which the parent molecules released by the nucleus decay with a mean lifetime τ_1 corresponding to a distance (called *scale length*) of $R_1 = v_1 \tau_1$. The daughter products then in turn decay with a lifetime τ_0 corresponding to $R_0 = v_0 \tau_0$. For such a two-component model, the number density takes the form

$$n(R) = \frac{C}{R^2} \left(e^{-\beta_0 R} - e^{-\beta_1 R} \right). \tag{6.16}$$

Here $\beta_0 (\equiv (\tau_0 v_0)^{-1})$ denotes the reciprocal of the mean distances travelled by or scale lengths of the observed molecules, β_1 denotes the same by the parent molecule before they are dissociated. The Eq. (6.16) can be written as

$$n(R) = \frac{C}{R^2} f(R) \tag{6.17}$$

where $f(R)$ is the function representing the quantity in brackets of Eq. (6.16). If $n(R_1)$ denotes the density at the distance R at which $f(R)$ has its maximum value, then

$$n(R) = n(R_1) \frac{R_1^2}{R^2} \frac{\beta_1}{\beta_1 - \beta_0} \left[e^{-\beta_0 (R - R_1)} - \frac{\beta_0}{\beta_1} e^{-\beta_1 (R - R_1)} \right]. \tag{6.18}$$

The integration of the above equation along the line of sight at the projected distance ρ from the nucleus gives the column density and hence the surface brightness at that point for an optically thin case. The resulting expression is of the form

$$N(\rho) = 2n(R_1) R_1^2 \frac{\beta_0 \beta_1}{\beta_1 - \beta_0} e^{+\beta_0 R_1} \frac{1}{\beta_0 \rho} [B(\beta_0 \rho) - B(\beta_1 \rho)] \tag{6.19}$$

where
$$B(z) = \frac{\pi}{2} - \int_0^z K_0(y) dy. \tag{6.19a}$$

Here $K_0(y)$ is the modified Bessel function of the second kind of order zero. The Eq. (6.19) can be written as

$$N(\rho) = (constant) s(x) = \frac{(constant)}{x} \left[B(x) - B\left(\frac{\beta_1 x}{\beta_0}\right) \right] \tag{6.20}$$

where $x = \beta_0\rho$. The equation shows that the shape of the surface brightness distribution $s(x)$, depends essentially upon the relative values β_1 and β_0. The Fig. 6.1 shows a plot of brightness profiles predicted for various values of the ratio β_1/β_0. The curves become rather flat in the central region for small values of β_1/β_0, and this means that the two scale lengths do not differ from each other very much. One can also see the deviation from the simple model in which the dependence is given by $1/\rho$.

The total number of molecules $M(\rho)$ within a cylinder of radius ρ centred on the nucleus can be obtained by the integration of Eq. (6.19) over ρ, i.e.,

$$M(\rho) = \int_0^\rho 2\pi\sigma N(\sigma)d\sigma$$
$$= 4\pi\, n(R_1) R_1^2 \frac{\beta_1}{\beta_0(\beta_1 - \beta_0)} e^{\beta_0 R_1} \beta_0 \rho [G(\beta_0\rho) - G(\beta_1\rho)] \quad (6.21)$$

where

$$G(z) = \frac{\pi}{2} - \int_0^z K_0(y)dy + \frac{1}{z} - K_1(z).$$

Here K_0 and K_1 are the modified Bessel functions of the second kind of order 0 and 1 respectively. The above expression can also be expressed more conveniently in terms of the production rate as

$$M(\rho) = \frac{Q\rho}{v}\left[\int_x^{\mu x} K_0(y)dy + \frac{1}{x}\left(1 - \frac{1}{\mu}\right) + K_1(\mu x) - K_1(x)\right] \quad (6.21a)$$

where $x = \beta_0\rho$ and $\mu = \beta_1/\beta_0$. The quantity occurring in bracket depends only on the parameters μ and x which could be tabulated for various values of x and μ.

The above two-stage model usually known as *Haser's model*, can be further extended to three stages in which the daughter products themselves dissociate. Let τ_1, τ_2 and τ_3 denote the lifetime of the parent, daughter and the granddaughter species respectively. If the corresponding velocities are v_1, v_2 and v_3 then $\beta_i = (\tau_i v_i)$ with $i = 1, 2$ and 3. The densities of the three corresponding species can be expressed conveniently in terms of the production rate Q_1 of the parent molecule as follows.

$$n_1(R) = \frac{Q_1}{4\pi v_1 R^2}\exp(-\beta_1 R) \quad (6.22)$$

$$n_2(R) = \frac{Q_1}{4\pi v_2 R^2}\frac{\beta_1}{(\beta_2 - \beta_1)}[\exp(-\beta_1 R) - \exp(-\beta_2 R)] \quad (6.23)$$

and

$$n_3(R) = \frac{Q_1}{4\pi v_3 R^2}[A\exp(-\beta_1 R) + B\exp(-\beta_3 R) + C\exp(-\beta_2 R)] \quad (6.24)$$

where

$$A = \frac{\beta_1 \beta_2}{(\beta_1 - \beta_2)(\beta_1 - \beta_3)},$$

$$B = \frac{-A(\beta_1 - \beta_3)}{(\beta_2 - \beta_3)}$$

and

$$C = \frac{-B(\beta_1 - \beta_2)}{(\beta_1 - \beta_3)}.$$

The corresponding column density of the species over the field of view of linear radius ρ is given by

$$N_1(\rho) = \frac{Q_1}{4\pi v_1}\frac{2}{\rho} B(\beta_1 \rho), \quad (6.25)$$

$$N_2(\rho) = \frac{Q_1}{4\pi v_2}\frac{2}{\rho}\left(\frac{\beta_1}{\beta_2 - \beta_1}\right)[B(\beta_1\rho) - B(\beta_2\rho)] \quad (6.26)$$

and

$$N_3(\rho) = \frac{Q_1}{4\pi v_3}\frac{2}{\rho}[AB(\beta_1\rho) + BB(\beta_2\rho) + CB(B_3\rho)] \quad (6.27)$$

where $B(Z)$ is given by Eq. (6.19a). The above procedure can easily be extended to cascade processes involving more than three species.

6.1.3. *From number densities*

The relation between the emission coefficient j and the number density n of the species for the resonance fluorescence excitation process is given by

$$\frac{j}{h\nu_{em}} = \frac{\pi e^2 f}{h\nu_{abs}m}\rho_\nu(r)nP. \quad (6.28)$$

Here ν_{em} and ν_{abs} are the emission and absorption frequencies, $\rho_\nu(r)$ the solar radiation density at the heliocentric distance r and P is the probability that a particular transition takes place in comparison with other decays. All other constants have their usual meanings. The integration of Eq. (6.28) along the line of sight through the comet and up to a projected radius ρ gives the relation

$$\frac{L(\rho)}{h\nu_{em}} = \frac{\pi e^2 f}{h\nu_{abs} m} \rho_\nu(r) \, PM(\rho) \tag{6.29}$$

where $L(\rho)$ is the observed luminosity in the emission line and $M(\rho)$ is the total number of the species in the field of view. The above equation can be written as

$$M(\rho) = \left(\frac{\nu_{abs}}{\nu_{em}}\right) \frac{mL(\rho)}{\pi e^2 f \rho_\nu(r) P} \tag{6.30}$$

or

$$\log M(\rho) = \log C + \log L(\rho) + 2\log r \tag{6.31}$$

where

$$C = \left(\frac{\nu_{abs}}{\nu_{em}}\right) \frac{m}{\pi e^2 f \rho_\nu(r) P}.$$

The last term in equation (6.31) arises as solar radiation varies as r^{-2} where r is the heliocentric distance of the comet. The constant C depends upon the transition of interest.

For C_2 molecule, as one generally observes band sequence fluxes with $\Delta V = +1$ or 0, it is necessary to include in the calculation of constant C, the absorptions arising from various vibrational levels from the lower electronic state through the term $(f\rho_\nu(r)/\nu)$. In a similar manner the population of vibrational levels in the upper electronic state enters through the probability P. Therefore, a mean value for the above ratio has to be used, i.e.,

$$\overline{\left(\frac{f\rho}{\nu}\right)} = \sum_{v''} x_{v''} \sum f_{v'',\Delta v}\, \rho(\nu_{v'',\Delta v})/\nu_{v'',\Delta v}. \tag{6.32}$$

Here v'' represents the vibrational level in the lower electronic state and Δv the various band sequences being considered. $x_{v''}$ represents the weight factors for the vibrational levels of the lower electronic state. Similarly the transition probability P for the downward transition should also take into account the population distribution in the upper vibrational levels. The Eq. (6.31) gives the abundance of the species of interest within the field of view of the comet as a function of r.

The comparison of the observed surface brightness distribution with the expected distribution of Fig. 6.1, fixes the values of β_0 and β_1 as well as R_1 from its definition. These can then be used to compute $M(\rho)$ from Eq. (6.21) for comparing with the observed distribution [Eq. (6.31)]. This fixes the value of $n(R_1)$ and hence the density distribution in the coma through Eq. (6.18). The observed distribution of $M(\rho)$ can also be used to derive the production rate Q directly from Eq. (6.21a).

The radial outflow decay model has been used extensively in the literature with a high degree of success. Basically from a fit of the expected surface brightness distribution with the observed distribution, it is possible to derive the production rates and the two scale lengths corresponding to the parent and the dissociated product. From a knowledge of the scale lengths, the lifetime of the species can be determined provided the velocity is known. However, it is found that the scale lengths derived from the Haser's model are generally smaller than the values calculated using the estimated outflow speed of the gas and the photodissociation lifetime computed from the solar UV flux and the measured photoabsorption cross-sections. There are some limitations of this model. First of all it is applicable only to photodestruction products. It does not allow for situations where they could be produced through chemical reactions. Other effects like non-radial flow, the effect of solar radiation pressure and the velocity distribution of the species have to be taken into account in a more realistic model. Some attempts have been made in this direction. Mention may be made of the method based on the *vectorial formalism*. In this method, the molecular fragments of a dissociated parent molecule are ejected isotropically in a reference frame attached to the parent molecule. Also, unlike in the case of Haser's model, the effects of velocity and lifetimes are separated which allow the study of velocity dependent phenomena. These are the two main improvements over the Haser's model. Another formalism has been developed based on an average random walk model which is the *Monte Carlo approach*. The calculations are done in 3-dimensions centred on the comet nucleus. The results are then reduced to 2-dimensional maps for comparing with the observation of molecules. It allows the asymmetric ejection of the parent molecules from the molecules such as jets etc. It can also take into account the acceleration due to solar radiation pressure. It includes several dissociation steps for each specie. More physical effects could be incorporated into the model. These two approaches are more complicated than Haser's model. However, these models yield the surface distributions which resemble rather closely with those of Haser's model.

6.1.4. *Semi-empirical photometric theory*

The gas production rate can, in principle, be determined from the knowledge of the observed light curve. The light curve of a comet basically gives the variation of apparent brightness as a function of the heliocentric distance. In general, the observed brightness in the visual region is mainly due to the continuum and the Swan bands of the C_2 molecule. The continuum is made up of scattering by the dust particles in the coma as well as the reflection from the nucleus. Therefore, a simple photometric equation can be written relating the observed light to the light contributed by the dust grains and the nucleus in the same visual band pass. It can be written as

$$A\phi_n(\alpha) + Bf_1(r) + C\phi_d(\alpha)f_2(r) = r^2 10^{0.4[m_\odot - m_1(r) - 5\log \Delta]}. \quad (6.33)$$

The three terms on the left hand side of the above equation represent the contribution from the nucleus, gas and dust respectively. Here $m_1(r)$ is the apparent visual magnitude of the comet, $m_\odot$ the apparent magnitude of the Sun, $\phi_n(\alpha)$ and $\phi_d(\alpha)$ are the phase functions of the nucleus and the dust, at phase angle α, $f_1(r)$ and $f_2(r)$ are the functional behavior of the gas and dust production with r. The equation (6.33) can be simplified with the assumption that the cometary activity is basically given by the production rate of hydrogen Q_H. This is a reasonable assumption as H_2O is the most abundant molecule in a comet and H is the dissociated product of H_2O. If the gas and dust are well mixed and remain almost the same with several apparitions, then one can write

$$Cf_2(r) \simeq C(r)Q_H(r). \quad (6.34)$$

The observed production rate of C_2 of several comets appears to vary roughly as a quadratic function of Q_H. Therefore, one can write to a first approximation,

$$Bf_1(r) \approx B(r)Q_H^2(r). \quad (6.35)$$

Therefore, Eq. (6.33) can be written in the form

$$P_n R_n^2 \phi_n(\alpha) + B(r)Q_H^2(r) + C(r)Q_H(r)\phi_d(\alpha) = r^2 10^{0.4[m_\odot - m_1(r) - 5\log \Delta]} \quad (6.36)$$

where R_n and P_n represent the radius of the nucleus and geometric albedo respectively. The Eq. (6.36) can be solved as a quadratic equation in $Q_H(r)$. For further simplification one can write the ratio of the dust scatter to that of the gas in the same band pass as

$$\delta = \frac{C(r)Q_H(r)\phi_d(\alpha)}{B(r)Q_H^2(r)}. \quad (6.37)$$

The term $B(r)$ should be proportional to the local lifetime of the molecule and so it can be written as

$$B(r) = R\tau_{c_2}(r/r_0)^2 \tag{6.38}$$

with the C_2 lifetime referring to $r = r_0 = 1 au$. Here R refers to the resonance fluorescence efficiency. The Eq. (6.36) can be written as

$$P_n R_n^2 \phi_n(\alpha) + (1+\delta) R\tau_{c_2} Q_H^2(r) = r^2 10^{0.4[m_\odot - m_1(r) + 5\log\Delta]}. \tag{6.39}$$

This gives the solution for the hydrogen production rate as

$$Q_H(r) = \left[\frac{r^2 10^{0.4(m_\odot - m_1(r) + 5\log\Delta)} - P_n R_n^2 \phi_n(\alpha)}{(1+\delta) R\tau_{c_2}}\right]^{1/2}. \tag{6.40}$$

If the second term involving $P_n R_n^2$ is smaller than the first term, which is the case at shorter distances, it is easier to calculate Q_H from the observed light curve.

6.2. Results

6.2.1. *OH and H*

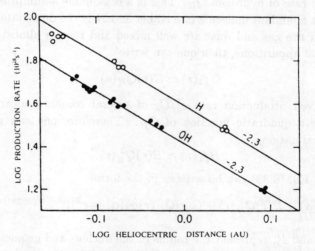

Fig. 6.2. A plot of the production rates of OH and H as a function of heliocentric distance for Comet Bennett (Keller, H. U. and Lillie, C. F. 1974. *Astr. Ap.* **34** 187.)

Extensive observations of the Lyman α of hydrogen as well as of OH emission at 3090 Å have been carried out for various comets using rockets and satellites. These observations have been used to extract the production rate of OH and H. These studies were essential in giving a clue as to whether H_2O is the parent molecule of OH and H, as the direct detection of H_2O was made only in 1986 on Comet Halley. If OH and H arise from the dissociation of H_2O, the ratio is $Q(H)/Q(OH) \approx 2$. The analysis of the brightness measurements for the Comet Bennett showed that the dependence of the production rates of H and OH with heliocentric distance had almost the same variation as $r^{-2.3}$ (Fig. 6.2). This implies that both H and OH must have come out of the same parent molecule, which presumably is water. The derived production rates for OH and H at 1au were $(2.0 \pm 0.8) \times 10^{29}$ molecules/sec and $(5.4 \pm 2.7) \times 10^{29}$ atoms/sec respectively. This gives a ratio of $Q(H)/Q(OH) \approx 2.7$. The International Ultraviolet Explorer Satellite (IUE) is well suited for this study as it

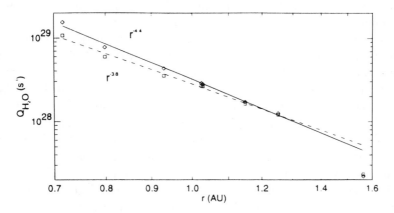

Fig. 6.3. Production rate of H_2O with heliocentric distance for Comet Bradfield (Budzien, S. A., Frestou, M. C. and Feldman, P. D. 1994. Icarus **107**, 164).

covers the wavelength region from 1150 to 3400Å wherein the emissions from H, OH and O, all presumably the decay products of H_2O, lie (Fig. 4.4). Therefore, it is possible to make simultaneous brightness measurements of all these three decay products as a function of the radial distance from the nucleus. Such spectra covering a wide range of heliocentric distances give a homogeneous set of observations. From a fit with the expected brightness distribution with those of observations, the production rate of H_2O can be determined for various heliocentric distances. The results for Comet Bradfield as shown in Fig. 6.3 indicate the dependence of the production rate with heliocentric distance as $r^{-3.8}$ to $r^{-4.4}$. For Comet Austin (1990 V) the variation is around $r^{-1.8}$ to $r^{-2.8}$. The ranges in the power of r for a given comet come about due to the fact whether the effect of solar activity is included or not in the model calculations. The derived production rates of water in Comet Bradfield as shown in Fig. 6.3 can also explain reasonably well the observed brightness distribution of hydrogen and oxygen atoms with two velocity distributions for H atoms of 8 and 20 km/sec (Sec 6.3.1). Therefore, the observed brightness of OH, H and O are consistent with a common source of all the three species, which most probably is H_2O. The heliocentric variation of the production rate of OH for Comet Stephan-Oterma is found to have a dependence, as $r^{-4.8\pm0.6}$ for the pre-perihelion distance. The variations seen in several comets are much steeper than that observed for the Comet Bennett. The typical number densities of H_2O and OH at a distance of $10^4 km$ from the nucleus are of the order of $2 \times 10^5/cm^3$ and $4 \times 10^4/cm^3$ respectively.

The study of the 18 cm line of OH in the radio region should also give information about the production rates and the velocity fields in the coma. This should supplement the information obtained from the study of the lines in the ultraviolet region. A large amount of work has been carried out on various comets based on the radio frequency lines of the OH radical. Since the radio frequency lines are very narrow and since it is also possible to make high resolution radio observations, it is possible to study the velocity distribution of OH molecule in the coma. In fact, the precision of the velocity field is limited only by the quality of the data. The observed profile is generally asymmetric about the centre. In addition, there is an asymmetry in the spatial brightness distribution in the East-West of the comet's centre. These asymmetries arise due to the fact that the velocities of the OH molecules in the coma are not uniform throughout, i.e., the differential velocity effect is present. The observed asymmetry implies that the OH radicals are not isotropically distributed around the nucleus.

The calculation of the total number of OH molecules from the observed emission is not so simple as the lines arise due to the excitation by ultraviolet radiation where Swings effect is very effective. The expression relating the total flux density emitted by the entire comet to the total number of OH molecules in the ground state is given by

$$F = \frac{A_{ul}\, ikT_{BG}}{4\pi\Delta^2} \frac{2F_u + 1}{8} N_{OH} \qquad (6.41)$$

where T_{BG} is the background emission, F_u is the total angular momentum quantum number of the upper state of the transition, i is the 'inversion' of the lambda doublet (Fig. 5.11) and Δ is the geocentric distance of the comet.

The OH production rate can be calculated from the relation $Q_P = N_{OH}/\tau_{OH}$ where τ_{OH} is the lifetime of the OH molecules. In general, for a better estimate it is necessary to take into account the collisional quenching as well as the distribution of molecules and the molecular outflow velocities in the coma in the interpretation of the observed flux in terms of the total number of molecules and hence the production rate. To specify the distribution of OH in the coma requires a model for the production of OH from H_2O and the destruction of OH due to photodissociation and photoionization. The Haser's model is most commonly used for the analysis of OH observations, i.e.,

$$n(OH) = \frac{Q_P}{4\pi r^2 v_{OH}} \frac{l_{OH}}{l_{OH} - l_P} \left(e^{-r/l_{OH}} - e^{-r/l_P}\right) \qquad (6.42)$$

where $n(OH)$ is the density of the OH molecule. l_{OH} and l_P are the scale lengths (same as β^{-1} of Sec. 6.1.2) for OH and the parent molecule respectively. v_{OH} is the radial velocity of the OH molecule. The Fig. 6.4 shows a plot of the OH-parent production rate as a function of the total visual brightness for a number of comets, both referred to $r = \Delta = 1 au$.

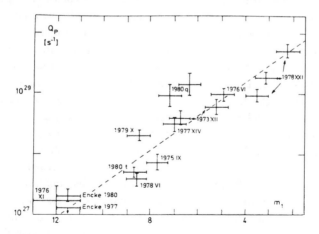

Fig. 6.4. Shows the production rate of OH as a function of the total visual brightness reduced to $r = \Delta = 1 au$ for various comets. The dashed line drawn is for $\log Q_p = 30.33 - 0.28 m_1$ (Despois, D., Gerard, E., Crovisier, J. and I. Kazes, I. 1981. *Astr. Ap.* **99** 320.)

The dashed line represents the relation

$$\log Q_P = 30.33 - 0.28 m_1. \quad (6.43)$$

It is of interest to compare the production rates of OH obtained from both the radio and the ultraviolet observations for the same comet, to see whether there is any discrepancy between the two determinations. It is found that the ultraviolet production rates are systematically higher compared to those of radio values.

There are several causes for this observed discrepancy. First of all the calculation of the production rate of OH is model dependent. It requires the knowledge of lifetime of OH and the velocity distribution. The model also depends upon the heliocentric distance, radial velocity, anisotropic outgassing and the solar cycle. Owing to all these factors, the problem is a complex one. Therefore, the discrepancy appears to be associated with the use of different model parameters which enter in the interpretation of the UV and radio observations. For example, the lifetime of OH used in the two cases are different. In addition, Λ-doublet OH population is sensitive to collisions in the inner coma and so it has to be taken into account. Therefore, the derived production rate of OH from radio observations depends upon the beam width of the radio telescope and so on. Basically, the physics of the problem is complicated and it involves the specification of too many parameters. So far a comparison of the production rate of OH from UV and radio methods were carried out for different dates of observations, radio telescopes and so on. Therefore, in order to understand better this discrepancy, if it is real, is to use simultaneous UV and radio observations so that the physical characteristics of the coma are roughly the same. Through this procedure it may be possible to put tight constraints on the model parameters. Such an exercise was carried out for Comet Halley in 1985. The results of these studies showed that there was reasonable agreement between the radio and the UV OH production rates. Therefore, it appears that the observed difference is present more due to poorly determined model parameters than a real discrepancy.

The study of $18 cm$ line of OH has been carried out routinely on comets. The results for Comet Halley showed a variation of the production rate with the heliocentric distance as r^{-2} for $r < 2AU$. The results based on $O[^1D]$ measurements for Comet Austin (1990 V) showed a variation of $r^{-2.7 \pm 0.2}$.

The production rate of H_2O can also be determined independently of the study of $[OI]$ $6300\text{\AA}$ and $H\alpha(6565\text{\AA})$ observations of comets. This will provide an independent check on the consistency of O and H arising out of H_2O. The Table 6.9 shows the derived values for the production rate of H_2O for Comet Halley, by the two methods based on the observations carried out on the same night and with the same instrument. The agreement between the two is good. Therefore, the production rates of H_2O derived from the $H\alpha$ brightness measurements are consistent with the H_2O production rates derived from the $O[^1D]$ measurements, both of which are based on the coma containing OH, H and O arising out of the dissociation of H_2O. Therefore, at the present time, it is an excellent approximation that H_2O is the source of observed species OH, H and O in comets.

The heliocentric variation of the production rate of H_2O is of particular interest. As already mentioned, if the evaporation of the gases from the nuclei is due to the absorption of the total solar energy, the variation of the production rate of H_2O should also behave as r^{-2}. The results presented earlier showed that for some comets it is in reasonable accord with this dependence. However, for some other comets the variation is vastly different. The reason for such a variation is not clear at the present time. There is an indication that the effect of the presence of dust in the coma is to steepen the heliocentric variation of the gas production. In addition, the evaporation of the gaseous material from the nucleus is dependent upon the nature and the composition of the nucleus. Therefore, more complicated models might have to be considered for explaining the observations.

6.2.2. CN, C_2, C_3, CH, NH_2, CO

(a) CN, C_2, C_3: It is easier to study the spectral lines of the molecules like C_2, CN, etc in comets using the standard techniques as their lines lie in the visual region. It is also possible to get a homogeneous set of data on many comets which are of prime importance for investigating the possible physical correlations like the composition variations or the similarities among the comets of various types. Unfortunately, the observations on a single comet covering a large range of heliocentric distances are still scarce at the present time.

The method based on Eq. (6.31) is particularly suitable for the study of molecules like C_2, CN, etc., as the total flux of a band or a band sequence in a certain field of view can easily be measured using the standard methods. The total number of the molecules can then be calculated from Eq. (6.31) as the constant can be evaluated using the spectroscopic data for the molecules of interest. The production rate can be calculated from Eq. (6.21a) from

Table 6.3. Scale lengths of molecules at 1 au.

Species	Parent(km)[a]	Daughter(km)[a]	Parent(km)[b]	Daughter(km)[b]
$CN(\Delta v = 0)$	1.7(4)	3.0(5)	2.19(4)	3.0(5)
$C_2(\Delta v = 0)$	2.5(4)	1.2(5)	1.6(4)	1.1(5)
$C_3(\lambda 4050)$	3.1(3)	1.45(5)	1.0(3)	6.0(4)
$NH(\Delta v = 0)$	-	-	5.8(3)	4.34(5)
$OH(0,0)$	-	-	4.1(4)	1.16(5)

[a] Cochran, A. L. 1985. *AJ* **90**, 2609.
[b] Schleicher, D. G. *et al.* 1987. *Astr. Ap.* **187**, 531.

Table 6.4. Observed column densities $(/cm^2)$ of CN, C_2 and C_3.

Comet	$r(au)$	$\rho(10^4 km)$	$N(CN)$	$N(C_2)$	$N(C_3)$
	$\bar{r} \simeq 0.7 au$				
Bennett	0.79	2.5	4.1(12)	-	-
Bradfield	0.70	3.7	1.7(11)	4.9(11)	2.2(11)
West	0.65	5.5	1.4(12)	2.5(12)	1.9(12)
Kohoutek	0.66	5.7	3.5(11)	7.0(11)	4.1(11)
Encke	0.71	6.9	5.8(9)	7.1(9)	-
	$\bar{r} \simeq 1.0 au$				
Bradfield	1.0	0.96	1.1(11)	2.5(11)	5.0(11)
West	1.04	5.1	5.9(11)	1.5(12)	1.5(12)
Kohoutek	1.03	6.9	5.9(10)	1.5(11)	9.7(11)
Kobayashi-Berger-Milon	1.07	whole coma	3.0(10)	-	-
Kohler	1.00	2.4	2.0(11)	4.8(11)	4.4(11)
Encke	1.07	0.66	4.7(8)	5.7(9)	6.1(10)

Swift, M. B. and Mitchell, G. F. 1981. *Icarus* **47** 412.

molecular column densities knowing the scale length of the two species and for an assumed outflow velocity, $\sim 1 km/sec$. The production rates can easily be scaled for any other value of the outflow velocity. There is some difference in the scale lengths used by different investigators. It is not clear which of the best values is to be used in the calculation of the production rates. As an example, the scale lengths derived for some molecules of interest are given in Table 6.3. The derived column densities of CN, C_2 and C_3 for several comets are given in Table 6.4 around a mean value of r. The results for the heliocentric variation of the production rates of C_2, CN and C_3 for Comet Bradfield show a dependence as $r^{-3.2}$ (Fig. 6.5) compared to $r^{-1.7\pm0.3}$ for CN observed for Comet West. For the dusty comet

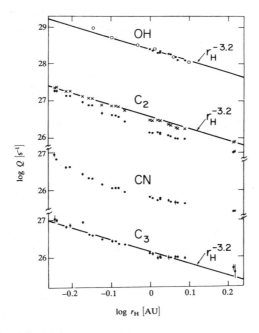

Fig. 6.5. The variation of the production rates of OH, C_2, CN and C_3 with heliocentric distance for Comet Bradfield. Dots are based on ground based data while open circles are from the International Ultraviolet Explorer Satellite. Crosses denote the production rate which is obtained by using for the parent scale length a value which is three times larger than the standard value. The heliocentric variation is also shown by straight lines (A'Hearn, M. F., Mills, R. L. and Birch, P. V. 1981. *Ap. J.* **86** 1559.)

Stephán-Oterma, the production rate of CN and C_2 varied approximately as $r^{-4.2\pm0.3}$ and $r^{-5.6\pm0.4}$ respectively. The curves become even steeper after the perihelion passage. For Comet Halley the pre-and post-perihelion production rate of C_2 varies as $r^{-3.25}$ and $r^{-2.39}$ respectively. Therefore, the heliocentric variation of the production rate of C_2 and CN appears to vary among comets. However, the observed dependences depend to some extent on the scale lengths used in the calculation of production rates. Still the observed dependences are steeper than the r^{-2} dependence required for the equilibrium vaporization model of the nucleus. Therefore, these results cannot be explained by the simple model as was also found earlier for the case of H_2O production rates. It is necessary to invoke other factors for the gas production.

Table 6.5. Average relative molecular production rates for comets.*

$\frac{Q_{OH}}{Q_{CN}} = 330 \pm 160$		
$\frac{Q_{OH}}{Q_{C2}} = 480 \pm 240$	$\frac{Q_{CN}}{Q_{C2}} = 1.5 \pm 0.3$	
$\frac{Q_{OH}}{Q_{C3}} = 5100 \pm 2000$	$\frac{Q_{CN}}{Q_{C3}} = 17 \pm 5$	$\frac{Q_{C2}}{Q_{C3}} = 11 \pm 2$
$\frac{Q_{OH}}{Q_{NH}} = 3800 \pm 1500$	$\frac{Q_{CN}}{Q_{NH}} = 13 \pm 4$	$\frac{Q_{C2}}{Q_{NH}} = 9 \pm 3$

* Results based on six comets. They were reduced using the same model parameters: Schleicher, D. G., Mills, R. L. and Birch, P. V. 1987. *Astr. Ap.* **187**, 531.

It is of interest to look at the relative abundance of the species among comets. The average relative molecular production rates of several species derived from six comets are shown in Table 6.5. There are some comets for which the values deviates from the above average values like comet Giacobini-Zinner. From the results of study of several comets it is possible to come to some general conclusions as the above results refer to comets of various types, dynamical ages, morphologies, continuum to line ratios and so on. Even with such a wide variation in the cometary properties, they all seem to show very little variation in the chemical abundances. This appears to indicate that comets may have been formed out of the same type of material. The non-correlation with the dynamical age of the comet indicates the homogeneous nature of the material with the depth of the nucleus of a comet.This implies indirectly that the differentiation process may not be very important.

The typical production rates at 1 au for CN, C_2, H, OH and H_2O expressed in the unit of $Q(H_2O) = 1$ are roughly the following:

$$Q(H_2O) \approx Q(OH) \approx 1;$$
$$Q(H) \approx 2;$$
$$Q(C_2), Q(CN) \approx 0.01$$

These values indicate that the production rates of C_2 and CN and other species seen in the visual spectral region are less by a factor of about 100 or so compared to that of H_2O or H. Thus, even though the lines of C_2, CN, etc., dominate the visual spectral region, they are only minor constituents of the cometary material. Therefore, the total amount of gas ejected from the nucleus can be represented to a very good approximation by the production rate of H_2O. Also, the dominant source of H from H_2O is an excellent approximation.

With regard to molecular hydrogen, the general feeling is that it should be present in comets. However, the Lyman bands of H_2 arising out of ($B\ ^1\Sigma^+ - X\ ^1\Sigma^+$) transitions occur around $\lambda \sim 1100$Å, whose intensities may be weak due to the lack of solar fluxes available for excitation in this spectral region. Therefore, the most possible way to detect the molecular hydrogen in comets is through resonance fluorescence of Lyman α absorption from the vibrationally excited states of $H_2(v'' \geq 2)$. The emission of the Lyman bands arising out of this process lies in the wavelength region of about 1400 to 1700 Å with a maximum around 1600 Å. So far H_2 lines have not been seen. The future observations of bright comets in this spectral region might help the detection of the lines. Based on the presently available observations around the 1600 Å wavelength region, it is estimated that the value of $Q(H_2)$ is less a factor of 10 or more compared to that of $Q(H)$.

(b) CH, NH_2: There are a few determinations of the column densities of the molecules CH and NH_2 in comets. The column density of CH for the Comet Kohoutek at $r = 0.5$ au is about $4 \times 10^{10}/cm^2$ averaged over $5 \times 10^4\ km$. Similarly the column density of NH_2 at $r = 1au$ is about $6 \times 10^{10}/cm^2$ which is the average value over $3 \times 10^4 km$. These two molecules have also been observed in the Comet West for the heliocentric distances from 0.60 to 1.58 au and only the relative values with respect to C_2 are known. The mean relative value of CH and NH_2 with respect to C_2 integrated over the whole coma is determined to be 0.013 ± 0.003 and 0.030 ± 0.013 respectively for distances $r \leq 1au$. These values show that the abundance of CH is approximately 1 to 2% of C_2 and that of NH_2 about 3% of C_2. The derived production rate of CH and NH_2 for Comet Halley at $r = 0.89$ au is $(4.1 \pm 1.2) \times 10^{27}/sec$ and $(1.6 \pm 0.3) \times 10^{27}/sec$ respectively. This gives a ratio for the production rate of CH and NH_2 with respect to H_2O of about 0.007 and 0.003 respectively.

(c) CO: The ultraviolet observations made on the Comet West at the heliocentric distance of $0.385 au$ showed the strong bands of $(A - X)$ transitions of the CO molecule. The deduced production rate of CO from this observation based on Eq. (6.4) with $\tau(at\ 1\ au) = 1.4 \times 10^6 sec$ comes out to be

$$Q(CO) = 2.6 \times 10^{29}/sec.$$

There is some uncertainty in the above value of $Q(CO)$ due to the uncertainty in the value of the lifetime. The above bands of CO have been seen from several comets, like Bradfield, Halley, Levy and so on.

The derived production rate of CO relative to H_2O in Comet West was around 0.3 and in Comet Bradfield the ratio was about 0.02. Comet Levy gave a ratio of 0.11 ± 0.02. Comet Halley gave a ratio of ≈ 0.17 but showed an interesting result. The results derived from NMS measurements on Giotto spacecraft showed that $\lesssim 7\%$ of CO in the coma of Comet Halley comes from the nucleus and the rest is apparently produced from a distributed source. This result came from a comparison of the observed distribution of molecules with $M/q = 28(CO)$ with $m/q = 18(H_2O)$. The observations of $M/q = 18$ showed a R^{-2} variation consistent with simple radial expansion from the nucleus, while the distribution of $M/q = 28$ required an additional source with a distribution peaking around $10^4 km$ from the nucleus. This could possibly arise from the evaporation of CHON particles.

6.2.3. CS, S_2

The other minor species whose production rates are of interest are the sulphur containing molecules, CS and S_2. The (0,0) band of CS at 2576Å has been observed in many comets mainly with the IUE satellite. The derived column densities for a few comets are given in Table 6.6. If the observed CS comes from CS_2, then $CS_2/H_2O \approx CS/H_2O \approx 10^{-3}$. Therefore, CS_2 is also a trace constituent of the nucleus.

The derived production rate of S_2 for the Comet IRAS-Araki-Alcock observed at $\Delta = 0.032 au$ is about $2 \times 10^{25}/sec$. This is roughly about 10^{-3} times of the H_2O production rate for the same distance.

6.2.4. Ions

The plasma tail is dominated by molecular ions of various kinds. The most dominant among them are the ions CO^+ and H_2O^+ whose lines lie in the near blue and in the visible regions respectively. The early estimates

Table 6.6. Comparison of column densities of some species in comets.[a]

Comet	Date	$r(au)$	$\Delta(au)$	N(O) ($10^{13}/cm^2$)	N(CS)[b] ($10^{12}/cm^2$)	N(C) ($10^{12}/cm^2$)	N(S) ($10^{12}/cm^2$)
Bradfield	16 Jan 1980	0.80	0.40	2.5	0.8	1.7	1.7
	31 Jan 1980	1.03	0.29	0.8	-	0.68	1.4
Encke	5 Nov 1980	0.81	0.32	0.40	0.48	0.27	0.59
	24 Oct 1980	1.01	0.29	-	0.09	-	-
Tuttle	7 Dec 1980	1.02	0.50	1.2	0.34	1.0	1.3
Borrelly	6 Mar 1981	1.33	1.61	-	0.13	-	-
Stephen-Oterma	7 Dec 1980	1.58	0.59	-	≤ 0.14	-	-
Meier	6 Dec 1980	1.52	1.89	-	≤ 0.17	-	-
Panther	6 Mar 1981	1.73	1.39	-	≤ 0.17	-	-

[a] Weaver, H. A., Feldman, P. D., Festou, M. C., A'Hearn, M. F. and Keller, H. U. 1981. *Icarus* **47**, 449.
[b] Sanzovo, G. C., Singh, P. D. and Huebner, W. F. 1993, *AJ* **106**, 1237.

for the column densities of various ions were made based on the observations of Comets Kohoutek and Bradfield for a projected distance of about 10^4 km from the nucleus and for $r = 0.5au$. The deduced column density ($/cm^2$) of CH^+, CO^+ and N_2^+ for Comet Kohoutek are of the order of $10^{10.9 \pm 1.1}$, $10^{12.6 \pm 0.9}$ and $10^{10.7 \pm 1}$ respectively. The above column densities for Comet Kohoutek roughly correspond to density of about 1 to 10^3 particles/cm^3 at a distance of $10^4 km$ from the nucleus.

An analysis of the surface brightness profiles of H_2O^+ in Comet Bennett and CO^+ in Comet West has been investigated. The calculated production rate of CO^+ for $r = 0.440$ and 0.842 au is $Q(CO^+) = 2.4 \times 10^{28}/sec$ and $0.22 \times 10^{28}/sec$ respectively with the heliocentric distance variation of $r^{-4.6 \pm 1.0}$. The estimated production rate of H_2O^+ for Comet Bennett is about $Q(H_2O^+) = 2 \times 10^{26}/sec$ at $r = 0.841au$. This value may be compared with the early estimates of $Q(H_2O^+) = 3.7 \times 10^{24}/sec$ for Comet Kohoutek at $r = 0.9au$ and $Q(H_2O^+) = 2 \times 10^{23}/sec$ for Comet Bradfield at $r = 0.6au$.

Table 6.7. Observed and theoretical values of $Q_{H_2O^+}/Q_{H_2O}$.

Comet	$r(au)$	Solar min	Solar max
Austin 1990V	0.514-1.190	-	2.6×10^{-3}
Halley 1986 III	0.828-1.709	1.2×10^{-3}	-
Halley 1986 III	0.76	3.5×10^{-3}	-
Bennett 1970 II	0.481	-	7×10^{-4}
Liller 1988 V	1.16	-	6.7×10^{-3}
Levy 1990 XX	1.46	-	5×10^{-3}
Theoretical Values		5.9×10^{-2}	6.9×10^{-2}
		3.2×10^{-2}	3.9×10^{-2}

(Adapted from Schultz, D., Scherb, F. and Roesler, F. L. 1994, *Icarus* **104**, 185. The references to individual values are given in this paper).

The production rate of H_2O^+ can also be determined from the Doppler shifted emission lines of H_2O^+ 6159Å and 6147Å emission doublets in the (0, 8, 0) band based on the Fabry Perot technique. The observed ratio $Q(H_2O^+)/Q(H_2O)$ for several comets is shown in Table 6.7. The theoretical value for the same ratio based on the photoionization reaction

$$H_2O + h\nu \to H_2O^+ + e$$

is also given in Table 6.7. The derived ratio of $Q(H_2O^+)/Q(H_2O)$ for Comet Halley is 1.2×10^{-3} and for Comet Austin is 2.6×10^{-3}. These values are significantly smaller than the theoretical values. The destruction of H_2O through the reaction

$$H_2O + H_2O^+ \to H_3O^+ + OH$$

does not seem to account for the low values of observed H_2O^+. There is the possibility that there could be uncertainties in the theoretical parameters or it is possible that H_2O^+ ions produced outside the collision zone may have been accelerated to high velocities by the solar wind so that the Doppler shifted lines went beyond the range of the instrument. There is however no satisfactory explanation for the observed discrepancy.

The presence of CO_2^+ ion in comets implied the presence of CO_2. But CO_2 was observed for the first time in Comet Halley from the infrared spectrometer on board the Vega spacecraft which detected ν_3 fundamental band near 4.3μm. The abundance of CO_2 relative to that of H_2O is around

4% for Comet Halley. Therefore, CO_2 is also a trace constituent of the nucleus.

6.2.5. Complex molecules

Despite an earnest effort, only a few of the molecules have definitely been identified from the observations in the radio frequency region. It is, therefore, interesting to calculate the expected intensities of rotational lines from molecules to see whether they could be observable.

The observations in the radio frequency region referring to source intensity are generally expressed in terms of equivalent brightness temperature, T_B. This is related to the line optical depth τ through the relation, which can be derived from the radiative transfer, as

$$T_B = (1 - e^{-\tau}) \frac{h\nu}{k} \left[\frac{1}{e^{h\nu/kT_{ex}} - 1} - \frac{1}{e^{h\nu/kT_{bg}} - 1} \right]. \qquad (6.44)$$

Here T_{ex} is the excitation temperature, T_{bg} is the background brightness temperature which can be taken to be 2.7K and ν is the line frequency. The optical depth at the line centre is given by

$$\tau = \frac{c^2}{8\pi\nu^2} \frac{2J+3}{2J+1} A_{J+1 \to J} \frac{<N_J>}{\Delta\nu} [1 - exp(-h\nu/kT_{ex})] \qquad (6.45)$$

where $\Delta\nu$ is the line width in frequency units and $<N_J>$ is the mean volume density within the observed field in the rotational state J. For most of the molecules, the lines are optically thin and therefore $(1 - e^{-\tau}) \approx \tau$ in Eq. (6.44). The average excitation temperature T_{ex} can be calculated from the relation

$$\frac{<N_{J+1}>}{<N_J>} = \frac{2J+3}{2J+1} \exp\left(-\frac{h\nu}{kT_{ex}}\right). \qquad (6.46)$$

In order to proceed ahead, the excitation mechanism of the molecules in a cometary atmosphere is required. To model the radio emission from a molecule, it is necessary to consider the excitation scheme of the molecules by taking into account the rotational, the vibrational and the electronic states of the molecule. However, for most of the molecules in the coma, the main excitation process is the pumping of the fundamental bands of vibration by the solar infrared radiation field and the thermal excitation by collisions in the inner coma. Such species do not have a significant electronic excitation because they are generally predissociative and also they lead to destruction rather than fluorescence. An important exception is the very stable CO molecule. In the case of CO molecule, the excitation rate of the ground state $X^1\Sigma^+$ vibrational levels by solar infrared radiation field at 1au is about $2.6 \times 10^{-4}/sec$ for $v'' = 0$ to 1 and $3 \times 10^{-6}/sec$ for $v'' = 0$ to 2. The rate is $\sim 10^{-6}$ to $10^{-9}/sec$ for excitation from the ground level to higher electronic states. Hence, the excitation rates are too low to populate electronic states as well as $v'' = 1$ level. Therefore, it is a very good approximation to consider the excitation starting from the $v'' = 0$ ground vibrational state.

To a very good approximation, the radiative excitation is dominated by one or two vibrational bands. The balance between the infrared excitation and spontaneous decay completely determines the rotational distribution at fluorescence equilibrium. The fluorescence equilibrium can be evaluated without much difficulty as it depends on the molecular constants and the solar radiation field. However, the collisional excitation due to collisions with H_2O molecules is more complicated and is also rather uncertain due to the lack of knowledge about cross-sections. The kinetic temperature of the coma gas has also to be known. One assumes spherical symmetry with constant expansion velocity, v_{exp} of the gas and a reasonable value for collision cross-section between the molecule and H_2O and the coma temperature distribution derived from hydrodynamic models. The molecular mean column densities can be derived by volume integration within the instrument beam. The excitation temperature, T_{ex} can therefore be estimated from Eq. (6.46). The expected rotational line intensities can be calculated for reasonable values of the relevant parameters in Eq. (6.44). On the other hand, from the observed intensities, it is possible to calculate the mean column density of the molecule. The production rate of the molecule can then be calculated from the relation (6.25).

As an example the relative population distribution of linear molecules CO and HCN is shown in Fig. 6.6. For such molecules, the rotational population distribution is determined by the ratio of total infrared excitation rate to the spontaneous decay rate of the rotational levels. Therefore, the molecules with large rotational Einstein A coefficients like that of HCN molecule, will have population concentrated mostly in the lowest rotational levels as compared to molecules like CO and HC_3N with small rotational Einstein A values which have populations over a wider range of rotational levels. Therefore, this explains the observational result as to why the line $J = 1 \rightarrow 0$ of HCN is seen in comets but not the case of the CO molecule

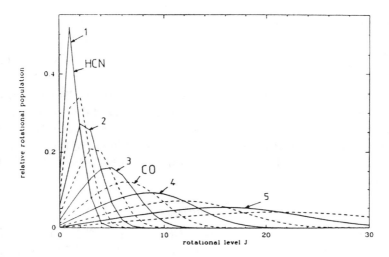

Fig. 6.6. The rotational population distribution under resonance fluorescence equilibrium for CO and HCN is shown as a function of rotational quantum number and for logarithmic of ratio of the infrared excitation rate to the rate of rotational de-excitation from the $J = 1$ level (Bockelee-Movan, D. and Crovisier, J. 1985. *Astr. Ap* **151**, 90.)

even though CO is more abundant than HCN. The production rate of HCN is around 10^{-3} times that of H_2O indicating that HCN is a minor constituent of the cometary nucleus.

In the list of molecules searched for in Comet Halley, HCN and H_2CO were detected. Other molecules such as CH_3CN and NH_3 whose detections were reported in an earlier comet were not detected in Comet Halley. The nondetection of most of the molecules of Table 4.1 is in general agreement with what is known about the molecular excitation of the lines and also their abundances in comets.

6.2.6. O, C, N, S

(a) Oxygen: The abundances or the production rates of oxygen atoms can be calculated from the observation of red doublet lines of oxygen atom occurring at 6300 and 6364Å. Since the line at 6300Å is quite strong in comets, it has been studied extensively for deriving the production rates of oxygen atoms. Since these lines arise due to the dissociation of some parent molecule which populate the ^1D state (Fig. 4.9), the column densities of the parent oxygen atoms can be easily obtained. The observed intensity of the line is therefore given by

$$I \propto \frac{\alpha \beta N}{\tau_P} \qquad (6.47)$$

where α and β are the yields for the particular state for dissociation and branching ratio respectively. N is the column density of the parents and τ_P is the dissociation lifetime of the parent. The oxygen ^1D production rate of the comet can be calculated using the observed fluxes of the oxygen line that has been corrected for various blending and other effects (F_{corr}). The photon rate is given by

$$P = \frac{F_{corr}\, 4\pi \Delta^2}{(h\nu)_{6300}}. \qquad (6.48)$$

Here Δ is the comet-earth distance. If the lifetime of the ^1D state is τ sec for the 6300Å line ($\sim 100\,sec$), multiplying P by the lifetime will give the total number of oxygen atoms within a cylindrical column of radius ρ, $M(\rho)$, centred on the nucleus, i.e.,

$$M(\rho) = \frac{F_{corr}\, 4\pi \Delta^2 \tau}{(h\nu)_{6300}}\; atoms. \qquad (6.49)$$

The production rate is generally calculated using the Haser's model with H_2O as the parent molecule and O as the daughter-product with corresponding scale lengths l_{H_2O} and l_0 respectively.

Fabry-Perot observations:

The main difficulty in the measurement of neutral oxygen from 6300Å emission is the contamination due to NH_2 emission lines. High resolution spectroscopy allows the separation of spectrally NH_2 lines from 6300Å line thereby providing an uncontaminated 6300Å line for quantitative studies. High resolution observations also help to avoid the airglow effects. The Fabry-Perot spectrometer can be used for this purpose. An example of the scan of 6300Å line is shown in Fig. 6.7. For the calculation of the production rate of $O[^1D]$ atoms, it is necessary to know the photodissociation of molecules responsible for the production of $O[^1D]$ atoms. $O[^1D]$ may be produced by the dissociation of H_2O and OH through the reactions

$$H_2O + h\nu \rightarrow H_2 + O[^1D]$$
$$H_2O + h\nu \rightarrow H + OH$$
$$OH + h\nu \rightarrow H + O[^1D]$$

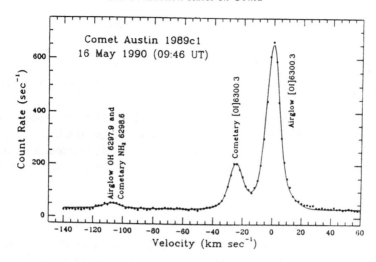

Fig. 6.7. The observation of the cometary $[OI]$ 6300 Å line. The zero point for the velocity scale is set at the position of the terrestrial airglow line. The solid curve represents a Gaussian least square fit to the data (Schultz, D., Li, G. S. H., Scherb, F. and Roesler, F. L. 1992. *Icarus* **96**, 190).

with respective branching ratios $BR1, BR2$ and $BR3$. $O[^1D]$ can also be produced from the reaction

$$CO + h\nu \rightarrow C + O[^1D]$$
$$CO_2 + h\nu \rightarrow CO + O[^1D]$$

with branching ratios BR4 and BR5. However, as CO and CO_2 are minor constituents of comets and their photodissociation lifetimes are much larger than those of H_2O and OH, the contribution from these sources to 6300Å emission is relatively small within the field of view of observations. Therefore, the contribution arising out of CO and CO_2 is generally neglected. The intensity of 6300Å emission at the projected distance ρ from the nucleus is given by

$$I_{6300}(\rho) = \frac{N_{H_2O}(\rho)}{4\pi\tau_1} + \frac{N_{OH}(\rho)}{4\pi\tau_3}. \quad (6.50)$$

Here τ_1 and τ_3 are the photodissociation lifetime of H_2O and OH leading to $O[^1D]$ channels. N_{H_2O} and N_{OH} are the column densities of H_2O and OH respectively at ρ. Using azimuthally symmetric parent-daughter Haser model for representing N_{H_2O} and N_{OH}, the intensity $I_{6300}(\rho)$ can be written as

$$I_{6300}(\rho) \alpha \frac{1}{\rho} \int_0^{\pi/2} \exp\left(\frac{-\rho \sec\theta}{l_1}\right) d\theta$$

$$+ \frac{BR_2 \cdot BR3}{BR1 \cdot \rho} \left[\frac{l_1}{l_1 - l_2}\right] \int_0^{\pi/2} \left\{\exp\left(\frac{-\rho\sec\theta}{l_1}\right) - \exp\left(\frac{\rho\sec\theta}{l_2}\right)\right\} d\theta. \quad (6.51)$$

Here l_1 and l_2 are the Haser parent-daughter photodissociative scale length for the (H_2O, H) and (OH, H) respectively. By integrating Eq. (6.51) with respect to ρ, it is possible to calculate the ratio of the total flux in the 6300Å line to the flux within the field of view of the observation. This ratio is given by

$$AC = \frac{\int_0^\alpha I_{6300}(\rho)\rho d\rho}{\int_0^{\rho_1} I_{6300}(\rho)\rho d\rho} \quad (6.52)$$

where ρ_1 is the projected radius of the field of view of the observation. Hence AC can be calculated, based on the values of the parameters and theoretical branching ratios $BR1$, $BR2$ and $BR3$. The photodissociation scale length can be derived from a comparison of Eq. (6.51) with several high spectral resolution images of 6300Å emission. The production rate $Q(O^1D)$ can be derived from the relation

$$Q(O^1D) = (\frac{4}{3})(4\pi\Delta^2 \Omega I_{6300}) AC. \quad (6.53)$$

Here Ω is the solid angle of the field of view, I_{6300} is the average intensity of the 6300Å emission within the field of view (in photons $/cm^2/sec/ster$). The factor (4/3) takes into account the fact that 1/4 of the O^1D atoms radiate to the ground state via 6364Å channel.

Table 6.8. Branching ratios.

Reaction	Theoretical values
$H_2O + h\nu \rightarrow H_2 + O(^1D)$	0.044, 0.054
$H_2O + h\nu \rightarrow H + OH$	0.840, 0.905
$OH + h\nu \rightarrow H + O(^1D)$	0.103

(Schultz, D., Li, G. S. H., Scherb, F. and Roesler, F. L. 1993, *Icarus* **101**, 95 and is for solar minimum conditions. The individual references to the above values are given in the paper.)

To convert $Q(O^1D)$ to $Q(H_2O)$, the ratio of $Q(H_2O)/Q(O^1D)$ has to be known. This ratio can be determined from the photochemical branching ratios. It can also be determined by comparing the production of $Q(^1D)$ with the results of $Q(H_2O)$ derived from other observations. There is a variation in the derived ratios of $Q(^1D)$ to $Q(H_2O)$. The estimated factor for converting $O[^1D]$ production rate to $Q(H_2O)$ from the branching ratios is given in Table 6.8. It is around 11.7 for pre-perihelion observations and 13.7 for post-perihelion observations for Comet Halley. The conversion factor for pre-and post-perihelion observations are different as the branching ratio $BR3$ depends upon the heliocentric velocity of OH. The resulting production rate of H_2O is compared with those derived from $H\alpha$ observations in Table 6.9, which shows a general agreement between the two determinations.

Table 6.9. H_2O Production rate for Comet Halley from $H\alpha$ and [OI]6300Å observations.

Date of observation	H_2O Production rate ($10^{29} s^{-1}$)	
	$H\alpha$	$[OI]6300Å$
Dec 13, 1985	4.3 ± 0.5	4.42 ± 0.82
— 15, 1985	3.2 ± 0.4	3.71 ± 0.85
— 16, 1985	2.8 ± 0.3	3.22 ± 0.61
Jan 4, 1986	6.3 ± 0.8	7.65 ± 1.5
— 7, 1986	8.3 ± 0.9	9.49 ± 1.8
— 9, 1986	11.6 ± 1.5	12.1 ± 2.3
— 12, 1986	7.8 ± 0.9	11.8 ± 2.3
— 13, 1986	17.5 ± 2.0	18.5 ± 2.6

(Adapted from Smyth, W. H., Marconi, M. L., Scherb, F. and Roesler, F. 1993. *Ap. J.* **413**, 756).

Green line of oxygen at 5577.3Å: The forbidden line of oxygen arising out of the ^1S level which lies at 5577.3Å is highly blended with the Swan band sequence of C_2 corresponding to $\Delta v = -1$. The study of this line can also help in clarifying the problem of the parent molecule of oxygen atoms. The H_2O parent molecule model calculations have indicated the intensity of the green line to be weaker by a factor of 10 or so compared to that of the red line. Therefore, if the green line is detected, the intensity ratio of green to red line could help in this regard. One way to extract the 5577.3Å line of oxygen based on low dispersion spectra is through the comparative spectral synthesis of $\Delta v = -1 C_2$ Swan band sequence. The calculated profile which is a superposition of all the rotational lines arising out of bands (0,1) to (5,6) and which has been corrected for the instrumental effect show a slight excess of flux at $\lambda \sim 5577\text{Å}$ compared to the expected flux which could be attributed to the contribution of the oxygen line of 5577Å. These results indicate that the intensity of the oxygen line at 5577.3Å should be roughly about 3 to 5% of the oxygen line at 6300Å. A similar ratio is obtained from the recent measurements on Comet IRAS-Araki-Alcock wherein the line 5577.3Å could be resolved. These results are, therefore, consistent with H_2O being the parent molecule for excited oxygen atoms.

The ultraviolet lines occurring at 1304Å have also been used to derive the production rates of oxygen atoms (Table 6.6).

(b) Carbon: The brightness of the carbon line at 1657Å arising out of the ^3P state has been used to derive the column densities of carbon for several comets which are given in Table 6.6.

The detection of the carbon line at 1931Å arising from the metastable level ^1D state is of particular interest. This is because the parent molecule should be able to explain the presence of carbon both in the $^3P(1657\text{Å}$ line) and ^1D states.

There are several molecules which can give rise to carbon atoms such as CO and several minor species like CO_2 and so on. The *in situ* measurements of Comet Halley showed that carbon is contained in CHON grains. Therefore, this is also source of carbon atoms. In fact, the inclusion of this carbon component from the grains to the observed component from the fluorescent lines has removed the problem of carbon deficiency in comets.

(c) Nitrogen, Sulphur: It is not possible to make an estimate directly of the production rate of nitrogen as the spectral line has not been observed so far. This is because it has no fluorescence bands above 1200Å. It could only be observable around 1000Å. Also due to Swings effect it is rather difficult to observe the line. Therefore, the estimates have to be made indirectly, based on the nitrogen bearing molecules like CN, NH, N_2^+ etc., and they give a rough value for the ratio $(N/O) \approx 0.1$. The observed brightness of the well known sulphur lines observed near 1812Å has been used to deduce the column densities for sulphur. The results for some comets are given in Table 6.6.

The striking result that comes out of the observations made on several comets in the ultraviolet region is that most of the comets are very similar in nature, in the sense that the relative abundances of minor species with respect to that of water are almost the same.

Table 6.10 gives a rough representation of the elemental abundances of H, C, N, O and S for comets. The table shows that the observed ratio of C/O in comets is similar to the solar value. However, the ratio H/O

Table 6.10. Average abundance ratios.

Ratios	Comets	Sun	CI chondrites
H/O	2.0	1175	0.17
C/O	0.40	0.43	0.10
N/O	0.09	0.13	0.007
S/O	0.03	0.02	0.07

Delsemme, A. H. 1991. In *Comets in the Post-Halley Era* eds. R. L. Newburn, Jr. *et al.* Kluwer Academic Publishers, Vol. 1, p. 377.

is very much smaller in comets compared to the solar value. This implies that hydrogen is very much depleted in comets. This in turn means that either the hydrogen has escaped from the system or it was not trapped at the time of formation. A plausible physical explanation could be that in the case of the sun it is the gravitational binding while in comets it is the chemical binding which is important. Hence, in the case of the Sun, gravity essentially prevents hydrogen from escaping, while in the case of comets only a small fraction of hydrogen can be bound in molecules. Therefore, most of the hydrogen as well as H_2 disappear due to their high volatility. On the other hand, the heavy elements like Na, Ca, Fe, Mn, etc are assumed to have sublimated out of the refractory grains and have relative abundances similar to the solar value (Chap. 5).

6.3. Analysis of Hydrogen Observations

6.3.1. *Analysis of Lyman α measurements*

The presence of a huge halo of hydrogen gas around comets was predicted by Biermann in 1968 based on the dissociation of the H_2O molecule. In general, the photon energy available for dissociation is much larger than the dissociation energy. Therefore, this excess energy is usually carried away by the dissociated products. For example, in the case of H_2O the excess energy gives a velocity $\sim 15 km/$sec to the hydrogen atom, and this is much larger than the gaseous outflow velocity which is of the order of $0.5 km/\text{sec}$. Since the lifetime for the ionization of the molecule is $\sim 10^6$ sec, the observed extent of the size of the halo which is given by $v\tau$ is about 10^6 to $10^7 km$. Based on this simple physical argument and with a hydrogen gas production rate of 10^{30} to 10^{31} molecule/sec, it was concluded that the comets should be bright in the Lyman α line. The first ultraviolet observations made in Lyman α with $OAO-2$ on Comets Tago-Sato-Kosaka and Bennett confirmed the presence of the hydrogen halo extending up to about 10^6 to 10^7 km around the nucleus. This result was further confirmed through a series of rocket and satellite observations made on various other comets. In recent years, with the availability of better instruments and spatial resolution, it has been possible to get good and high quality isophotes of the Lyman α region (Fig. 6.9). These high quality isophotes have shown that they extend farther in the direction away from the Sun compared to that in the direction toward the Sun. The isophotes also become more and more elongated in the antisolar direction as the comet goes nearer the Sun. These observations clearly show the effect of intense Lyman α radiation pressure effects. The high quality isophotes also show that the axis of the isophotes is not along the direction of the Sun-Comet line, but is inclined at an angle to this line. The theoretical models have to explain some of these observed effects.

In order to make a comparison with the observed isophotes, the emission of the hydrogen atoms has to be calculated. The collisions could be important in the inner coma region $\lesssim 10^4 km$ for the gas production rate of about 10^{30} molecules/sec. However, the dissociation takes place at distances $\gtrsim 10^5 km$ and therefore the collisions are not important. For an optically thin case the emission is given by

$$B = gN \ (photons/sec/cm^2) \qquad (6.54)$$

where g is the emission rate factor and N is the column density of hydrogen atoms. Therefore, the calculation of emission intensities essentially depends upon N, which is model dependent.

To start with, one can assume an isotropic radial outflow without collisions from a point source. The above assumption is reasonable since the observed distribution of molecules is nearly spherically symmetric around the nucleus. This is borne out by the shape of the isophotes in the visible region. Therefore, the assumption of a point source with the radial outflow of material usually termed as the *Fountain model* is reasonable for the interpretation of the hydrogen coma. In this steady state model the hydrogen atoms are released at the rate of Q_H from the nucleus and are pushed away due to solar radiation pressure in the direction away from the Sun with a lifetime t_H.

The lifetime t_H for hydrogen atoms is the sum total of lifetimes due to photoionization (t_{ph}) and to charge exchange with protons (t_{pr}). The calculated photoionization lifetime at $1 au$ is $t_{ph} = 1.4 \times 10^7 \text{sec}$. The lifetime for charge exchange with protons of flux $F_w \approx 2 \times 10^8 \text{ proton}/cm^2/sec$ at $1 au$ and for an effective cross-section $\sigma = 2 \times 10^{-15} \, cm^2$ is equal to

$$t_{pr} = \frac{1}{\sigma F_w} = 2.5 \times 10^6 \text{sec}.$$

Therefore, the total lifetime, t_H is given by

$$t_H = \left[t_{ph}^{-1} + t_{pr}^{-1}\right]^{-1} = 2.1 \times 10^6 \text{sec}. \tag{6.55}$$

The acceleration due to the radiation pressure force is given by

$$b = \frac{h\nu}{m_H c} \left(\frac{\pi e^2}{m_e c}\right) f \frac{F_\odot}{r^2}. \tag{6.56}$$

The velocity distribution of the hydrogen atoms is assumed to be Maxwellian and is given by

$$f(v)dv = \frac{4}{\sqrt{\pi}} \frac{v^2}{v_0^2} e^{-(v/v_0)^2} \, dv. \tag{6.57}$$

The most probable speed v_0 is related to the mean velocity v_H by the relation

$$v_H = \left(\frac{2}{\sqrt{\pi}}\right) v_0. \tag{6.58}$$

The density of hydrogen atoms at a distance r from the nucleus and for a time of travel t is given by

$$n_H = \frac{Q_H}{4\pi v_H r^2} e^{-t/t_H}. \tag{6.59}$$

The distance r has now to be expressed in terms of the orbit of the hydrogen atoms after they are released from the nucleus. The co-ordinate system is chosen such that the nucleus is the origin, the Sun is in the negative z direction and the positive x-axis is in the direction of the Earth. The hydrogen atoms leaving the nucleus follow the parabolic orbits and they can reach any point (x, y, z) with a velocity v through two different trajectories. The total density which is made up of these two components is given by the relation

$$n_H(x,y,z) = \frac{Q_H}{4\pi v} [(a \pm \sqrt{x_0^2 - x^2})\sqrt{x_0^2 - x^2}]^{-1}$$
$$\times \exp\left[-\frac{1}{t_H}\left(\frac{2}{b}\right)^{1/2}\sqrt{a \pm (x_0^2 - x^2)^{1/2}}\right]. \quad (6.60)$$

Here $a = z + (v^2/b)$ and $x_0 = \sqrt{x^2 - (z^2 + y^2)}$. The parameter x_0 is a function of y and z and defines a surface of a rotational paraboloid about the z-axis containing all the atoms within the maximum velocity v. The total column density along the line of sight s, including the velocity distribution of atoms, is given by

$$N(s) = \int_0^\alpha \frac{v}{v_H} f(v) 2 \int_0^{x_0} n_H(x,y,z) \, ds \, dv. \quad (6.61)$$

The model-based isophotes of Lyman-α can, therefore be calculated for different values of Q_H, t_H and v_H. The calculated isophotes for Comet Bennett are shown in Fig. 6.8. The number marked on each isophote gives

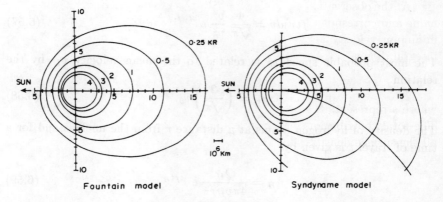

Fig. 6.8. Calculated Lyman α isophotes based on the Fountain and Syndyname models for Comet Bennett for April 1, 1970. The parameters used are: production rate $Q_H = 1.2 \times 10^{30}/sec$, outflow velocity $\nu_H = 7.9 km/sec$, lifetime $t_H = 2.5 \times 10^6 sec$ at 1 au, solar Lyman α flux = 3.2×10^{11} photons cm^2/sec/Å. (Adapted from Keller, H. U. 1976. *Space Sci. Rev.* **18** 641.)

the intensity expressed in Kilo Rayleighs (KR) where 1 Rayleigh = 10^6 $photons/cm^2/sec$. The calculated Lyman−α isophotes based on the above model were in qualitative agreement with the observed isophotes; however in details, there were differences. For example, the calculated isophotes were symmetrical about the sun-comet line while the observed isophotes were inclined at an angle with respect to this line. This reflects the limitations of the model. The main assumption of the model is that the average parameters do not change with time, and this implies a steady state. But this is far from true in a real situation, wherein one has to consider the non-steady situations.

The non-steady state models were then developed and are usually referred to as *Syndyname models*. The formalism is very similar to the case of the formation of the dust tail of comets, wherein one finds dust particles in the tail emitted at different earlier epochs (Chap. 7). The hydrogen atoms coming out of the nuclear region are being acted upon by the forces of radiation pressure and gravitation. The resulting trajectories of the hydrogen atoms in the orbital plane of the comet depend upon the ratio of these two forces generally denoted as $(1-\mu)$. The particles ejected continuously and having the same value of $(1-\mu)$, describe a curve called *Syndyname*. The resultant effect of these two forces is to give curvature to the path of the hydrogen atoms. The total column density of the hydrogen atoms is the integral over all these paths corresponding to all the fictitious source points.

At the observation time t_0, all the points are arranged along the Syndyname and correspond to the same emission time t_e. Let v_P represent some minimum velocity above which the contribution reaches the line-of-sight integral. Then

$$v_P = \frac{s}{(t_0 - t_e)}$$

where s represents the minimum distance from the fictitious source point. The total contribution to the column density at time t is the integral over all the surface densities and is given by

$$N_t(s) = \frac{1}{2\pi} \int_{v_P}^{\alpha} \frac{P_t(v)\, dv}{v^2 t^2 [1 - (v_P/v)^2]^{1/2}} \qquad (6.62)$$

where $P_t(v)dv$ represents the production rate of hydrogen atoms with velocities between v and $v + dv$. Here the subscript t is used to denote the production rate as a function of time, which in turn depends on the heliocentric distance. The total column density arising out of all the fictitious source points on the Syndyname is given by

$$N = \int N_t(s)dt = \frac{1}{2\pi} \int_0^\infty dt \int_{v_P}^\infty \frac{P_t(v)dv}{v^2 t^2 [1-(v_P/v)^2]^{1/2}}. \quad (6.63)$$

If the hydrogen atoms decay with a lifetime t_H which may be a function of t, then the exponential factor $\exp[-\int_0^t t_H^{-1}(t')\,dt']$ has to be included in the expression for $N_t(S)$. Therefore, the expression for the total column density is given by

$$N = \frac{1}{2\pi} \int_0^\infty dt \, \exp\left[-\int_0^t t_H^{-1}(t')dt'\right] \int_{v_P}^\infty \frac{P_t(v)dv}{v^2 t^2 [1-(v_P/v)^2]^{1/2}}. \quad (6.64)$$

The velocity distribution is related to the production rate of hydrogen atoms through the relation

$$P_t(v)dv = \left(\frac{Q_H}{v_H}\right) v f(v) dv. \quad (6.65)$$

The Fig. 6.8 shows the calculated isophotes for Comet Bennett based on the above model for the same parameters of v_H, t_H, Q_H and $F_\odot$ used in the Fountain model. The figure clearly shows that the isophotes are now inclined at a certain angle to the Sun-Comet line and are symmetrical about the Syndyname, and this is consistent with the observations.

The model described above can be improved further by assuming the parent molecules to be released by the nucleus and allowing for the decay of the molecules as they flow outwards. This eliminates the assumption of the hydrogen atoms being emitted from a point source. In this model the parent molecules are released isotropically from the nucleus. These parent molecules could give rise to two or more decays along their paths with a certain value for the scale length for each decay. This in turn is dependent on the lifetime and the velocity of the molecules. The formalism developed in Sec. 6.1.2 can easily be applied to the present situation.

The models described above has been applied to the Lyman$-\alpha$ observations of many comets with great success. A comparison of the observed and the calculated isophotes allows one to determine quantitatively the production rate, the lifetime and the outflow velocity of hydrogen atoms. The detailed observations of Comet Bennett showed that two Maxwellian velocity distributions in the model calculation corresponding to $v_H = 7$ and 21 km/sec with 50:50 of each are required to fit the observations. The high velocity distribution is required mainly to fit isophotes at the outer

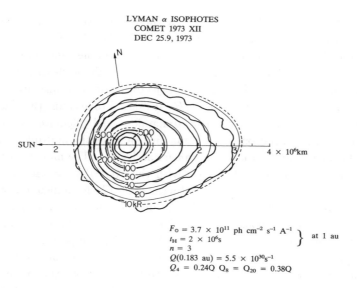

Fig. 6.9. Comparison of the observed (wavy line) and the model isophotes (thin line) for Lyman α in Comet Kohoutek for the parameters given in the insert (Meier, R. R., Opal, C. B., Keller, H. U., Page, T. L. and Carruthers, G. R. 1976. *Astr. Ap.* **52** 283.)

boundaries. The Fig. 6.9 shows a comparison between the expected and the observed Lyman–α isophotes for Comet Kohoutek. The hydrogen production rate for Comet Bennett had a value of about 5×10^{29}/sec at 1 au, an outflow velocity of about $9 \pm 2 km$/sec and $t_H = 2.2 \times 10^6$ sec.

Monte-Carlo approach:

In the formulation outlined above, the source of hydrogen atoms was considered to be an isotropic point source. The model used the calculation of the syndyname on the sky plane to approximate the combined effects of radiation pressure and gravity on the atom trajectories on a large spatially extended hydrogen coma. The model could reproduce the two-dimensional sky-images of Comet Kohoutek, which required three Maxwellian contributions with probable speeds of 4, 8 and 20 km/sec for the H atoms. These are generally consistent with the results expected from the dissociation of H_2O. However, many details are still to be understood. This, therefore, led to the consideration of more realistic models which take into account the isotropic ejection of daughter radicals or atoms at the time of photodissociation of the parent molecule. Therefore, the Monte-Carlo approach has been developed. The model developed is a three-dimensional time dependent Monte-Carlo particle trajectory model (MCPTM). It takes into account the physical processes which are important in the inner and outer coma of comets. In particular, it includes a physically realistic description of the detailed production mechanism and trajectories of H atoms produced by the photodissociation of H_2O and OH. It takes into account the solar radiation pressure acceleration. The gas dynamic model calculates the velocity and temperature variation in the coma and includes the multiple collisions, photochemical heating, radiative transfer and so on. The model couples the simple gas-dynamic flow of the gas with the MCPTM formalism. Therefore, the gas-dynamic model coupled with MCPTM formalism considers in a consistent way and also in a realistic manner the actual trajectories of as many as $\sim 10^5$ radicals or atoms. The space and column densities and hence the emission rates are then calculated. For comparing with the observation, the calculated results have to be referred to the projection on the sky-plane which in turn depend upon the relative geometry of the Earth, the Sun and the comet at the time of the observations. These are then compared with observations. The calculated Lyman-α isophotes are compared to the observed Lyman-α coma images of Comet Kohoutek in Fig. 6.10 and give a good match. The derived production rates of H_2O at heliocentric distance of 0.43au and 0.18au are 9.6×10^{29} and 8.4×10^{30}/sec respectively, as against the values of 6.2×10^{29} and 5.5×10^{30}/sec respectively obtained from the Syndyname model. The difference in the production rates between the two methods arises mainly due to the different solar Lyman α flux used in the two calculations. Therefore, the two methods give basically the same production rate. However, there are fundamental differences in the two methods as pointed out earlier. The velocity distributions obtained from MCPTM results are shown to be consistent with the three Maxwellian velocity distributions required to explain the observed isophotes based on the Syndyname models. Therefore, the MCPTM model can account in a natural way the two or three Maxwellian distributions which are required to explain the observed isophotes based on the Syndyname model. The physically based MCPTM model also appears to show that the Syndyname model is applicable to comets with high production rates at small heliocentric distances.

As is evident from the above brief description of the MCPTM formalism which takes into account in detail the important physical processes in the inner and outer extended coma region, is a very general method of

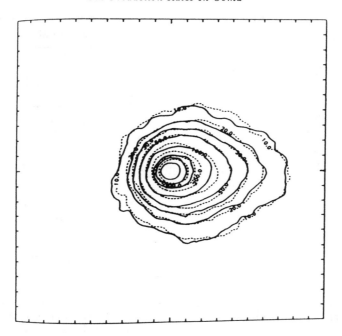

Fig. 6.10. Comparison of the $Ly\alpha$ isophote contours (solid lines) of Comet Kohoutek at $r = 0.18 au$ with the calculated isophotes based on fully time-dependent three-dimensional self-consistent MCPTM result (dashed lines). The contours are in units of kilo-Rayleighs and the distance between the two tick marks in the circumscribed box is $4.0 \times 10^5 km$ (Combi, M. R. and Smyth, W. H. 1988 Ap. J. **327**, 1044).

studying the observations of Lyman α in comets. The additional complications in terms of physics and dynamics can be incorporated as and when the need arises. However, this is a numerical simulation approach and the limitation is related to the computing power.

6.3.2. Analysis of $H\alpha$ observations:

The $H\alpha$ line at $6563 Å$ can also be used to study the daughter species and could in turn be used to monitor the production rate of H_2O. Unfortunately, this line is usually faint and in addition could be contaminated with geocoronal $H\alpha$ emission, diffuse galactic $H\alpha$ emission particularly when the comet is located near the galactic plane and cometary $H_2O^+ (0, 7, 0)$ emission at $6562.8 Å$. However, the observations carried out with the Fabry-Perot spectrometer, the Doppler shift of the cometary $H\alpha$

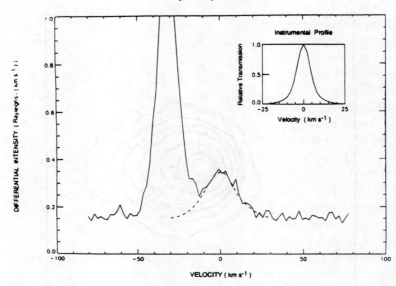

Fig. 6.11. Fabry-Perot observation of $H\alpha(6563A)$ line of Comet Halley taken on January 4, 1986. The smaller feature which is centered at zero velocity is the cometary $H\alpha$ emission while the larger feature centered around $-32km/sec$ is the geocorona $H\alpha$ emission. The model fit to the observed cometary $H\alpha$ is shown by the dashed line (Smyth, W. H., Marconi, M. L., Seherb, F. and Roesler, F. 1993. *Ap. J.* **413**, 756).

may be sufficient to avoid the geocoronal $H\alpha$ emission (Fig. 6.11). The contamination of galactic $H\alpha$ emission and H_2O^+ emission can be minimized by taking proper precautions in making the observations. The commonly used Haser's model is not adequate for the analysis of $H\alpha$ emission as the hydrogen atoms have large dispersion in velocities. These had limited the study of $H\alpha$ emission in comets.

As discussed earlier, the Monte-Carlo Particle Trajectory Model (MCPTM) has been successfully applied to Lyman α observations of comets. This formalism can be modified and used for the analysis of observations of $H\alpha$ emission in comets. The main quantity to be determined in the application of MCPTM model is the g-factor for the $H\alpha$ emission. This has to be calculated taking into account the contributions from all the transitions which can give rise to this emission. This involves the knowledge of the transition probabilities of all the relevant transitions which contribute to this emission and the solar radiation which excites these levels, i.e., through excitation and de-excitation. The principal source of $H\alpha$

emission is through the excitation of H atoms by solar Lyman β photons. The actual profile of Lyman β has to be used since H atoms are in motion. For solar minimum conditions the total flux of Lyman β line at 1 au is $3.5 \times 10^9\,photons/cm^2/sec$ while the corresponding Lyman α flux is $3.1 \times 10^{11}\,photons/cm^2/sec$. For an accurate calculation of g-factor for $H\alpha$ emission, in addition to principal contribution from the $3P$ to $3S$ transitions, several other secondary contributions have also been considered. The derived g-factor is velocity dependent, as in the case of OH (see Fig. 5.11). The MCPTM can be used to simulate observations of $H\alpha$ emission. The calculated and the observed line profile for Comet Halley is shown in Fig 6.11. The derived velocities for such a fit require a broad velocity peak from 5 to 9 km/sec representing the bulk of ($\sim$ 70%) of H atoms released and a velocity $\sim 8km/sec$ due to the dissociation of OH. Additional smaller peaks also appear at around $11km/sec$ ($\sim$ 7%) and $22-26km/sec$ ($\sim$ 20%) due to the formation of H from OH and at 15 to 19 km/sec due to photodissociation of H_2O. A comparison of the derived production rate of H_2O from $H\alpha$ emission with those derived from 6300Å emission observations should help in checking the photochemical processes which give rise to atomic oxygen and hydrogen from the parent molecule H_2O. Fortunately such a comparison can be made for Comet Halley, as measurements of $H\alpha$ and 6300Å emissions were carried out on the same dates, with the same instrument, same calibration procedure and so on. The results derived from such an analysis are shown in Table 6.9. The agreement is reasonable in view of the various uncertainties involved in the calculation of parameters such as, g factor for $H\alpha$ emission and the branching ratios for the production of $O[^1D]$ atoms from photodissociation of H_2O and OH.

6.4. Gas-Phase Chemistry in the Coma

The understanding of the formation of various observed species in the coma has attracted much attention in recent years (Chap. 4). They could be produced either through the break-up of complex molecules or by some other physical process, as the gas expands outwards in the coma. Therefore, the distribution of the observed species in the coma does not necessarily reflect the original gaseous material ejected from the nucleus. In addition to the various physical processes that could modify this gas, the gas-phase chemical reactions among various constituents (neutral and ions) in the inner coma could also alter the original gas. The chemical reactions can take place due to the fact that the timescales involved for many of the reactions are much faster than the solar photodestruction rates at a heliocentric distance of 1 au. In fact, the early investigations did indicate that ion-molecule reactions could be quite important and various reaction sequences were considered. They did produce substantial amounts of various types of species that have been observed in comets. However, the reaction schemes considered were quite simple so that the calculations could be carried out without much difficulty. But in a real physical situation various networks of chemical reactions have to be considered between neutrals and ions of different atoms and molecules. This has been made possible with the availability of computers in recent years. They have also been helpful in calculating more realistic and time dependent models which take into account a large number of physical effects.

The nucleus is assumed to be spherical in shape containing the volatile material in certain proportions. The gases evaporate from the nucleus as a result of the solar heating. The vaporization theory (Chap. 11) could be used to calculate the gas production rate, outstream velocity, temperature, etc., which is consistent with the assumed mixture. The total number density present in the expanding coma gas at a distance r from the nucleus is related to the density at the nucleus surface $n(R_0)$ through the conservation of mass by

$$n(r) = n(R_0) \left(\frac{R_0}{r}\right)^2. \tag{6.66}$$

Here R_0 is the radius of the nucleus. To take into account the various physical processes and the chemical reactions which modify the particular species, it is necessary to have the particle conservation equation. The equation for a general case for a species i can be written in the form

$$\frac{\partial n_i}{\partial t} + \nabla \cdot n_i v_i = \sum_i R_{ji} - n_i \sum_k R_{ik}, \tag{6.67}$$

where n_i and v_i denote the number density and the flow velocity respectively. The first and the second terms on the right-hand side of the equation denote the rate of formation of species i, by processes involving species j and the rate of destruction of species i through processes involving species k. The whole set of equations given by Eq. (6.67) are coupled to each other through sources and sinks which are on the right-hand side of the equation. The assumption of spherical symmetry and a constant outflow velocity for all the species will simplify the equations considerably. Photodissociation and photoionization processes as well as the opacity effects

Table 6.11. Gas phase chemical reactions with examples.

Reaction type	Example
Photodissociation	$h\nu + H_2O \rightarrow H + OH$
Photoionization	$h\nu + CO \rightarrow CO^+ + e$
Photodissociative ionization	$h\nu + CO_2 \rightarrow O + CO^+ + e$
Electron impact dissociation	$e + N_2 \rightarrow N + N + e$
Electron impact ionization	$e + CO \rightarrow CO^+ + e + e$
Electron impact dissociative ionization	$e + CO_2 \rightarrow O + CO^+ + e + e$
Positive ion-atom interchange	$CO^+ + H_2O \rightarrow HCO^+ + OH$
Positive ion charge transfer	$CO^+ + H_2O \rightarrow H_2O^+ + CO$
Electron dissociative recombination	$C_2H^+ + e \rightarrow C_2 + H$
Three-body positive ion-neutral association	$C_2H_2^+ + H_2 + M \rightarrow C_2H_4^+ + M$
Neutral rearrangement	$N + CH \rightarrow CN + H$
Three-body neutral recombination	$C_2H_2 + H + M \rightarrow C_2H_3 + M$
Radiative electronic state deexcitation	$O(^1D) \rightarrow O(^3P) + h\nu$
Radiative recombination	$e + H^+ \rightarrow H + h\nu$
Radiation stabilized positive ion-neutral association	$C^+ + H \rightarrow CH^+ + h\nu$
Radiation stabilized neutral recombination	$C + C \rightarrow C_2 + h\nu$
Neutral-neutral associative ionization	$CH + O \rightarrow HCO^+ + e$
Neutral impact electronic state quenching	$O(^1D) + CO_2 \rightarrow O(^3P) + CO_2$
Electron impact electronic state excitation	$CO(^1\Sigma) + e \rightarrow CO(^1\Pi) + e$

(Huebner, W. F., Boice, D. C., Schmidt, H. U. and Wegmann, R. 1991. In *Comets in the Post-Halley Era*, Eds. R L Newburn, Jr. et al., Kluwer Academic Publishers, p. 907).

for the incident solar radiation are taken into account. The various chemical reactions between ions, electrons, neutrals and free radicals have been considered (Table 6.11). A step-by-step solution of the coupled Eq. (6.67) gives the number density of the species and its variation with r. Figures 6.12 and 6.13 show the typical results of such calculations for various neutrals and ions. Generally, the calculations are carried out for an assumed mixture of the molecules present in the nucleus. The resulting number densities of the various species are then compared with the observed number densities. The process is repeated by modifying the initial composition till an overall agreement is obtained with the observations. In one set of calculations the component of the nucleus was assumed to be composed of

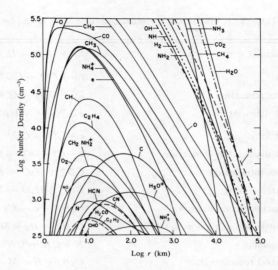

Fig. 6.12. Typical variation of the number density of various species plotted as a function of the distance from the centre of the nucleus. The radius of the nuclues = 1 km and r = 1 au (Huebner, W. F. and Giguere, P. T. 1980. *Ap. J.* **238** 753.)

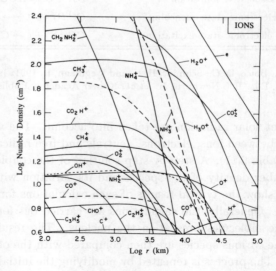

Fig. 6.13. Typical variation of number density of ions plotted as a function of the distance from the centre of the nucleus. (Huebner, W. F. and Giguere, P. T., 1980, *op. cit.*.)

Table 6.12. Comparison of observed and calculated densities.

Species	Comet	$r(au)$	log(column density)$[/cm^2]$		
			Observed	Computed	
				Comp (a)	Comp (b)
CN	West	0.6	12.3	12.8	12.9
		1.0	11.8	12.4	12.4
C_2	West	0.6	12.3	12.4	12.3
	Kohoutek	1.0	12.2	11.6	11.6
C_3	West	0.6	12.0	12.1	12.1
NH_2	Kohoutek	1.0	10.8	10.7	10.7
CH	Kohoutek	0.6	11.5	10.7	10.5

Biermann, L., Giguere, P. T. and Huebner, W. F. 1982. *Astr. Ap.* **108**, 221.

H_2O, CO_2, NH_3 and CH_4 corresponding to the situation of comets being formed in the neighbourhood of giant planets (Table 12.2). These results were not in good agreement with the observed abundances of many species. Some of the discrepancies could be removed by using a mixture composed of some interstellar molecules for the initial composition such as given in Table 6.14. Such a model is one among the various other possible combinations that have been considered for the initial composition of the nucleus. This is due to the fact that there is no unique way of fixing the chemical composition of the nucleus.

The solution of the equations gives the number density of each species. In practice, it is possible to observe only column densities. These can be calculated by integrating the number density along the line of sight. However, the column density being an integrated effect, has the disadvantage of washing away the fine structure. The results of model calculations show that for a molecule H_2O, the variation with the distance from the nucleus is very steep, while for others like OH, CN, NH etc., the curve is flat up to a certain distance and then falls off. The abundances of various species converge for large values of R. The variation of column density of H_2O with R, essentially goes as $1/R^2$, arising out of the dilution effect. For larger distances, the curve falls more steeply due to the increase in the photodestruction rates. The column densities of C_2 and O are flat up to projected distances, $\log \rho \sim 5$, while for CN it is not so. That is basically related to the formation and destruction mechanisms of the molecule. If the molecule is continuously being formed by different processes, then the column density will not change very much with R. On the other hand, if the molecule is only destroyed, then the column density has to decrease with R. The calculations also show that with a decrease in the heliocentric distance, the maximum of the density profiles of radicals are found to move away from the nucleus, which is a consequence of the opacity effect. This arises due to the fact that for smaller heliocentric distances the ultraviolet opacity increases, shielding the solar radiation penetrating inside, and this results in the reduction of the photodestruction process in the inner coma. A comparison of the results for some of the observed species with model calculations for Comets Kohoutek and West is shown in Table 6.12. The agreement is quite satisfactory considering the fact that the uncertainties in the observed abundances are around +0.3 to +0.5. The calculated heliocentric variation of several neutral species is also in rough agreement with the observed variations.

It is interesting to note from the model calculations (Fig. 6.13) that the most abundant ions in the inner coma should be H_3O^+, NH_4^+ and others. Therefore, in a H_2O dominant comet, the most abundant ion in the inner coma should be H_3O^+. The dominant H_3O^+ in the inner coma arises as a result of the production of H_3O^+ through photoionization of H_2O followed by ion-neutral reaction and destruction by dissociative recombinations,

i.e., $\quad\quad H_2O \to H_2O^+ + e$,
$\quad\quad\quad\quad H_2O^+ + H_2O \to H_3O^+ + OH$
and $\quad\quad H_3O^+ + e \to H_2O + H$.

The general conclusion that comes out of several independent model calculations is that the icy nucleus containing some observed interstellar molecules gives better agreement with the abundances of observed radicals. These results are interesting because they show the possible similarity between the composition of the material present in comets and in interstellar clouds (see Chaps. 9, 13).

The results discussed so far, based on the model calculation of the gas-phase chemistry of the coma, should be taken as an indication of the trend of the expected abundances of various neutrals and ions rather than the actual numbers themselves. This is due to the fact that various assumptions have been made in the model calculations as well as due to the inadequate knowledge of the various phenomena that could take place in the inner coma. For example, the nucleus is assumed to be spherical in shape and homogeneous in composition. The effect of dust on the chemical reaction has not been taken into account. The outgassing from the nucleus is assumed to be symmetrical and the radiation pressure effects are neglected. In addition, the effect of the interaction of the solar wind with the ions might have to be considered as it will have a direct effect on the ions and their flow. Even with these simplifying assumptions, the model calculations are highly complex in nature. It is, therefore, encouraging that even with simple models, it is possible to account for the observed abundances of various species reasonably well. However, many attempts have been made to take into account several of the above effects in the model calculations.

Another approach that has been followed is to take into account several of the physical effects mentioned above but, restrict the chemistry to a few reactions.

6.4.1. *In situ mass spectrometer for ions*

The *in situ* measurements carried out with spacecrafts on Comet Halley gave confirming evidence for some of the basic ideas of gas-phase chemistry discussed earlier. In fact the gas-phase chemistry has been used extensively in the interpretation of ions observed in the inner coma of Comet Halley to extract the production rates of ions.

The ion mass spectrometer on board the Giotto spacecraft has provided important information about the ions present in the coma of Comet Halley to a distance $\sim 1000 km$. The mass resolution is excellent. This makes it easier to calculate the ion densities. The instrument measures the distribution of m/q. Therefore, a careful analysis of the data with the theoretical modeling is required before the peaks can be identified with certainty. However, the singly ionized peaks can easily be identified. The peaks at $m/q = 18, 17$ and 16 gave the direct detection of H_2O as these peaks belong to the H_2O group of ions, i.e., H_2O^+, OH^+ and O^+. The peak at $m/q = 12$ is certainly C^+. The peak at $m/q = 19$ is almost certainly H_3O^+ which supported the chemical models. Molecular ions have been detected ranging beyond 100 amu. The major ion species and also some of the secondary ones which can contribute to the observed m/q peaks are shown in Table 6.13. The important reactions for forming some of these ions are shown in Fig. 6.14. As can be seen from Table 6.13, NH_4^+ can

Table 6.13. Dominant ion species contributing to mass range detected in the ion mass spectrometer.

M/q	Ion Species	M/q	Ion Species
12	C^+	34	$^{34}S^+$, H_2S^+
13	CH^+, $^{13}C^+$	35	H_3S^+
14	CH_2^+, N^+	36	C_3^+
15	CH^+_3, NH^+	37	C_3H^+, $H_3O^+ \cdot H_2O$
16	O^+, CH_4^+, NH_2^+	38	C_2N^+, $C_3H_2^+$
17	OH^+, NH_3^+, CH_5^+	39	$C_3H_3^+$
18	H_2O^+, NH_4^+	40	$C_3H_4^+$, CH_2CN^+
19	H_3O^+	41	$C_3H_5^+$
20	$H_2^{18}O^+$	42	-
21	$H_3^{18}O^+$	43	CH_3CO^+
22	-	44	CO_2^+, CS^+, $C_3H_8^+$
23	-	45	HCS^+, HCO_2^+
24	C_2^+	46	$H_2CO_2^+$, NS^+, H_2CS^+
25	C_2H^+	47	H_3CS^+, HNS^+
26	$C_2H_2^+$, CN^+	48	SO^+
27	$C_2H_3^+$, HCN^+	49	HSO^+
28	H_2CN^+, CO^+, N_2^+, $C_2H_4^+$	50	$C_4H_2^+$
29	HCO^+, $C_2H_5^+$, N_2H^+	51	$C_4H_3^+$
30	H_2CO^+, CH_4N^+, NO^+, $C_2H_6^+$	52	$C_3H_2N^+$
31	CH_2OH^+, HNO^+	53	$C_4H_5^+$
32	S^+, O_2^+	54	-
33	HS^+, O_2H^+	55	$C_3H_3O^+$, $C_4H_7^+$

(Huebner, W. F., Boice, D. C., Schmidt, H. U. and Wegmann, R. 1991. In *Comets in the Post-Halley Era*, Eds. R L Newburn, Jr. *et al.*, Kluwer Academic Publishers, p. 907).

also make contribution to the peak at $q/m = 18$. Similarly, NH_3^+ could contribute to the peak at $q/m = 17$. Several peaks have been seen with the positive ion cluster analyzer on Giotto spacecraft with an alternate mass difference of 14 and 16amu as can be seen from Fig. 6.15. A suggestion has been made that it could arise due to ion fragments CH_2 and O breaking

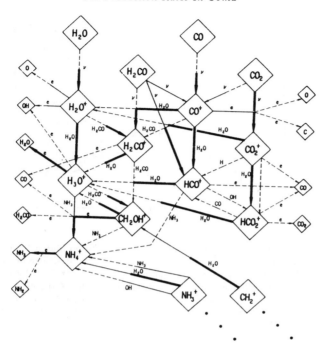

Fig. 6.14. For illustrative purpose, this figure shows the network of major chemical reactions involving the formation and destructive processes that have to be considered in the calculation of abundances of some of the species. The arrow indicates the direction of the reaction. ν represents photodissociation or photoionization and e indicates electron dissociative recombination (Huebner, W. F., Boice, D. C., Schmidt, H. U. and Wegmann, R. 1991. In *Comets in the Post-Halley Era*, eds R. L. Newburn, Jr. *et al.*, Kluwer Academic Publishers, p. 907).

off from the large parent molecule formaldehyde $(H_2CO)_n$, also known as Polyoxymethylene (POM).

To derive the abundances of neutral molecules from the measured densities of the different ion species, a detailed modelling of the ion chemistry in the cometary coma is needed. The elaborate models have taken into account around 1000 reactions involving 59 neutrals and 76 ionized species. A critical parameter is the electron temperature as the ions are produced inside the collision zone. A reasonable temperature profile for the gas is used along with the measured expansion velocity of the gas and ions by the Giotto spacecraft in the analysis of the NMS data. Based on the realistic representation of the relative abundances of the major parent molecules in

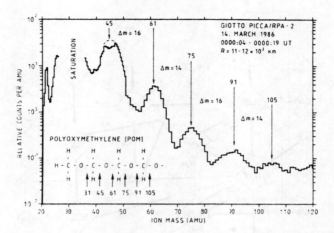

Fig. 6.15. The ion mass spectrum of Comet Halley observed on March 14, 1986 with the Giotto spacecraft at a distance $11 - 12 \times 10^3$ km from the nucleus. The peaks at regular intervals with alternate difference in mass number of 14 and 16 have been interpreted as dissciation production of the molecule polyoxymethylene. (Huebner, W. F. et al. 1987. *Symposium on the diversity and similarity of comets*, ESA-SP 278, p. 163; see also Vanysek, V and Wickramasinghe, N. C. 1975. Astrophys. Space Sci., **33**, L19).

the coma close to the nucleus, the reaction scheme is then used to calculate the expected distribution of ions. This can then be compared with the observed distribution. From such studies, the production rate of the species relative to water can be derived. The analysis has been used extensively for interpreting the observed distribution of ions in Comet Halley. Such studies not only identify the species and give the relative production rate relative to water, but also give rise to the identification of new species. The derived abundances of some of the species are given in Table 6.16. The production rate of CH_3OH and H_2S relative to water is around 1.7% and 0.4% respectively. The Fig. 6.16 shows a model fit to the observed distribution of ion mass spectrum with the composition of the primary gas as given in Table 6.14. The model calculations based on two primary compositions give the same fit to the observed distribution. Therefore, the primary composition cannot be predicted for abundances of the order of 1% or smaller with any realistic accuracy. In addition to various uncertainties involved in the model calculations, the reaction rates of many of the reactions are not known accurately. They also introduce some uncertainties in the derived production rates.

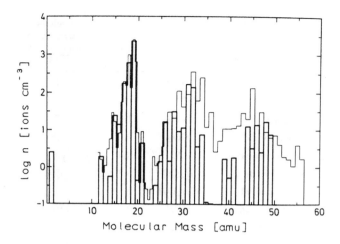

Fig. 6.16. The comparison of the model ion mass spectra with the observed ion mass spectra for Comet Halley. The thin lines show the results obtained from ion spectrometer of the Giotto spacecraft at a distance of 1500 km from the nucleus. The model curve is the best fit for the assumed mixture of primary molecules (Wegman, R., Schmidt, H. U., Huebner, W. F. and Boice, D. C. 1987, *Astr. Ap.*, **187**, 339).

6.5. Temperature and Velocity of the Coma Gas:

The knowledge of the temperature and velocity of the coma gas as a function of distance from the nucleus is required for the calculation of the excitation process as well as for the derivation of molecular abundances. This can, in principle, be derived from the model calculations which take into account the thermodynamical and hydrodynamical processes in the coma. For the description of the state of the gas, the equations of mass, momentum and energy have to be considered. The corresponding fluid dynamic equations with velocity u, mass density ρ and pressure P can be written as

$$\frac{\partial \rho}{\partial t} + \nabla \cdot (\rho u) = \dot{\rho}, \tag{6.68}$$

$$\frac{\partial}{\partial t}(\rho u) + (u \cdot \nabla)\rho u + \rho u(\nabla \cdot u) + \nabla P = \dot{q}, \tag{6.69}$$

$$\frac{\partial}{\partial t}(\rho \frac{u^2}{2} + \rho \epsilon) + \nabla \cdot \left[\rho u(\frac{u^2}{2} + h)\right] = \dot{E}. \tag{6.70}$$

Table 6.14. Initial chemical composition of the nucleus.

Species		Relative Number (%)	
		(a)	(b)
Water	H_2O	100	100
Methane	CH_4	1	2.5
Ammonia	NH_3	1.5	2.5
Carbon dioxide	CO_2	2.0	3.75
Carbon monoxide	CO	15.0	10.0
Formaldehyde	H_2CO	3.8	2.5
Methyl alcohol	CH_3OH	0.8	-
Hydrogen cyanide	HCN	0.1	0.07
Methyl cyanide	CH_3CN	-	0.2
Nitrogen	N_2	0.1	0.08
Methylamine	CH_3NH_2	-	0.1
Acetylene	C_2H_2	1.0	0.84
Nitric oxide	NO	0.2	-
Ethylene	C_2H_4	0.3	-
Ethane	C_2H_6	0.3	-
Hydrogen sulphide	H_2S	0.1	-
Carbon disulphide	CS_2	-	1.25
Formic acid	H_2CO_2	0.37	-

[a] Geiss, J. et al. 1991 Astr. Ap. **247**, 226 (Model G 1991).
[b] Schmidt, H. U., Wegmann, R., Huebner, W. F. and Boice, D. C. 1988 Comp Phys Comm **49**, 17 (Model S 1988).

Here ϵ and h are the specific internal energy and enthalpy respectively. The quantities on the right hand side represent the sources and sinks for the mass, momentum and energy respectively. These equations have to be solved with the appropriate initial boundary conditions. The equations are generally solved in a one-dimensional form with radial variation. Such solutions are sufficient in the inner coma, but in the outer coma where the solar wind interactions occur, the description of the state of the ionic gas may require two or three-dimensional solutions. The corresponding equations for a spherically symmetric coma for the one dimensional case are given by

$$\frac{1}{r^2}\frac{d}{dr}(r^2 u\rho) = \dot{\rho}, \tag{6.71}$$

$$\rho u\frac{du}{dr} + \frac{dP}{dr} = \dot{q} \tag{6.72}$$

and

$$\frac{1}{r^2}\frac{d}{dr}\left[r^2\rho u(h + \frac{u^2}{2})\right] + \left[\frac{\overline{\gamma}}{\overline{\gamma}-1}\rho u\right] = \dot{Q} - \dot{L}. \tag{6.73}$$

Here $\overline{\gamma}$ is the specific heat ratio for the gas. $\dot{Q}$ and $\dot{L}$ are the energy input and energy loss rates respectively. The above equations can easily be modified to include the dust component. The method is to write the analogous equations for the interaction of dust and the gas which are therefore treated as separate fluids. They also include the dust drag force and dust-gas energy exchange terms. The solution of pure single-fluid equations has a singularity at the sonic point, typically within 1 km from the surface of the nucleus. Taking the surface of the nucleus having a vaporization temperature of water $\sim 190K$ and an outflow speed equal to the sonic velocity $\sim 0.34 km/sec$ as the initial boundary simplified the problem enormously. The calculations become complicated for getting the solution of the equations for the case of gas and dust together. It involves an iterative procedure. But the calculations have shown that the time dependent gas-dust model seems to eliminate the singularity.

The results of the solution of the equations depend on the various physical processes that have been considered in the terms on the right hand side of the equations. In particular, it depends on the source of energy input, energy loss, chemistry and so on. Therefore, the emphasis on the solution depends upon the problem of interest. For example, several studies have been carried out which mainly emphasize certain aspects like nucleus-coma interface, gas-dust interaction, detailed study of the chemistry or the solar wind interaction and so on.

The inner part of the coma is the collision dominated region (LTE). It decreases in importance moving away from the nucleus. The hydrodynamic flow which is a good approximation in the inner region becomes free molecular flow in the outer region of the coma, with a transition region in between. Therefore, the solutions of the hydrodynamic equations can represent approximately the real solution provided proper quantities are used referring either to LTE or NLTE situations and a suitable form for in-between regions. The photolytic heating is given by the thermalisation of the excess kinetic energy carried off by the photolytic products due to photodissociation or exothermic chemical reactions. The dominant contribution to the heating process comes from the dissociation of H_2O through various channels. This means that collisions have to be important which imply that the time scale for collision must be smaller than the time scale for the expansion of the gas. The studies have shown that at least 20 collisions are necessary for the complete thermalisation of the hydrogen atoms. Therefore, the efficiency of the heating rate depends upon the collision parameter which defines the extent of the collision zone. The efficiency factor for heating is expressed in terms of the dimensionless parameter r/r^* where

$$r^\star = \frac{\alpha_H \, \sigma_H \, Q(H_2O)}{4\pi v}, \tag{6.74}$$

where α_H represents the mean relative energy transfer per collision in a $H - H_2O$ collisions, σ_H the cross-section for $H - H_2O$ collisions, $Q(H_2O)$ is the production rate of H_2O and v is the velocity of the gas. The Fig. 6.17 shows the heating efficiency deduced from several studies.

The cooling process of the coma gas comes mainly due to H_2O molecules because of its large abundance. The cooling takes place mainly through their strong rotational transitions. An approximate relation valid under LTE conditions and which is commonly used is given by

$$L = \frac{8.5 \times 10^{-19} \, T^2 \, N^2(H_2O)}{N(H_2O) + 2.7 \times 10^7 \, T} \; ergs/cm^3/sec. \tag{6.75}$$

The detailed energy level diagram and Einstein A coefficients for para and ortho H_2O are now available. Based on this, there exists several studies which make a better estimate for the cooling rate. The cooling rate of H_2O can be represented as

$$L_0 = \frac{L_{LT}^0 \, n_{H_2O}^2}{n_{H_2O} + L_{LT}^0/L_{NLT}^0} \tag{6.76}$$

where L_{LT}^0 and L_{NLT}^0 are respectively LTE and non-LTE cooling rates and $n(H_2O)$ is the density of water molecules. The analytical expression for L_{NLT}^0 is given by

$$\begin{aligned} L_{NLT}^0 &= 1.561 \times 10^{26} \, T^{0.9045} \, \exp[-89.362/T] \; \text{for } T < 100, \\ &= 1.8116 \times 10^{-27} \, T^{1.307} \, \exp[-68.34/T] \; \text{for } T > 100 \end{aligned} \tag{6.77}$$

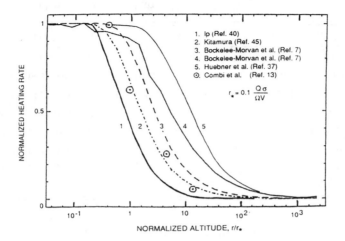

Fig. 6.17. The efficiency factor for photolytic heating as a function of the dimensionless parameter (r/r^*) deduced by various authors. This function controls the water temperature and outflow velocity (Crifo, J. F. 1991. In *Comets in the post-Halley Era*, Eds. R. L. Newburn, Jr., et al. Kluwer Academic Publishers, p. 937).

which assumes that $H_2O - H_2$ collisional cooling to be true for $H_2O - H_2O$ collisions. L_{LT}^0 can be written as

$$L_{LT}^0 = (3L^0 + L^P)/4 \tag{6.78}$$

where L^0 and L^P are respectively cooling rates for ortho and para-$H_2O \cdot L^0$ and L^P are determined from detailed energy level diagrams and Einstein A values and can be expressed as

$$L^i = 1.9862 \times 10^{-16} \sum_{I=1}^{23} Y_i(I) \exp\left[-1.43879 X_i(I)/T\right]. \tag{6.79}$$

Here i stands for O or P. The values of $Y_i(I)$ and $X_i(I)$ are given in Table 6.15. The cooling rate as given by Eq. (6.76) is compared with the cooling rate of Eq. (6.75) for different temperatures and H_2O densities (Fig. 6.18). It can be seen that the Eq. (6.76) gives larger values compared to earlier values and the difference is density dependent.

Table 6.15. Values of $X_P(I)$ and $Y_P(I)$ for Para-H_2O.

I	$X_P(I)$	$Y_P(I)$	I	$X_P(I)$	$Y_P(I)$	I	$X_P(I)$	$Y_P(I)$
1	37.1391	6.8392-1	9	285.2200	1.9275+2	17	542.907	7.0861+1
2	70.0907	1.9231-1	10	315.7792	6.6844+1	18	586.4800	2.4825+2
3	95.1757	1.7715-1	11	326.2656	4.6635+1	19	602.7742	1.3158+2
4	136.1641	2.6580+1	12	383.8427	2.6736+2	20	610.1151	8.9653+2
5	142.2783	9.0271+0	13	416.2088	1.0481+2	21	661.5492	4.7864+2
6	206.3013	4.2525+1	14	446.6972	8.5629+1	22	709.6093	2.5306+2
7	222.0529	1.3775+1	15	488.1349	7.0866+2	23	742.0729	1.8336+3
8	275.4971	4.2854+0	16	503.9682	3.5993+2			

Values of $X_O(I)$ and $Y_O(I)$ for Ortho-H_2O.

I	$X_O(I)$	$Y_O(I)$	I	$X_O(I)$	$Y_O(I)$	I	$X_O(I)$	$Y_O(I)$
1	42.3717	6.4241-2	9	300.3621	6.7548+1	17	552.9119	1.6258+2
2	79.4963	3.1153+0	10	325.3483	3.9217+1	18	586.2445	1.6074+2
3	134.9018	2.5416+1	11	382.5171	2.6605+2	19	610.3418	8.9672+2
4	136.7617	2.8907+0	12	399.4581	2.1137+1	20	648.9791	4.7538+2
5	173.3656	7.0452-1	13	446.5107	9.2758+1	21	704.2158	2.6980+2
6	212.1561	4.4905+1	14	447.2528	9.1327+1	22	742.0761	1.8337+3
7	224.8383	2.1689+1	15	488.1084	7.0875+2			
8	285.4192	1.9327+2	16	508.8121	3.5261+2			

(Courtesy S. P. Tarafdar)

The results of model calculations for the velocity and temperature variation in the coma is given in Fig. 6.19. The velocity of the gas changes very slowly for distances $< 10^3 km$ and then increases more rapidly. However, the temperature distribution falls steeply to a minimum temperature $\sim 20K$ at a distance $\sim 10^2 km$ and then increases again. The general shape of the temperature variation is the resultant effect of the relative importance of heating and cooling rates of H_2O and the expansion cooling. The model results are in reasonable agreement with the observation of gas velocity deduced from Giotto observations for distances between 10^3 and 3×10^4 km for Comet Halley (Fig. 6.19). The observed velocities are deduced from the

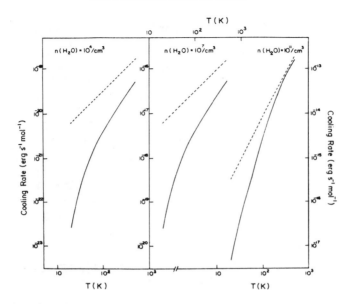

Fig. 6.18. The comparison of cooling rates due to H_2O molecule. The continuous lines from Eq. (6.76) and dashed lines from Eq. (6.75) (Shimizu, M. 1976. *Astro. Space Sci.* **40**, 149).

energy shifts of peaks in the neutral mass spectrometer energy spectra and show a variation from a value of about 0.8 to 1.05 km/sec. There are several other determinations of the mean velocity of the gas of the coma of Comet Halley which show scattering around a mean velocity $\sim 0.85 km/sec$. The radial outflow velocities have also been inferred from the study of the expansion velocity of the halos of the coma of comets. These values can be fitted well with a relation of the type

$$V = 0.535\, r^{-0.6}\ km/sec. \tag{6.80}$$

With regard to the temperature, a low temperature of around $60 K$ for $r > 10^3 km$ is suggestive from the study of the $2.7 \mu m$ band of H_2O from Comet Halley.

The problem of division of the coma into various regions can be avoided if the Monte-Carlo simulation procedure is adopted. This takes into account the actual properties of the coma. In fact, the results based on Monte-Carlo methods may be used to specify the approximate region of transition from hydrodynamic to free molecular flow. In practical terms, the region

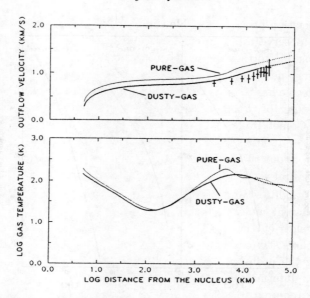

Fig. 6.19. The calculated variation of expansion velocity with distance from the nucleus is compared with the observed variation for Comet Halley deduced from the Giotto measurements (top figure). The bottom figure shows the calculated variation of gas temperature as a function of distance from the nuclues. (Combi, M. R. 1989. *Icarus* **81**, 41).

of transition from hydrodynamic to free molecular flow depends to some extent on the physical assumption made or criteria used in the problem, like the mean free path length of H_2O to the hydrodynamic sphere or to the number of collisions required for thermalisation etc. The results of calculations indicate that the hydrodynamic model for Comet Halley is valid for nuclear distances upto $\sim 10^4\ km$ or so and are not applicable for $r \gtrsim 5 \times 10^4\ km$ or so. Another advantage of the Monte-Carlo method is that it is possible to consider, in detail, various physical processes including the detailed chemical processes and their evolution with time. Though such calculations have been carried out, the approach is more of a computer simulation and hence may be restrictive.

6.6. Parent Molecules

Most of the observed species in comets (Chap. 4) cannot exist as such in the nucleus. Therefore, it is suggested that they could arise out of the photodissociation process of stable complex molecules. However, it is very difficult to observe these parent molecules directly from the Earth-based observations due to their short lifetime, as they are dissociated in most of the cases very near the nucleus. Hence the information about the possible parent molecule has to come from indirect means and, in particular, mainly from the study of the decay products. However, the spacecrafts which passed by Comet Halley in 1986 and made *in situ* measurements of the inner coma $\sim 1000 km$ from the nucleus have given important and direct information about the possible nature of the parent molecules. In general, there are several ways to approach this problem.

The straightforward approach is to look for the dissociated products of simple parent molecules as in the case of H_2O molecules. There are not many clear-cut cases of this type. The laboratory investigations of the photochemistry of various types of molecules can help a great deal in elucidating the problem of the parent molecules. The laboratory studies have to give information with regard to the products of photodissociation, the energy required to initiate the reaction and also the branching ratios for various paths of the reactions. At present, the data is limited in extent. Also such studies have to be performed for all molecules of astrophysical interest. To start with, the various types of molecules identified in the interstellar space could be used. The calculated photochemical lifetimes of several molecules based on the laboratory data is given in Table 6.2 and they show that for most molecules the lifetime is below 2×10^4 sec at 1 au. The results based on laboratory photochemical studies should be used only as a guideline when applying the same to a cometary atmosphere. This is due to the fact that the astrophysical situations are vastly different from laboratory conditions.

The surface brightness distribution as a function of the radial distance from the nucleus can give information about the parent molecule (Sec. 6.1.2). The comparison of the observed brightness profiles with those of calculated profiles gives the scale lengths of the species. The lifetime of the species, can therefore, be obtained by dividing the scale length by the expansion velocity. These lifetimes in turn can give information with regard to the possible types of the parent molecule.

Another approach is through the study of gas-phase reactions in the coma. If an agreement can be obtained between the observed and the calculated abundances of various species, one can hopefully trace backwards through the network of reactions and this could help in identifying the plausible parent molecule of various species. However, the procedure may not be very practical in most of the cases.

Another possible method is through the spectral study of the observed lines and particularly from those molecules which give lines arising out of singlet-singlet and triplet-triplet transitions. The idea is to see whether the observed intensities of lines arise preferentially from the singlet-singlet or the triplet-triplet states of the molecule. This could in turn be related to the products arising out of the dissociation of the parent molecule.

The study of forbidden lines can also help in identifying the parent molecule of the species which gives rise to these lines. By combining different types of studies it is possible to infer the possible nature of the parent molecule of some of the observed species. We may illustrate them with a few examples.

Until the direct detection of H_2O from Comet Halley in 1986, all the earlier observations had built up a strong case for the H_2O as the dominant volatile component of the nuclei of comets. This is basically related to the fact that all the dissociated products of H_2O are seen and they dominate the ultraviolet spectra of comets. But the direct detection of H_2O came from the observation made from the Kuiper Airborne Observatory of the ν_3 band near $\lambda \sim 2.65 \mu m$ in Comet Halley. The same band was also observed by the infrared spectrometer on the Vega spacecraft. The results based on Neutral Mass Spectrometer (NMS) measurements on the Giotto spacecraft showed that H_2O comprised around 80% of the volatile of Comet Halley.

The presence of CO in comets was indicated by the detection of fourth positive bands at $\lambda \sim 1500 \text{Å}$. Its abundance relative to water varies from a few percent to around 30%. For Comet Halley, the value derived from various observations is $\sim 17\%$, while NMS data gave a value $\sim 7\%$. The lower value is attributed to CO coming from the nucleus and the rest is believed to arise from an extended source. CO is the second most abundant molecule in comets.

The observation of the ν_3 band of CH_4 around $3.3 \mu m$ has yielded an estimate for the production rate of CH_4 which is $\sim 1\%$. The determination of the abundance of CH_4 from the NMS on Giotto is rather difficult as it could be contaminated with the dissociated fragments from H_2O, possibly NH_3 and others, but is $\sim 2\%$.

Although CO_2 was believed to exist due to the presence of CO_2^+ ion in comets, it was first detected in Comet Halley from the observation of ν_3 band at $4.3 \mu m$ from the IKS instrument in the Vega spacecraft and the

NMS instrument on Giotto which gave a value ∼ 3% relative to water. The observed distribution of the column density of CO_2 is in good agreement with the ρ^{-1} law. This is consistent with the distribution of R^{-2} expected for parent molecules in the vicinity of the nucleus.

The presence of NH_2 and NH radicals in comets indicate strongly that NH_3 must be present. The *in situ* measurements and the study of NH_2 bands indicate roughly a ratio ∼ 1% for the production of NH_3 relative to water. There is little doubt that NH_3 is present in comets although its abundance is rather uncertain.

The photodissociation of HCN, namely

$$HCN + h\nu \rightarrow CN + H$$

can explain the CN radical both from the point of view of production rates as well as from the kinematic considerations. HCN has been seen in Comet Halley and the derived abundance is about 0.1% relative to water.

Extensive observations of Comet Halley have been carried out with the 3880Å line of CN with a three-channel spectrometer on board the Vega 2 spacecraft. From a thousand spectra taken at different geometrical configurations and distances from the nucleus, an estimate of the physical parameters such as the lifetime τ_P, the mean velocity of the daughter molecule during photolysis with respect to that of parent molecule and the distance from the nucleus where the radius optical depth of the coma for the daughter molecule is unity, have been derived. The distance in turn is used to extract the absorption cross-section for solar radiation of species produced by the photolysis of the parent. The derived values of $\tau_P = 1600 \pm 200 secs$ at $1au, V = 0.7 \pm 0.3 km/sec$ and $d = 100 \pm 50km$ are interpreted in terms of a parent mixture composed of $HCN(\sim 50\%)$, $C_4N_2(\sim 35\%)$ and $HC_3N(\sim 15\%)$.

The source of S_2 (lifetime ∼ $350 secs$) has been the subject of considerable debate as the presence of S_2 in the cometary nuclei has great significance for the origin of cometary material. CS_2 hardly photodissociates into S_2. It is highly unlikely that solid S_2 has formed from gas phase condensation, which primarily produces other compounds. So is also with the irradiation of the nucleus by cosmic rays or ultraviolet radiation and so on. Therefore, S_2 could be a parent molecule.

The most plausible parent of CS is the short-lived (scale length ≈

300km) CS_2 molecule,

i.e.,
$$CS_2 + h\nu \rightarrow CS + S(^3P),$$
$$CS_2 + h\nu \rightarrow CS + S(^1D).$$

However, CS_2 has not yet been observed.

The S produced from CS_2 and S_2 does not appear to account for the observed amount of sulphur. The molecule H_2S which has been detected in Comet Austin (1990 V) and Levy (1990 XX) may contribute to the observed sulphur. The molecule OCS has been invoked to explain the observed spatial distribution of S atoms in the coma. Here again, the OCS molecule has not been detected in comets.

The analysis of the C_2 brightness profile seems to suggest that it could be formed in a three-step photodissociation process of some parent molecule. The laboratory studies indicate that C_2 can be produced in a two-step process as follows:

$$C_2H_2 + h\nu \rightarrow C_2H + H$$

followed by

$$C_2H + h\nu \rightarrow C_2 + H.$$

The molecule propynal (C_3H_2O) is also suggested for the parent of C_2.

The C_3 molecule is a difficult case for which there are some suggestions. Laboratory studies have shown that C_3 molecule can be formed from C_3H_4. This molecule photodecomposes to give C_3H_3 and C_3H_2 as intermediate products, which can then absorb a photon to give C_3 molecule.

As in the case of CN, the extensive spacecraft data obtained from Vega 2 for the 4050Å emission of C_3 from Comet Halley gave the values for $\tau_P = (2 \pm 0.1) \times 10^4$ sec at 1au, $v = 2.5 \pm 0.5$ km/sec and $d = 40 \pm 20$ km. These are interpreted in terms of propynal (C_3H_2O) and allene (C_3H_4) as parent molecule of C_3.

Table 6.16 gives a limited list of probable parent molecules and their abundances of some species observed in comets. Some of the identifications are to be taken with caution as they are still preliminary in nature.

Table 6.16. Possible parent molecules and abundances of some of them.

Observed species	Possible parent molecule	Abundance	Method
H_2O, H_2O^+, OH, H, O	H_2O	1.0	
CO_2^+, CO^+, CO	CO_2	≤ 0.035	a
		0.027	c
CO, CO^+, C	CO	0.07	a
		0.17 - 0.20	b
CH	H_2CO	0.045	c
		0.015	f
CN	HCN, C_4N_2, H_3CN		
HCN	HCN	~ 0.001	g
NH, NH_2	NH_3	≤ 0.1	a
		0.01 - 0.02	d
		0.004 - 0.008	e
C_2	C_2H_2, C_3H_2O		
C_3	C_3H_2, C_3H_4, C_3H_2O		
CS	CS_2		
S_2	S_2		
S	S_2, OCS, CS_2, H_2S		
	CH_4	≤ 0.07	a
		~ 0.02	d
	N_2	≤ 0.02	d

Adapted from: Delsemme, A. H. 1981. In *Comets and the origin of life*, ed. C. Ponnamperuma. D. Reidel Publication Company. p. 141; Krankowsky, D. 1991. In *Comets in the Post-Halley Era*. eds. R L Newburn, Jr., *et al*. Kluwer Academic publishers. Vol. 2. p. 855, (a) Giotto NMS, gas spectra (b) Rocket UV experiment (c) Vega IKS, IR spectra (d) Giotto IMS, ion spectra (e) optical spectroscopy (f) VLA, radiowave (g) millimeter spectra.

6.7. Summary

We may briefly summarize the results of the discussion presented so far. The study of several comets has shown a variation of the production rate of H_2O with the heliocentric distance vastly different from that of r^{-2} dependence. In order to explain such a variation, it may be necessary to invoke more complicated models than those have been considered so far. The results for the production rates of CN, C_2, C_3, CH, etc., indicate to a first approximation that comets of various types, dynamical ages and morphologies have very little variation in their chemical composition. In addition, even though the lines of C_2, CN and others dominate the visual spectral region, their production rates are less by a factor of 100 or so compared to that of H_2O or H. These molecules are, therefore, the minor constituents of the cometary material. Most of the elements have abundance ratio very similar to the solar atmospheric value except hydrogen which is very much depleted in comets.

The *in situ* measurements of Comet Halley made by spacecrafts have shown the complex nature of ions present in the coma. The production rates of these species derived for the first time making use of the detailed gasphase chemistry show them to be a minor contributor of the gas. There is a strong evidence to show that grains can also act as suppliers of volatiles in the coma. The expected velocity distribution of the coma gas based on the detailed modeling agrees with the Giotto results. Lastly, the spacecraft observations have detected several possible parent molecules. All the earlier studies indicating H_2O as the parent of OH, H and O have been further reinforced.

Problems

1. Calculate the surface brightenss profiles for a few values of β_1/β_0 as shown in Fig. 6.1.

2. Deduce the values of β_1 and β_0 for the following observed surface brightness distribution: at $\rho = 10^3, 4 \times 10^4$ and $7 \times 10^4 km$, the observed values of surface brightness are 350, 35 and 12 kilo Rayleighs respectively. What is the lifetime of the species for an assumed velocity of 1 km/sec?

3. Assume that the diatomic molecule NH is formed due to the collision process of N and H with α as the rate constant for the reaction. What is the resulting concentration of NH for $n_H = 10/cm^3$, $\alpha = 10^{-10}$ cm^3/sec and for a time scale of 4×10^9 years?

4. Write down the expression for the equilibrium abundances of CH and CH^+ in terms of other abundances for the following reactions: $C + H \to CH + h\nu$; $C^+ + H \to CH^+ + h\nu$; $CH + h\nu \to CH^+ + e$; $CH + h\nu \to C + H$; $CH^+ + h\nu \to C^+ + H$. The rate constants for these reactions are k_1, k_2, β_1, β_2 and β_3 respectively.

References

The calculations of surface brightness distribution for the two component models are discussed in these two papers.
1. Haser, L. 1957. *Bull. Acad. Belg.cl. Sc.,* 5e Ser. **43** 740.
2. O'Dell, C. R. and Osterbrock, D. E. 1962. *Ap. J.* **136** 559.

The values required for the calculation of surface brightness distribution are given in
3. A'Hearn, M. F. and Cowan, J. J. 1975. *Ap. J.* **80** 852.

The three component model is given in the paper.
4. O'Dell, C. R., Robinson, R. R., Krishna Swamy, K. S., Spinrad, H. and McCarthy, P. J. 1988. *Ap. J.* **334**, 476.

The vectorial formalism model is presented in
5. Festou, M C. 1981. *Astr. Ap.* **95**, 69.

The Monte Carlo approach is discussed in
6. Combi, M. R. and Delsemme, A. H. 1980 *Ap. J.* **237**, 633.
7. Ferro, A. J. 1990. In *Proc of the workshop on observation of recent comets* (1990). eds. W. F.Huebner et al. South West Research Institute, San Antonio, Texas, p. 160.
8. Hodges, Jr. R. R. 1990. Icarus, **83**, 410.

The semi-empirical photometry theory is discussed in the following papers.
9. Newburn, R. L. 1981. In *The Comet Halley Dust and Gas Environment.* ESA SP-174, p.3.
10. Divine, N. *et al.* 1986. *Space Sci. Rev.* **43**, 1.

The discussion about complex molecules is drawn from
11. Crovisier, J. 1987. *Astr. Ap. Suppl.,* **68**, 223.

The following paper pertains to Sec. 6.2.6 of oxygen.
12. Schultz, D., Li, G. S. H., Scherb, F. and Roesler, F. L. 1992. *Icarus* **96**, 190; **101**, 95, 1993.

The following paper refers to discussion of Sec. 6.3.1
13. Keller, H. U. 1976. *Space Sci. Rev.* **18** 641.

Monte-Carlo approach is presented in
14. Combi, M. R. and Smyth, W. H. 1988. *Ap. J.* **327**, 1044.
15. Combi, M. R., Bos, B. J. and Smyth, W. H. 1993, *Ap. J.* **404**, 668.
16. Smyth, W. H., Marconi, M. L., Scherb, F. and Roesler, F. 1993, *Ap. J.* **413**, 756.

The basic formulation of gas phase chemistry in the coma can be found in
17. Huebner, W. F. and Giguere, P. T. 1980. *Ap. J.* **238** 753.

The following papers give some results based on in-situ measurements.
18. Balsiger, H. et al. 1986. *Nature*, **321**, 330.
19. Eberhardt, P. Meier, R., Krankowsky, D. and Hodges, R. 1994. *Astr. Ap.*, **288**, 315.
20. Meier, R., Eberhardt, P., Krankowsky, D. and Hodges, A. R. 1994. *Astr. Ap.* **287**, 268.
21. Krankowsky, D. 1991. In *Comets in the Post-Halley Era*, eds. R. L. Newburn, Jr., et al. Kluwer Academic Publishers, P. 855.

The following papers discuss the hydrodynamics of the coma gas.
22. Hodges, R. R. 1990. *Icarus* **83**, 410.
23. Combi, M. R. 1989. *Icarus* **81**, 41.
24. Crifo, J. F. 1990. *Icarus* **84**, 414.

The cooling of H_2O given in Section 6.5 is contributed by Professor S. P. Tarafdar.
25. Chandra, S. and Tarafdar, S. P. 1992. *Astr. Ap.* **253**, 537.

The following papers may be referred for parent molecules.
26. Krasnopoesky, V. A. 1991. *Astr. Ap.* **245**, 310.
27. Moroz, V. I. et al. 1987. *Astr. Ap.* **187**, 513.

CHAPTER 7

DUST TAIL

The gaseous material released by the nucleus was studied in the previous chapter. This material also carries with it a large amount of dust. The idea that the dust particles are dragged into the tail by the effect of radiation pressure was first suggested by Bessel in the 1830's. This concept was refined by Bredichin around 1900. These ideas were put on a firm basis by Finson and Probstein in the 1950's. These authors have worked out in great detail the dynamics of grains based on fluid and kinetic concepts and the resulting intensity distributions.

7.1. Dynamics

Consider the case of only pure dust tails. The dust emitted by the nucleus is subjected to two opposite forces. The solar radiation pressure acting on the dust particles tries to push it away from the Sun while the force of solar gravity tries to pull it towards the Sun. Since the two forces vary as $(1/r^2)$, one usually defines an effective gravity, essentially given by the difference of the two forces. The Keplerian orbit mechanics is then used for the study of the dynamics of the dust particles. The ratio of radiation pressure to gravitational force is denoted by $(1 - \mu)$ where

$$(1 - \mu) = \frac{F_{rad}}{F_{grav}}. \tag{7.1}$$

The expressions for F_{rad} and F_{grav} for a spherical particle can be written as

$$F_{rad} = \frac{\pi d^2}{4} \frac{Q_{pr}}{c} \left(\frac{F_\odot}{4\pi r^2} \right) \tag{7.2}$$

and

$$F_{grav} = \frac{GM_\odot}{r^2} \left(\frac{\rho_d \pi d^3}{6}\right). \tag{7.3}$$

Here $F_\odot$ is the mean solar radiation field impinging on the grain of diameter d and density ρ_d. Q_{pr} is the efficiency factor for radiation pressure (Chap.8). Therefore Eq. (7.1) becomes

$$(1 - \mu) = C(\rho_d d)^{-1} \tag{7.4}$$

where $C = \dfrac{3Q_{pr}F_\odot}{8\pi c G M_\odot} = 1.2 \times 10^{-4} Q_{pr}$.

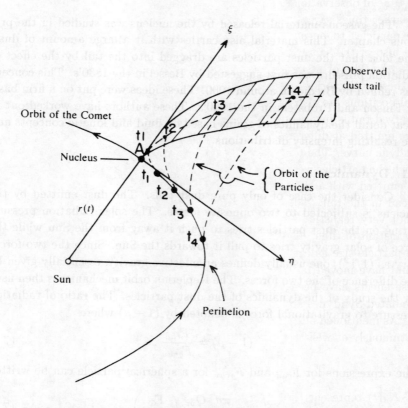

Fig. 7.1. Trajectories of the particles and the formation of the dust tail.

The nature of the orbit of the particle is decided by the value of $(1-\mu)$. There are two cases of interest that have to be considered. Firstly, particles are ejected continuously as a function of time by the nucleus. The locus of the particles which have the same value of $(1-\mu)$ is called the *Syndyname* or *Syndyne* curve (Chap. 6). Physically this means that particles of the same size are emitted by the nucleus. The other case involves the distribution of particles emitted at any one particular time, as in the case of an outburst. These particles will have varying $(1-\mu)$ values and the locus of the curve which describes this situation is called *Synchrone*.

The cometocentric coordinate system used to describe a comet tail is shown in Fig. 7.1. Here ξ and η represent the coordinates in the orbit plane of the comet. The third coordinate ζ is in the direction perpendicular to the plane of the orbit. Let the position of the comet be at A corresponding to the time of observation t. Consider the position of the comet corresponding to earlier times say t_1, t_2, t_3 and t_4 along its orbit. For simplicity assume the particles to be emitted from the nucleus with zero relative velocity and also of the same size. This implies that they have the same value of $(1-\mu)$. The orbit of the particles can be worked out and the position on these trajectories for the time of observation t can be calculated. The final positions are also marked as t_1, t_2, t_3 and t_4 in the tail. Therefore the observed tail is just the locus of the dust particles in the (ξ, η) plane, that were emitted at different earlier times. This represents the Syndyname or Syndyne curve. The farther one goes along the tail, the earlier the time at which the particles were emitted. However, in a real situation, the particles are emitted with a certain velocity spread, which gives rise to the observed spread in the tail. The assumption that the particles have the same size is not very realistic as generally particles of various sizes are emitted at any instant of time. The actual trajectories of the particles are sensitive to the particle size as can be seen from Eq. (7.4). The smaller size particles would have moved farther compared to those of larger size particles. All these effects have been incorporated in the dynamic model. Some of the calculated Syndynes and Synchrones are shown in Fig. 7.2.

As mentioned already, in a real situation the dust particles are emitted continuously as well as with varying sizes from the nucleus for which

$$(1-\mu) \propto (\rho_d d)^{-1}. \tag{7.5}$$

Let $\dot{N}_d(t)$ represent the dust production rate of all sizes in particles per second. For simplicity it is assumed that the distribution of particle sizes

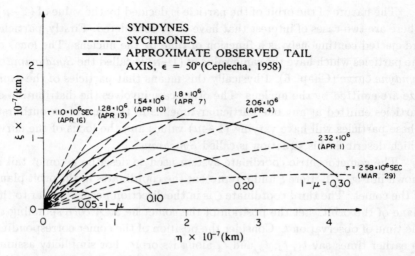

Fig. 7.2. Syndynes and Synchrones for Comet Arend-Roland on April 27, 1957. Time $\tau = 1.71 \times 10^6$ see at perihelion on April 8. The position of the observed tail of the comet is also shown. (This figure as well as all the other figures of this chapter are taken from Finson, M. L. and Probstein, R. F. 1968. *Ap. J..* **154** 327 and 353.)

remains constant along the orbit of the comet. The particle distribution function g is given by

$$g(\rho_d d)d(\rho_d d) \text{ such that } \int_0^\infty g(\rho_d d)d(\rho_d d) = 1. \quad (7.6)$$

The quantities $\dot{N}_d(t)$ and $g(\rho_d d)$ are usually obtained from the fitting of the model with the observations.

For the calculation of the orbit of the particle, it is necessary to know the initial velocities of the dust particles coming out of the nucleus. The model for the calculation of the velocity of the dust particles is based on the steady state fluid approach. The flow is assumed to be spherically symmetric with the same temperature for the gas and dust at the nucleus and dust particles having a uniform size. The gas and the dust coming out of the nucleus are coupled through drag forces, computation of which is based on molecular drag coefficient. With no relative velocity to start with, the particles are accelerated outwards. The solution of the conservation equations gives the velocity of the dust particles. These calculations which are quite involved show that the dust particles reach their terminal speed within about 20 radii of the nucleus. This is indeed a very short distance

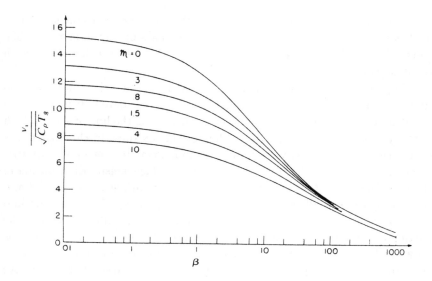

Fig. 7.3. Speed of the dust particles emitted from the inner head region (see text).

and hence one generally uses terminal speed for the dust particles in the tail calculations. The terminal speed denoted as v_i can be expressed in the functional form as

$$\frac{v_i}{(C_p T)^{1/2}} = g(\mathcal{M}, \beta) \tag{7.7}$$

$$\mathcal{M} = \frac{\dot{m}_d}{\dot{m}_g} \text{ and } \beta = \frac{16}{3}\pi \rho_d d R_0 (C_p T)^{1/2}/\dot{m}_g$$

where $\dot{m}_d$ is the mass flow rate for the dust and $\dot{m}_g$ that of the gas. Therefore $\mathcal{M}$ represents the dust-to-gas mass ratio. Here R_0 is the radius of the nucleus of the comet, T is the initial temperature of the gas and C_p the specific heat at constant pressure. Fig. 7.3 shows a plot of the velocity of the dust particles as a function of β. Therefore the velocity can be written in the form

$$v_i = v_i(\rho_d d, \dot{N}_d, \dot{m}_g) \tag{7.8}$$

where T and R_0 are known. Based on these results, it is generally assumed that the dust comes out of the nucleus in a spherically symmetric manner whose speeds are given by the results shown in Fig. 7.3. The results of more detailed and realistic dust particle size distribution calculations show that the derived velocities are similar to those in Fig. 7.3, except that the terminal dust velocity is smaller by about 20% with respect to the single size dust particles. One can also keep $\dot{m}_g$, the gas mass flow rate, as a variable. This value can be obtained by comparing the theoretical calculations with the observations.

In order to compare with the observed isophotes, the total emission from the comet along the line of sight has to be calculated. It is necessary to calculate the integral of the number density along the line of sight, which is termed as surface density. Since the optical thickness in the dust tail is small, the surface density is directly proportional to the light intensity. There are two approaches for calculating the surface density distribution in the tail. One procedure is to use the particles of various sizes emitted and then integrate over all the times (Synchrone approach). The other method is to consider one particle size and then integrate over all the sizes of the particles (Syndyne approach). The two approaches give the same result as they involve double integral involving $(1 - \mu)$ and time in both the cases. Only the order of integration in these two cases is reversed. For illustrative purposes, a brief discussion of the Syndyne approach will be presented.

Consider the position of the comet observed at a particular time t_c measured from the time of perihelion passage which is taken to be zero. For a given value of $(1 - \mu)$, the particles lying on a Syndyne curve would have been emitted at time

$$t = t_c - \tau \tag{7.9}$$

where τ increases along the Syndyne curve with zero value at the Syndyne origin. The number of particles emitted in the time τ and $\tau + d\tau$ and in the size range $(\rho_d d)$ and $(\rho_d d) + d(\rho_d d)$ is given by

$$\dot{N}_d(t) d\tau g(\rho_d d) d(\rho_d d). \tag{7.10}$$

This has to be multiplied by the light scattering function to get the total scattered intensity. The scattered intensity is proportional to the cross-sectional area of the particle and hence varies as $(\rho_d d)^2$. Working in terms of $(1 - \mu)$ rather than $(\rho_d d)$, the new weighted distribution function can be written as

$$f(1 - \mu) d(1 - \mu) \propto g(\rho_d d)(\rho_d d)^2 d(\rho_d d) \tag{7.11}$$

such that $\int_o^\infty f(1 - \mu) d(1 - \mu) = 1$. Therefore the Eq. (7.10) can also be expressed as

$$\dot{N}_d d\tau f(1 - \mu) d(1 - \mu). \tag{7.12}$$

These particles then follow well-defined orbits in the plane of the orbit of the comet. However what is actually seen is the projection of these orbits in the plane of the sky. Therefore, the expected intensities in the plane of the sky has to be calculated. The total modified surface density is obtained by integrating over various Syndyne tails for all values of $(1-\mu)$. The final expression is of the form

$$D = \int_{(1-\mu)_a}^{(1-\mu)_b} \dot{N}_d f(1-\mu) \left[2v_i\tau \frac{dx}{d\tau}(\tau; 1-\mu; t_c)\right]^{-1} d(1-\mu) \qquad (7.13)$$

Here $dx/d\tau$ represents the rate of change of length along the given Syndyne axis with respect to τ. Physically it represents the effect on the surface density due to dispersion in the longitudinal direction with respect to the tail axis. On the other hand the other term $v_i\tau$ arises due to the dispersion in the lateral direction. The Eq. (7.13) involves the functional parameters $\dot{N}_d(t), f(1-\mu)$ and $v_i(\tau; 1-\mu; t_c)$. These can be determined from a comparison of the expected and the observed intensity distributions. From a knowledge of these three parameters, all other quantities of interest can then be determined.

The above formalism has been applied very successfully to Comet Arend-Roland as can be seen from Fig. 7.4. Near perihelion, the gas and the dust emission rates are about 7.5×10^7 gm/sec and 6×10^7 gm/sec. This gas emission rate corresponds to a value of about 1.5×10^{30} molecule/sec. The dust to gas ratio for Comet Arend-Roland is about 0.8 and the average size of the particles in the tail is $\sim 1\mu$. The particles in the observed dust tail of Arend-Roland correspond to those which have been emitted about 10 to 15 days on either side of the perihelion passage.

Although Finson and Probstein formalism has been very successful in the studies of tail morphology, it has several limitations. All the grains are assumed to be of the same kind and it is difficult to distinguish between different kinds of grains. The observation of Comet Halley has clearly shown a wide variation in sizes and in the chemical and physical nature of grains. The value of Q_{pr} is also not constant but is dependent upon the sizes of the particles. If the grains of various sizes are emitted continuously, it is difficult to separate the effects depending on the time of ejection from those due to the size and properties of the grains. Therefore several studies have been carried out to take care of some of the above limitations. However, these studies with complicated numerical integration techniques, lose the great advantage of the simplicity of the Finson and Probstein approach.

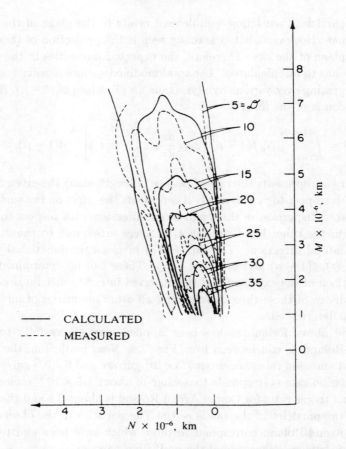

Fig. 7.4. Measured and calculated isophotes for Comet Arend-Roland for May 1, 1957. D is the density level.

7.2. Anti-Tail

Many comets generally show a short tail in the solar direction which is called the anti-tail (see Fig. 1.16). The theory developed for the dynamics of the grains of dust tails can also explain the presence of anti-tail in comets. The calculated Synchrones and Syndynes for Comet Arend-Roland are shown in Fig. 7.5 which shows the presence of Synchrones in the direction of the Sun. The results also show that only synchrones younger than about 30 days appear to produce the usual dust tail, while the older ones project in the direction of the Sun, which is seen as the anti-tail. The

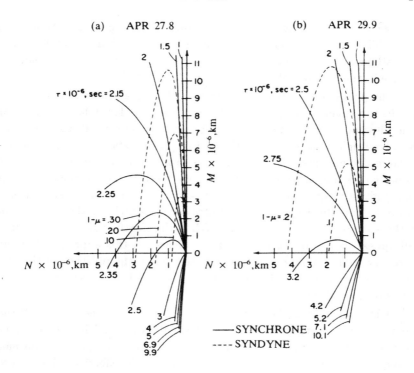

Fig. 7.5. Calculated Syndynes and Synchrones for tail and anti-tail for Comet Arend-Roland.

appearance of the anti-tail of Comet Arend-Roland changed with time within an interval of a few days. This arises due to the changing pattern of the positions of Synchrones. If the Synchrones are crowded together, then the anti-tail will be seen as a narrow tail, while if the Synchrones are spread out, the anti-tail will be a broad one. The visibility of an anti-tail depends mainly on the sun-comet-earth geometry. It can be seen clearly only when the orbit of the comet cuts the orbital plane of the Earth.

The anti-tail of Comet Kohoutek could be seen shortly after perihelion. Comet Halley also displayed anti-tail after its perihelion passage. In general, the anti-tail is more favourable for observation after perihelion.

The dynamical theory of grains has been applied very successfully to the observed anti-tail of many comets. In fact, the predictions of the anti-tail based on the theory have later been confirmed through observations. The size of the particles present in the anti-tail could be larger than about

10μ, which is much larger than those present in the dust tails which is $\lesssim 1\mu$. The large size grains present in the anti-tail is also consistent with the estimated size of the particles based on the absence of 10μ emission feature in the anti-tail of Comet Kohoutek (Chap.9).

The presence of anti-tail of comets is therefore generally explained as due to the effect of large size particles. However, large size particles are generally poor light scatterers. Therefore it is rather difficult to understand the large brightness observed in the anti-tail of Comets, if they are produced by large size particles. To overcome this problem, an attempt was made to properly take into account the focusing effect arising due to velocity effects. The study seems to indicate that even with the normal size distribution of grains it is possible to produce concentration of particles in the orbital plane at the second node of their orbits. When the Earth crosses this plane, it will appear as a spike to the observer. Hence, in this approach, it is not necessary to invoke large size particles to explain the presence of the anti-tail. This also eliminate the difficulty of observed brightness, as it involves grains of normal sizes only.

It is evident from the dynamical theory of grains [see Eq. (7.4)] that the size of the particles along the tail should vary and it depends upon the nature of Synchrone and Syndyne curves. Since the polarization measurements are sensitive to particle sizes, they could effectively be used for determining the sizes of the particles. The polarization measurements on comets do seem to indicate the variation of the particle size along the tail of the comet.

7.3. Dust Features

The high quality photographs near the nuclear region show highly complicated structure of various kinds as can be seen from Fig. 7.6. This arises due to the complicated flow pattern of the grains after they are released from the nucleus. To a first approximation it can be described in terms of the resulting force between the gravitational, F_{grav} and radiation F_{rad}, forces acting on the grains (Sec. 7.1). The trajectories of the grains under such a force in a cometocentric system are approximately parabolic in nature. Different patterns could be produced depending on the parameters of the grains like radius, density and the material property (Fig. 7.7). In a real situation the spin of the nucleus, its associated precession and nutation and the cometary activity has to be considered. These will give rise to complex behaviour of the flow patterns of the dust particles in the near nucleus

Dust Tail 203

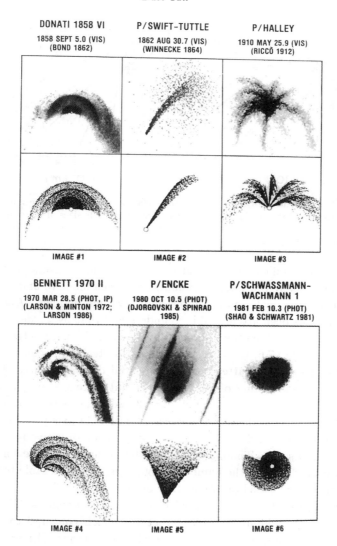

Fig. 7.6. The top rows show the various types of features observed in six comets in the vicinity of the nucleus. The bottom rows show the computer simulated images. There is a good similarity between the two. The Sun is at the top. (Sekanina, Z. 1991. In *Comets in the Post-Halley Era*, eds. R. L. Newburn, Jr., et al. Kluwer Academic Publishers, p. 769).

region. For example, arising from a single active area on the nucleus, the spiral jets unwind and develop into expanding parabolic envelopes and so

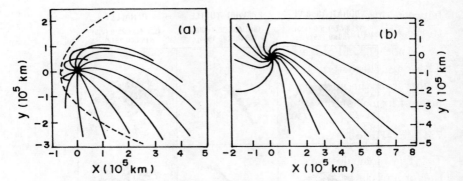

Fig. 7.7. Dust particle trajectories for Comet Halley released from the nucleus at the time of perihelion passage. (a) and (b) correspond to particle sizes of $10\mu m$ and 0.8 mm respectively. The dotted curve shows the resulting envelope (Fertig, J. and Schwehm, G. 1984. *Adv. Space Res.* **4**, 213).

on. It is rather difficult to take into account all these effects in the model calculations. Therefore several attempts have been made to understand the observed features in terms of simple hydrodynamic models. Another approach that has been attempted is to understand the flow patterns through computer generated synthetic images by varying the parameters. As can be seen from Fig. 7.6, the computer simulation technique can in principle reproduce the diverse patterns of the observed dust structures of comets. However the disadvantage of this method is that there are many parameters involved which have to be adjusted by trial and error method to match the observations, such as the angles associated with spin axis positions and its relation to Earth and Sun, rotation period, position of active regions, dust properties and so on. Therefore it is possible to have multiple solutions in this approach. To limit the number of solutions, it is preferable to fix some of the parameters based on some other observations.

Several other fine features have also been seen in comets. The streamers seen are straight and curved structures which converge towards the nucleus. They are generally associated with synchrones. For example the streamers seen in Comet West (1976 VI) have been attributed to particles of different sizes expelled simultaneously from the nucleus in the form of an outburst. Therefore long streamers are indicative of large dispersion of particle sizes, whereas a narrow width reflects a short duration outburst. Several well separated dust streamers has been seen in the images of Comet Halley. The analysis of these streamers is consistent with the interpretation that they represent diffuse synchrones of dust particles. The time interval between these streamers is found to be around 2.2 days which corresponds to the rotational period of the nucleus.

Another kind of feature, called striae has been seen in several comets. These are characterized by a series of parallel narrow bands at large distances from the nucleus. They do not start from the nucleus. The mechanism of formation of striae is not clear at the present time.

7.4. Icy-Halo

The molecules present in the nucleus of a comet are likely to be in the form of clathrates or hydrates wherein the molecules are loosely packed in the icy matrix (Chap. 11). These clathrates have successfully been produced in the laboratory under more or less cometary conditions and they have shown the presence of large size grains of diameter in the range of 0.1 to 1 mm which appear as powder snow. Therefore the evaporation of the gases from the nucleus could give rise to small size particles as well as larger sizes as seen in the laboratory experiments. The larger size grains also follow dynamics similar to those of smaller ones. One difference of course is that they travel farther out into the coma before they are vapourized. Therefore, this could give rise to a icy-halo around the nucleus. The radius of the halo for a heliocentric distance of 1 au could possibly be in the range of 10^3 to 10^5 km.

In Chap. 6, we have discussed the hypothesis wherein the parent molecules are assumed to be released by the nucleus and these then break up as they move outwards giving rise to the observed species. Another interesting possibility is that the large size grains emitted by the nucleus giving rise to the icy-halo as discussed above could be the source of the observed molecules in comets. The molecules stored in these grains could be released as they are vapourized. Strong evidence for such a process has come from Comet Halley Observations.

The dynamical theory developed to explain the shape of the dust tail of comets has been very successful in explaining the observations. In particular it can explain quantitatively the observed isophotes of dust tails of several comets. It has also been possible to get approximate production rates of the gas and dust and therefore the dust to gas ratio. The theory is also very successful in explaining the presence of anti-tails in comets. Several improvements have been carried out over the simple dynamical theory, which lead to loss of the simplicity of the approach.

Problems

1. A grain of size a and density ρ is moving in a medium of hydrogen density N_H due to the effect of radiation pressure of a star of temperature T and radius R. The grain is being slowed down by the drag force due to the hitting of the grain and sticking to it. Show that the terminal speed of the grain is given by

$$v_t = \left(\frac{R^2 \sigma T^4}{m_H c}\right)^{1/2} \frac{1}{N_H^{1/2} r}.$$

Hint: Use the conservation of momentum of the particle for calculating the drag force.

2. Why does the length of the dust tail change with heliocentric distance? At what distance from the Sun will the tail have the maximum extent.
3. How does the tail of a Comet look, when it passes by Jupiter at a close distance?

References

The theory of dynamics of dust tails is worked out in detail in these papers.
1. Finson, M. L. and Probstein, R. F. 1968 *Ap. J.* **154**, p. 327 (Paper I), p. 353 (Paper II).

Application of the dynamical theory to Comet Bennett can be seen from the following reference.
2. Sekanina, Z. and Miller, F. D. 1973 *Science* **179**, 565.

Some of the other relevant papers are the following:
3. Crifo, J. F. 1991, *In Comets in Post-Halley Era*, eds. R. L. Newburn, Jr. et al. Kluwer Academic Publishers, Vol. 2, p. 937.
4. Farnandez, J. A. and Jockers, K. 1983, *Rep. Progress in Phys.* **46**, 665.
5. Fulle, M. 1989, *Astr. Ap.* **217**, 283.
6. Grün, E. and Jessberger, E. K. 1990, *In physics and chemistry of comets*. ed. W. F. Huebner. Springer-Verlag. P. 113.
7. Richer, K. and Keller, H. U. 1987. *Astr. Ap.* **171**, 317.
8. Sekanina, Z. and Farrell. J. A. 1978. *A. J.* **87**, 1836.

The idea of clathrates was first proposed in
9. Delsemme, A. H. and Swings, p. 1952, *Ann. d'Astrophy.* **15**, 1.

Cometary dust features can be seen in
10. Rahe, J., Donn, B. and Wurm, K. 1969. *Atlas of Cometary Forms*, NASA SP-198 U.S. Govt. Printing Office, Washington D.C.

CHAPTER 8

LIGHT SCATTERING THEORY

The dust particles present in comets scatter and absorb the incident solar radiation. In fact, it is the scattered solar radiation which makes the dust tail of comets visible. The infrared radiation that has been seen from comets is also directly related to the amount of incident radiation absorbed by the dust particles. Therefore the theory of scattering of light by small particles is basic to the study of cometary grains. The aim is to determine, from theory, the distribution of intensity of the scattered radiation and the polarization as a function of the scattering angle. It is also of interest to know the cross sections for the absorption and scattering processes, which determine the albedo of the particles. The efficiency factor for the radiation pressure is also of interest, as was discussed in the last chapter. All these quantities depend upon the shape, structure and composition of the grain. The theories of scattering have been developed for well-defined particle shapes like spheres, concentric spheres, cylinders, spheroids and so on. However, grains in general are likely to be irregular in shape, inhomogeneous and fluffy in nature. The theory of scattering from such grains is highly complex. Hence, at the present time attempts are being made to study the scattering from such grains from the theoretical points of view with certain approximations. The study is also being pursued through experimental means. We will briefly discuss some of these aspects in this chapter.

8.1. Mie Scattering Theory

8.1.1. *Efficiency factors*

The theory of scattering by spherical particles of homogeneous composition involves the solution of Maxwell's equations with appropriate boundary conditions on the sphere. This was worked out by Mie in 1908 and independently by Debye in 1909. Since an excellent account of the solution of this scattering problem, generally known as Mie Theory, is discussed in many books, we will summarize here only the results with particular reference to the case of cometary dust.

The scattering properties of a particle depend upon the following quantities: (1) the property of the medium, usually specified by the complex refractive index, $m = n - ik$. where n and k are the refractive and absorptive indices respectively. (2) the wavelength of the incident radiation (λ) and (3) the size of the particle (a). As a result of radiation interacting with the particle, part of the radiation is absorbed and part of it is scattered. Therefore, the total amount of radiation lost from the incident beam (extinction) is the sum total of the absorbed and the scattered components. These are generally expressed in terms of the dimensionless efficiency factors, Q_{sca} and Q_{abs} for the scattering and absorption components. The efficiency factor for the total extinction is given by

$$Q_{ext} = Q_{sca} + Q_{abs}. \tag{8.1}$$

If C_{sca}, C_{abs} and C_{ext} denote the corresponding cross sections, then

$$C_{sca} = \pi a^2 Q_{sca} \tag{8.2}$$

$$C_{abs} = \pi a^2 Q_{abs} \tag{8.3}$$

and

$$C_{ext} = \pi a^2 Q_{ext}. \tag{8.4}$$

The main aim is to determine the efficiency factors Q_{sca} and Q_{abs} from the scattering theory. They are given by

$$Q_{sca} = \frac{2}{x^2} \sum_{n=1}^{\infty} (2n+1) \left\{ |a_n|^2 + |b_n|^2 \right\} \tag{8.5}$$

$$Q_{ext} = \frac{2}{x^2} \sum_{n=1}^{\infty} (2n+1) \{Re(a_n + b_n)\} \tag{8.6}$$

where the dimensionless parameter $x = (2\pi a/\lambda)$ and Re represents the real part. The scattering coefficients a_n and b_n are given by

$$a_n = \frac{\psi'_n(mx)\psi_n(x) - m\psi_n(mx)\psi'_n(x)}{\psi'_n(mx)\zeta_n(x) - m\psi_n(mx)\zeta'_n(x)} \tag{8.7}$$

$$b_n = \frac{m\psi'_n(mx)\psi_n(x) - \psi_n(mx)\psi'_n(x)}{m\psi'_n(mx)\zeta_n(x) - \psi_n(mx)\zeta'_n(x)} \tag{8.8}$$

ψ_n and ζ_n are the modified Bessel functions known as the Riccati-Bessel functions. The addition of a prime to the Riccati-Bessel functions denotes the differentiation with respect to their arguments. Riccati-Bessel functions can be expressed in terms of Bessel functions, J, as follows

$$\psi_n(y) = \left(\frac{\pi y}{2}\right)^{1/2} J_{n+1/2}(y) \tag{8.9}$$

and

$$\zeta_n(y) = \left(\frac{\pi y}{2}\right)^{1/2} [J_{n+1/2}(y) + i(-1)^n J_{-n-1/2}(y)] \tag{8.10}$$

where y represents either mx or x. The third Riccati-Bessel function can be defined as

$$\chi_n(y) = (-1)^n \left(\frac{\pi y}{2}\right)^{1/2} J_{-n-1/2}(y) \tag{8.11}$$

The functions $\psi_n(y)$ and $\chi_n(y)$ are connected through the identity

$$\zeta_n(y) = \psi_n(y) + i\chi_n(y). \tag{8.12}$$

In addition to energy, light carries momentum. Therefore a beam that interacts with the particle will exert a force on the particle called radiation pressure. With the assumption that all the photons absorbed by the particle transfer all their momentum and hence exert a force in the direction of propagation, the efficiency for radiation pressure is given by the difference between the total extinction and the amount of scattered light.

$$i.e.\ Q_{pr} = Q_{ext} - \overline{\cos\theta} Q_{sca} \tag{8.13}$$

where

$$\overline{\cos\theta}\, Q_{sca} = \frac{4}{x^2} \sum_{n=1}^{x} \left\{ \frac{n(n+2)}{n+1} [Re(a_n)Re(a_{n+1}) \right.$$
$$+ Im(a_n)Im(a_{n+1}) + Re(b_n)Re(b_{n+1})$$
$$\left. + Im(b_n)Im(b_{n+1})] + \frac{2n+1}{n(n+1)} [Re(a_n)Re(b_n) + Im(a_n)Im(b_n)] \right\}$$
$$\tag{8.14}$$

where Re and Im represent the real and imaginary quantities. The efficiency for radiation pressure is of interest in the study of dynamics of grains. The value of $\overline{cos\theta}$ is also sometimes known as the asymmetry factor.

$$i.e.\ g = \overline{cos\theta}. \quad (8.14a)$$

The value of $g = 0$ for a particle that scatters isotropically and $g > 0$ for preferentially scattering in the forward direction. For complete forward scattering $g = 1$. For $g < 0$ the scattered radiation is in the backward direction. The cross sections given by Eqs. (8.5) and (8.6) refer to a particle of a given size. However, in general, a medium will consist of particles of various sizes. If $n(a)$ represents the number of particles per unit volume in the size range a and $a + da$ then the total extinction is given by the expression

$$Q_{total}(\lambda) = \int \pi a^2 Q_{ext}(a,\lambda) n(a) da. \quad (8.15)$$

8.1.2. Albedo

Another quantity of interest is the amount of energy scattered from the incident beam, called the albedo of the particle. In the general case, when the scattered contributions from diffracted, refracted and reflected components are taken into account, the albedo of the particle is defined as

$$\gamma = \frac{Q_{sca}}{Q_{ext}}. \quad (8.16)$$

If the diffraction component is not taken into account, it is called the bond albedo. Therefore the bond albedo is defined as the ratio of the energy refracted and reflected by the particle in all directions to the energy incident on the geometric cross section.

$$i.e.\ A_B = \frac{1}{G} \int \sigma_r(\theta,\phi) d\omega \quad (8.16a)$$

Here $\sigma_r(\theta,\phi)$ is the differential scattering cross-section. For the case when there is symmetry with respect to the direction of incident radiation, then

$$\sigma_r(\theta,\phi) = \sigma_r(\theta).$$

The geometric albedo A_p is defined as the energy scattered by the particle at $\theta = 180°$ (backward scattering) to that scattered by a white disk of the same geometric cross section, scattering according to Lambert's Law $(\sigma_r(180°) = G/\pi)$.

$$i.e.\ A_p = \frac{\pi}{G}\sigma_r(180°). \tag{8.16b}$$

The bond albedo A_B is related to the geometrical albedo A_p by the relation

$$A_B = A_p \cdot q \tag{8.16c}$$

where

$$q = 2 \int_0^\pi j(\theta) \sin\theta d\theta.$$

Here $j(\theta) = \sigma_r(\theta)/\sigma_r(180°) \equiv F(\theta)/F(180°)$, the ratio of the phase functions and q is known as the phase integral. For isotropic scattering, $A_B = 1$ and $A_p = 0.25$ and therefore $q = 4$.

8.1.3. Scattered intensity

The major physical quantity of interest is the intensity of the scattered radiation as a function of the scattering angle. If I_0 is the original intensity impinging on the grain, the intensity of the light scattered into unit solid angle for the scattering angle θ defined with respect to the incident beam (Fig. 8.1) is given by $F(\theta)I_o$, where $F(\theta)$ denotes the phase function.

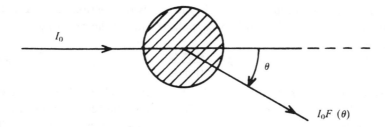

Fig. 8.1. Schematic diagram of a scattering process. θ is the scattering angle and $I_o(\theta)F(\theta)$ is the intensity of the scattered radiation from the original incident radiation of I_o.

Therefore a plot of $F(\theta)$ versus θ gives the angular distribution of the scattered radiation and is known as scattering diagram. The scattering phase function is related to the complex scattering amplitudes $s_1(\theta)$ and $s_2(\theta)$ as

$$\begin{aligned} F(\theta) &= \frac{1}{2k^2} \left[|s_1(\theta)|^2 + |s_2(\theta)|^2\right] \\ &= i_1 + i_2 \end{aligned} \tag{8.17}$$

where $k = (2\pi/\lambda)$. The quantity i_1 and i_2 are essentially the components of intensity in the direction perpendicular and parallel to the scattering plane. The scattering plane contains the incident radiation and the direction of the scattered wave. The expressions for $s_1(\theta)$ and $s_2(\theta)$ are given in terms of the scattering coefficients a_n and b_n as

$$s_1(\theta) = \sum_{n=1}^{\infty} \frac{2n+1}{n(n+1)} \{a_n \pi_n(\cos\theta) + b_n \tau_n(\cos\theta)\} \quad (8.18)$$

$$s_2(\theta) = \sum_{n=1}^{\infty} \frac{2n+1}{n(n+1)} \{b_n \pi_n(\cos\theta) + a_n \tau_n(\cos\theta)\} \quad (8.19)$$

where

$$\pi_n(\cos\theta) = \frac{1}{\sin\theta} P_n^1(\cos\theta) \quad (8.20)$$

and

$$\tau_n(\cos\theta) = \frac{d}{d\theta} P_n^1(\cos\theta). \quad (8.21)$$

Here P_n's are the Legendre polynomials.

8.1.4. *Polarization*

The scattered intensities i_1 and i_2 in the two directions are a function of the scattering angle. Therefore the polarization of the resulting radiation is defined as

$$P = \frac{i_1 - i_2}{i_1 + i_2}. \quad (8.22)$$

The value of $|P|$ varies from 0 to 1. The sign of P could be positive or negative. Positive and negative signs imply that the scattered light is polarized perpendicular or parallel to the scattering plane respectively. The value of P for the scattering angles, $\theta = 0°$ and $180°$, is equal to zero as $i_1 = i_2$. In addition to linear polarization, circular polarization also may be seen in certain favourable cases. The circular polarization arises if the refractive indices are different for the two states of polarization. In general, the circular polarization is one or two orders of magnitude smaller than that of linear polarization.

8.2. Approximate Expressions

So far we have been considering the exact expressions for the efficiency factors for the scattering and the extinction by the spherical particles. In many practical situations, like in the infrared and far infrared regions, where the condition that the size of the particle is very much less than the wavelength of the radiation, i.e., $x \ll 1$ is satisfied, the approximate analytical expressions of Eq. (8.5) and (8.6) may be used. Under the above condition, the Eqs. (8.5) and (8.6) reduce to

$$Q_{sca} = \frac{8}{3}x^4 Re\left\{\frac{m^2-1}{m^2+2}\right\}^2 \tag{8.23}$$

and

$$Q_{abs} = -4x Im\left\{\frac{m^2-1}{m^2+2}\right\} \tag{8.24}$$

If further m is real, i.e. no absorption, then

$$Q_{sca} = Q_{ext} \approx \frac{8}{3}x^4 \left|\frac{m^2-1}{m^2+2}\right|^2 \tag{8.25}$$

and $Q_{abs} = 0$.

8.3. Computation of Cross Sections

The actual evaluation of the efficiency factors has been considered in detail by various investigators. Various methods have been proposed for the calculation of efficiency factors starting from the simple expansion techniques. The main problem is to calculate the coefficients a_n and b_n and this is quite tedious. The calculation of a_n and b_n involves a knowledge of the quantities ψ_n, ψ'_n and others which in turn depend upon the Bessel functions with complex arguments. The computation of these quantities becomes complicated when the refractive index is complex. However they can be calculated easily using recurrence relations with the help of a computer. The following recurrence relations can be used to derive the Riccati-Bessel functions.

$$\psi_n(y) = \frac{2n-1}{y}\psi_{n-1}(y) - \psi_{n-2}(y) \tag{8.26}$$

$$\psi'_n(y) = -\frac{n}{y}\psi_n(y) + \psi_{n-1}(y) \tag{8.27}$$

$$\chi_n(y) = \frac{2n-1}{y}\chi_{n-1}(y) - \chi_{n-2}(y) \tag{8.28}$$

and

$$\chi'_n(y) = \frac{n+1}{y}\chi_n(y) - \chi_{n+1}(y) \tag{8.29}$$

with
$$\psi_0(y) = \sin y \tag{8.30}$$
$$\psi_1(y) = \frac{\sin y}{y} - \cos y \tag{8.31}$$
$$\chi_0(y) = \cos y \tag{8.32}$$
and
$$\chi_1(y) = \frac{\cos y}{y} + \sin y. \tag{8.33}$$

Therefore all the functions that enter in a_n and b_n can be generated. Hence the efficiency factors can be computed for any given value of m and x. The number of terms to be included in Eqs. (8.5) and (8.6) depend upon the parameters x and m. If x is large, more terms have to be included in the calculation of a_n and b_n. However with the availability of computers it is easy to sum over all the terms till the required accuracy is achieved in the calculated efficiency factors. Many investigators have developed efficient computer programs for the calculation of these quantities which are readily available in the literature.

8.4. Results

For astrophysical situations, one seems to deal with the real part of the refractive index of the particles, which is of the order of 1.3 to 1.6 (Chap.9). Therefore, for illustrative purposes, the extinction cross-section is plotted as a function of x for constant indices of refraction of 1.3 and 1.6 in Fig. 8.2. Since the indices of refraction are assumed to be constant, the nature of the curves remains the same when the size or wavelength is changed. But in a real situation, the refractive index of the material is a function of the wavelength which has to be incorporated. The results of such calculations give a better physical picture and show more structures in the curves arising due to the material property. The curves in Fig. 8.2 display several interesting features. A series of broad maxima and minima can be seen easily. Superposed on these band features are the small irregular fine structures generally called the ripple structure. The curve for small values of x refers to Rayleigh scattering region, where it varies as $(1/\lambda^4)$. For intermediate values of x, the variation is given by $(1/\lambda)$. For large values of x, the value of Q_{ext} approaches a value of 2. It is interesting to note that the limiting value 2 is twice as large as the cross section given by the geometrical optics. This is essentially related to the fact that the geometrical optics is strictly not valid in the neighbourhood of the edges of the object. So one

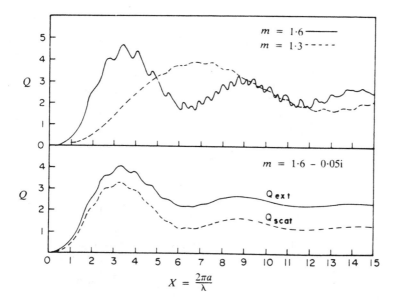

Fig. 8.2. The efficiency factors for extinction and scattering for spheres calculated from the Mie Theory is plotted as a function of the dimensionless parameter X. The results are shown for three sample cases. (Adapted from Wickramasinghe, N. C. 1973. *Light scattering functions for small particles*. London: Adam Hilger.)

has to use the diffraction theory. The additional factor of πa^2 comes due to this reason. The effect of absorption in the Q_{ext} is to reduce the strength of the resonances and the curve becomes smooth (Fig. 8.2). For very large values of x, the absorption and scattering contribute almost equally while for small values of x, the total extinction is mainly contributed by the absorption process. The effect of increase in the absorption coefficient is to reduce the albedo of the particle.

The scattering intensity for various angles is shown in Fig. 8.3. The values of the intensity at $0°$ and $180°$ are marked in the margin of the curve. The most important point to note from the curve is that the scattering is mostly in the forward direction ($\theta = 0°$) and it oscillates with the scattering angle. The scattering diagrams for a few cases are shown schematically in Fig. 8.4. For isotropic scattering the intensity is the same in all directions. For Rayleigh scattering i.e., particle size $\ll$ wavelength of the incident radiation, the phase function $\propto (1 + \cos^2 \theta)$ where θ is the scattering angle.

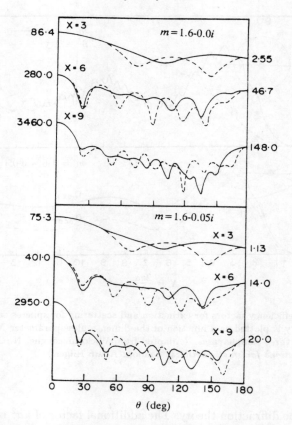

Fig. 8.3. The intensities I_1 (dashed curve) and I_2 (solid curve) are plotted as a function of the scattering angle for two representative values of the refractive index. Each curve is labeled by the parameter X. (Adapted from Wickramasinghe. N. C., 1973 op. cit.)

This dependence shows that the resultant diagram shown by the solid line is a superposition of two different components. One component which is isotropic, is shown by dashed lines and the second component is shown by two lobes. The Mie scattering diagram for $m = 1.3$, $x \approx 1$ and $x > 1$ is also shown in Fig. 8.4. Here as x increases, the scattering becomes more and more forward peaked.

However, in any real situation, there will be a distribution of particle sizes. Therefore the cross-sections have to be averaged over the size distribution function. Several analytical expressions have been used for averaging the cross-sections. For the purpose of illustration Fig. 8.5 shows

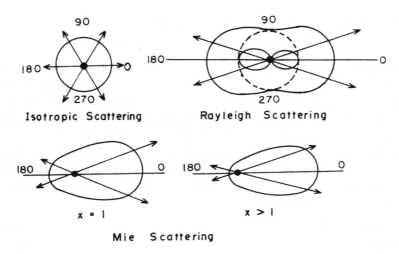

Fig. 8.4. Schematic scattering diagrams for various cases.

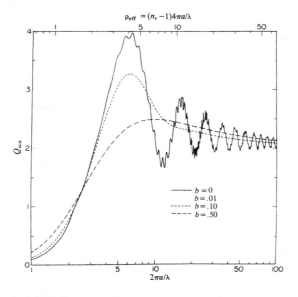

Fig. 8.5. A plot of the efficiency factor for scattering as a function of the dimensionless parameter $X = (2\pi a/\lambda)$. The curves represent the averaged value summed over the size distribution of the form $n(r) = \text{const.} \ r^{(1-3b)/b} e^{-r/ab}$. The refractive index is $n = 1.33$, $k \equiv n_i = 0$. (Hansen, J. E. and Travis, L. D. 1974. *Space Sci. Rev.* **16** 527.)

the results for the size distribution of the form

$$n(r) = \text{constant } r^{(1-3b)/b} e^{-r/ab} \tag{8.34}$$

where $a = r_{eff}$ and $b = v_{eff}$. Here r_{eff} and v_{eff} represent a mean radius for scattering and a measure of the width of the size distribution function respectively, $b = 0$ refers to the case of single particles. The effect of the size distribution function is to average out the ripple structure which smoothens it out. It also dampens the resonance effects. Therefore the curve becomes smooth. The results for phase function and linear polarization are shown in Fig. 8.6. Although the results refer to a particular size distribution function, the general trend and nature of the results are similar for any other size distribution function.

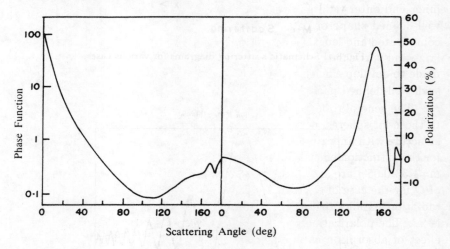

Fig. 8.6. Phase function and polarization for single scattering is plotted as a function of the scattering angle for wavelength of 5500 Å and $n = 1.33, k \equiv n_i = 0$ (Adapted from Hansen, J. E. and Travis, L. D. 1974 *op. cit.*)

8.5. Particles of Other Types

So far, the discussion has been limited to the case of spherical particles which is easier to solve mathematically. However in a real situation, the particles are far from being spherical in shape. Therefore considerable effort has been spent in the study of particles of other shapes through theoretical and numerical means. Several classical methods of solving scattering problems have been used. To name a few, separation of variables, perturbation methods, fields expanded in vector spherical harmonics among others have been used. The exact solutions for scattering from spheroids, infinite cylinders, concentric spheres have been worked out and are available in many books. The cases of cylinders with oblique incident, as well as concentric cylinders have also been investigated mathematically. The approximate formulae for the limiting cases are also available for concentric spheres, cylinders etc. However, even the well defined shape for the particles introduces many more additional parameters due to their shape, orientation etc. compared to that of spheres.

Two possibilities exist for the case of spheroids. They could be prolate or oblate. For infinite cylinders, the exact solution has been derived for the case when the particle diameter is much smaller than its length. In the case of concentric spheres there are various possibilities depending on the inner and outer radii and their relative refractive indices. Even from such well-defined shape of particles the computation of cross sections is quite complicated and lengthy. In addition, the computation time could be quite large and particularly so for larger size particles. Therefore efforts are being made to develop efficient methods for the calculation of cross sections from such well-defined particles.

The general results for nonspherical cases are qualitatively very similar to those for spheres, with a Rayleigh like increase, broad scale interference structure with finer ripples superposed over it. As a typical case, the results for the extinction efficiencies in the case of infinite cylinders for n=1.33 and m=1.33-0.05 i are shown in Fig. 8.7. Here Q_E and Q_H denote the value of Q for the case of electric vector and the magnetic vector of the incident radiation parallel to the axis of the cylinder. The radius of the cylinder is 'a'. The polarization $(Q_E - Q_H)$ is also shown in the same figure. The effect of absorption as pointed out before, is to dampen the polarization oscillations.

The discussion so far was limited to the ideal situations wherein the particles are assumed to have definite and regular shapes. However, in general, the dust particles in a real case are far from the above situations. It is more likely that they will be irregular in shape. They could also be fluffy, porous and could have surface roughness. Therefore several attempts have been made to consider the general case of scattering by an arbitrary particle. The general solutions involve the calculation of the effect on individual atoms by the incident field and also the combined fields of all other atoms. As the general case is almost impossible to solve, it necessarily involves making certain approximations and finally the cross sections have to be derived from numerical methods. Several studies have been carried out based

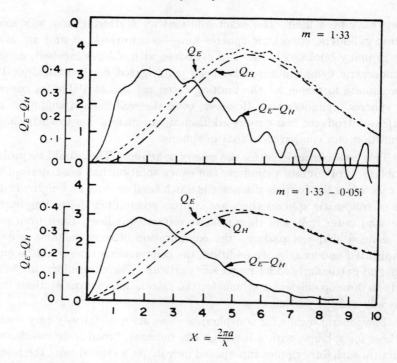

Fig. 8.7. A plot of the efficiency factors for extinction for cylinders oriented normal to the incident radiation for two values of refractive index. Q_E is for the electric vector E of the incident radiation parallel to the axis and Q_H is for the magnetic vector H of the incident radiation parallel to the axis. The polarization is also shown as $Q_E - Q_H$. (Adapted from Greenberg, J. M. 1978. In *Cosmic Dust.* ed. McDonnel, J. A. M. New York: John Wiley and Sons, p. 187.)

on perturbation approach, T-matrix method also known as the extended boundary condition method, study of statistical description of roughness in the frame work of the potential theory and so on.

However a method which has been successfully used is to subdivide the particle into several smaller identical elements, which itself is a collection of large number of atoms, and can be represented as a dipole oscillator. The average field due to the combined effect of all these individual oscillations is then calculated. This is the well known Discrete Dipole Approximation (DDA) method. Another approach that has been investigated is to make use of the integral representation of the macroscopic Maxwells equations. In this method, the particles are assumed to be an aggregate of small cubic volumes at the lattice site of which are located sub columns, of identical and

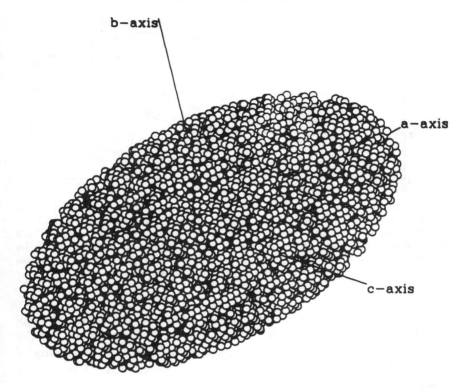

Fig. 8.8. The model of the dust particle (a prolate pseudospheroid with $2a = 60, 2b = 2c = 30$ in units of the lattice spacing d) used in the calculation of cross-section under Discrete-Dipole approximation. In this grain model 60% of the original 28256 dipoles have been removed (Wolff, M. J., Clayton, G. C., Martin, A. G. and Schulte-Ladbeck, R. E. 1994. Ap. J. Letters, **423**, 51).

homogeneous character. This is mathematically equivalent to that of DDA except that the computation is based on the calculation of volume element polarizations. A sample representation of a fluffy particle in terms of sub-particles which is considered in the computation of cross sections is shown in Fig. 8.8 The nature of fluffiness i.e. creating porosity in the grain, depends upon the number of dipoles, their sizes and the separation between them. The number of dipoles used in the calculation could be as large as 10^5. Each of the dipoles in the grain could be small enough to be thought of as Rayleigh-limit inclusions. Extensive calculations have been carried out for the study of scattering properties of such a grain under the DDA

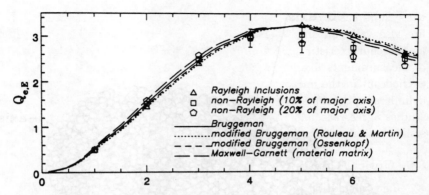

Fig. 8.9. The calculated efficiency factors for extinction with 40% porosity. The results are for two orientations of the incident electric vector E and H which are parallel and perpendicular to the major axis of a 2:1 prolate spheroid. The DDA results for both Rayleigh and non-Rayleigh vacuum inclusions are compared to the various mixing rules (continuous curves). The base grains have 3558, 8320 and 14440 dipoles for the ranges $x \leq 2, 3 \leq x \leq 4$ and $x \leq 5$ respectively (Wolff, M. J. et al. 1994., loc. cit)

formalism for examining the effects of porosity, size of the dipoles, topology of the grain and so on, on the derived results. A sample result for the extinction from a prolate spheroid is shown in Fig. 8.9. The validity of DDA results has to be tested by comparing the results with those derived from well-defined shapes.

In the past, the scattering from regular particles like spheres, spheroids etc., but with inhomogeneous composition, has been studied by replacing the inhomogeneous grain with a homogeneous grain by using an effective dielectric constant calculated either from the Maxwell-Garnett theory or the Bruggeman theory. It is therefore of interest to compare the results obtained from such calculations with those derived from DDA formalism, for the same grain parameters. Figure 8.9 shows such a comparison. The agreement in general is found to be better, as the number of dipoles in the DDA formalism is increased. Ultimately, how good are the calculations has to be tested by comparing with the observations. In general, the results based on these theories seem to represent astronomical observations better compared to the case of spherical particles.

It is easier to carry out laboratory measurements on larger size particles rather than on micron size particles. Therefore several studies have been carried out using the microwave techniques. If the optical constants of the material in the microwave and optical wavelength regions are the same, then the results based on the microwave studies carried on larger size particles are also valid for smaller size particles in the visible wavelength region, as the characteristic parameter x is the same. Therefore, various kinds of measurements like extinction, angular scattering and so on have been carried out in the microwave region on particles of various shapes such as cylinders, spheroids, spheres and irregular particles. The results of such studies agree reasonably well with those of calculated values, thus giving some credibility for the theories.

8.6. Optical Constants

The calculation of cross sections requires the knowledge of the gross material property of the grain. This is generally represented by the refractive index, $m = n + ik$, or the dielectric constant $\epsilon = \epsilon' + i\epsilon''$, of the medium. The optical constants represent a measure of the capacity of electrons in the material to oscillate for the incident radiation. m and ϵ are generally complex quantities. These two quantities are not independent of each other. Knowing one set of quantities, the other set can be calculated. The relations for ϵ' and ϵ'' are given by

$$\epsilon' = n^2 - k^2 \quad \text{and} \quad \epsilon'' = 2nk. \tag{8.35}$$

similarly the relations for n and k are given by

$$n = \left[\frac{(\epsilon'^2 + \epsilon''^2)^{1/2} + \epsilon'}{2} \right]^{1/2} \tag{8.36}$$

and

$$k = \left[\frac{(\epsilon'^2 + \epsilon''^2)^{1/2} - \epsilon'}{2} \right]^{1/2}. \tag{8.37}$$

The determination or the use of either refractive index or the dielectric constant depends upon the physical situation being considered.

The optical constants of materials are not directly measurable quantities. But they have to be determined from laboratory measurements of some property of light in combination with the suitable theory. Various types of measurements can be made for the determination of the optical constants of the material. They could be the measurement of the extinction coefficient from a set of thin slabs, transmittance and reflected light at near normal incidence or for various incidence angles, measurement of the phase shift of the reflected light and so on. The theoretical formulation for the interpretation of the above types of measurements are available. The choice of the method depends upon the material to be measured, experimental techniques available and so on. Various methods have been used to derive the optical constants of several types of material. However the most commonly used and probably the best method for the determination of optical constants is the use of dispersion relations, generally known as the Kramers-Kroning relations in combination with the reflectance measurements. Basically, the dispersion relation is an integral over the whole range of photon energies, which relate the dielectric function to the measured quantity like the extinction coefficients or the reflectance measurements. The real and imaginary parts of m or ϵ are not independent of each other but are interrelated through these dispersion relations.

The determination of optical constants of materials from the laboratory measurements are quite difficult and cumbersome. The sample has to be perfectly smooth to avoid scattering effects, it should be homogeneous and thin slab and so on. Most of the optical constants derived from laboratory measurements are of that of pure substances. In addition the wavelength coverage may not be extensive. Therefore one might have to make use of the derived optical constants over the limited wavelength regions available, to cover the wavelength region from ultraviolet to far infrared. The optical constants for several material like graphite, carbon, amorphous carbon, silicates and so on which are of interest for cometary studies are available. The optical constants for several terrestrial rock samples and moon samples have also been measured. Since these materials are not strictly homogeneous, the measurements in principle, should provide some sort of average dielectric constants of these materials. The cometary grains are inhomogeneous in character. Therefore the best way to get the refractive index of the material of interest is through laboratory studies. Since it is almost impossible to study all the cases of interest in the laboratory, a theoretical treatment of the problem becomes a necessity.

The dust grains present in the astrophysical environments are highly inhomogeneous in character. In general, it is not easy to calculate the average optical properties of inhomogeneous medium even if the property of the individual constituents are known. So one has to use various approximations. This has therefore given rise to several theories. The two theories that are most commonly used are the Maxwell-Garnett theory dating back to 1904 and the Bruggeman theory proposed in 1935. In the Maxwell-Garnett theory, the composite grain consists of matrix material in which small spherical inclusions are embedded having dielectric constants ϵ_m and ϵ respectively with f as the volume fraction of inclusions. The inclusions are assumed to be spherical in shape with sizes smaller than or nearly equal to the wavelength under consideration. The problem is to calculate the average electric field of the medium from the knowledge of the individual electric fields E and E_m of the two components arising out of the interactions, which are by themselves are averages over a smaller volume. With the assumption that the relations for the electric fields, polarization vector and the relations between E and E_m as well as between ϵ and ϵ_m also hold for the average fields, the following relation for the average dielectric constant is derived.

$$\epsilon_{av} = \epsilon_m \left[(2\epsilon_m + \epsilon_i - 2f(\epsilon_m - \epsilon_i))/(2\epsilon_m + \epsilon_i + f(\epsilon_m - \epsilon_i)) \right]. \qquad (8.38)$$

Here the average dielectric constant is independent of the position in the composite medium. This is generally known as the Maxwell-Garnett rule or simply as the M-G rule.

In the Bruggeman theory, there is no distinction between the matrix and the inclusions as in the case of Maxwell-Garnett theory. Here the grains are aggregate of small particles with their own dielectric constant (ϵ_i) and volume fraction (f_i). The effective dielectric constant for such a case is given by the expression

$$\sum_i \frac{f_i(\epsilon_i - \epsilon_{eff})}{(\epsilon_i + 2\epsilon_{eff})} = 0 \qquad (8.39)$$

The value of ϵ_{eff} is derived from the solution of the above equations. However there may be some convergence problem if there is a wide variation in optical constants. Maxwell-Garnett theory is more popular and is generally being used. The applicability and accuracy of the above relations has to be tested based on the comparison between the computed values and those of laboratory measurements of two component mixtures. The laboratory measured dielectric constants of several two component mixtures are in agreement with the effective dielectric constant values derived from Eq.(8.38). Therefore the M-G rule can be used to derive the average optical constants of a two component medium. It is possible to extend the method for a multicomponent mixture as well as for the shape distribution for the inclusions, which of course introduces several additional parameters.

The possibility of deriving the optical constants directly from laboratory measurements of extinction by small particles has also been considered.

However, the computed cross-sections based on the optical constants of bulk samples do not appear to represent fully the observations. This discrepancy could arise due to particle clustering in the laboratory studies. An approximate way of taking this into account is to consider the randomly oriented ellipsoids in the Rayleigh limit, which appears to give a better fit to the observations. Therefore this could be a method of taking into account the problem of clustering in the computation of cross-section.

The dust in comets are more complicated than the simple models considered in the theoretical treatments as summarized so far. For a real situation it is necessary to specify several parameters for the aggregate like the size, shape and the refractive index of the material and for the inclusions, the degree of porosity, size, shape, location and their orientation. To consider so many parameters in the actual calculation of cross-section becomes almost impossible from the practical point of view as well as enormous computing time is required. In view of the complexity of the theoretical models for arbitrary shape particles and also due to the excessive computing time required, such calculations are not in common use at the present time. Instead, the simplest of all the models, namely Mie theory of scattering for spheres is most commonly used. This is sometimes used in combination with the use of average dielectric constant based on Maxwell-Garnett theory. One should, however, keep in mind the limitations of this approach.

Problems

1. Show that for the condition $(2\pi a/\lambda) \ll 1$ and with refractive index $m = n - ik$, Q_{abs} is given by

$$Q_{abs} = \frac{a}{\lambda} \frac{48\pi nk}{(n^2 - k^2 + 2) + 4n^2 k^2}.$$

2. Express the above relation for Q_{abs} in terms of the dielectric function ϵ_1 and ϵ_2.
3. Calculate Q_{sca} and Q_{abs} for ice particles of $m = 1.3 - 0.05i$, $a = 2 \times 10^{-5}$ cm and $\lambda = 5\mu$.
4. For a particle size distribution of the form $n(a) \alpha a^{-3}$ for $a_1 < a < a_2$, calculate the constant of proportionality such that the total number is unity.
5. Assuming m=1, $a \ll \lambda$ and the size distribution of Problem 4, get an expression for the total Q_{abs}.

6. Discuss the physical reason for the occurrence of Kinks in the extinction curve. Why does it smooth out when integrated over the size distribution function?

References

The theory of scattering from spheres can be found in these classic works:
1. Debye, P. 1909, *Ann. Physik*, **30**, 59.
2. Hulst, Van de 1957. *Light Scattering by Small Particles*. New York: John Wiley and Sons.
3. Mie, G. 1908, *Ann. Physik*, **25**, 377.

A discussion of albedos can be found in
4. Hanner, M. S., Giese, R. H., Weiss, K. and Zerull, R. 1981 *Astr. Ap.*, **104**, 42.

The theory of core-mantle particles is worked out in
5. Guttler, A. 1952. *Ann. Phys.*, Lpz. 6 Folge, Bd. 11,5.

The case of core-mantle and cylindrical particles is included in
6. Wickramasinghe, N. C. 1973, *Light Scattering Functions for Small particles*, London: Adam Hilger Limited.

A good account is also given in the book
7. Bohren, C. F. and Huffman, D. R. 1983 *Absorption and Scattering of Light by Small particles*. John Wiley, New York.

The idea of Discrete-Dipole approximation was proposed in the following paper.
8. Purcell, E. M. and Pennypacker, C. R. 1973, *Ap. J.* **186**, 705.

More recent work can be found in the following papers.
9. Draine, B. T. and Goodman, J. J. 1993, *Ap. J.* **405**, 685.
10. Hage, J. I. and Greenberg, J. M. 1990, *Ap. J.* **361**, 251.
11. Wolff, M.J., Clayton, G. C., Martin, P. G. and Schulte-Ladbuck, R. E. 1994, *Ap. J. (letters)*, **423**, 51.

Maxwell-Garnett and Bruggeman rules are discussed in reference No. 7.

CHAPTER 9

THE NATURE OF DUST PARTICLES

As was pointed out in the previous chapter, when radiation is incident on a dust particle (also called the grain), it can scatter as well as polarize it. The absorbed energy by the dust particles can also be radiated in the infrared region and this has been seen in comets. There may be also spectral features in the infrared region depending on the composition of the dust particle. The radiation pressure acting on the particle can give rise to dynamical effects as discussed in Chap. 7. Each of these physical effects could be used to infer the nature and the composition of the cometary grains. The *in situ* measurements of Comet Halley performed by the instruments on board the spacecrafts Giotto and Vega, have given valuable information about the nature and the composition of dust particles. While it is known that comets do exhibit a great deal of variability, there still exists a general overall pattern of behaviour in their observed properties. The emphasis here is on this general pattern of the observed properties of comets. Here we would like to consider some of these effects.

9.1. Visible Continuum

Comets in general possess a continuum in the visible region of the spectrum. The strength of the continuum varies from comet to comet and with the heliocentric distance. The observed continuum is attributed to the scattering of the solar radiation by the dust particles. Therefore, as is wont to, the dusty comets should have a strong continuum. What is of interest, of course, is the dependence of the continuum as a function of the wavelength. The usual method adopted for getting the variation of continuum with wavelength is to compare the observed energy distribution of a comet with the energy distribution of the Sun or a similar star. But the cometary spectrum in the visible region is dominated by the emission features from the molecules (Chap. 4). Therefore, the continuum has to be corrected for these emission features or a spectral region has to be selected where the emission features are absent or minimal. From these observations, it is possible to determine the relative continuum intensities of comet with respect to the Sun or a star similar to the Sun. This yields the wavelength dependence of the observed continuum of the comet. Since the dependence arises due to the scattering of the dust particles in the comet, the interpretation of these observations should give information about the nature of the particles.

Let I_0 be the intensity of the incident solar radiation and r, Δ represent the solar-comet and the comet-Earth distances respectively. The intensity of the scattered radiation at the Earth at an angle θ is given by (Chap. 8).

$$I_{scat}(r, \Delta, \theta, \lambda) = \frac{I_0(\lambda)}{2k^2 r^2 \Delta^2} [i_1(\lambda, \theta) + i_2(\lambda, \theta)] \tag{9.1}$$

where $k = 2\pi/\lambda$. This has to be integrated in general with a size distribution function. Therefore, the problem reduces to the calculation of $I_{scat}(r, \Delta, \theta, \lambda)$ for a given size distribution function and for the assumed grain properties represented by the refractive index of the material, $m = n - ik$. The mean scattering intensities can be calculated from the Mie Theory or from other theories as discussed in the previous chapter. Therefore, the expected intensity distributions can be calculated for various input parameters which could then be compared with the observations.

There exists a large number of broad band observations of comets in the wavelength region from 0.5 to $20 \mu m$. To have an idea of the expected results from the scattering component of the cometary grains, some of the comets around a few heliocentric distance values and covering various scattering angles can be grouped together. Since the main interest is in the general shape of the wavelength dependence of the observed fluxes and not in the absolute amount, the shape of the curves could be shifted to obtain a mean dependence of the observed fluxes. This procedure enables one to extend the wavelength range of the scattered radiation for comets up to $\lambda \approx 4 \mu m$ as can be seen from Fig. 9.1. There exists a close relation in the various observations. The shape of the observed solar flux variation with the wavelength is nearly the same as that of the observed shape. This

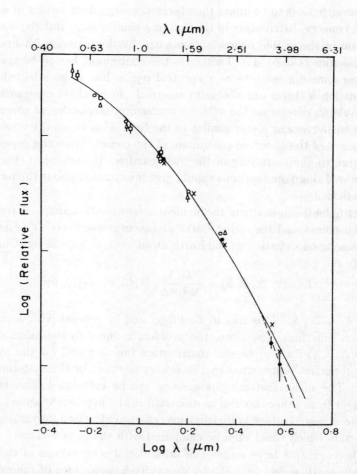

Fig. 9.1. Superposed plot of relative observed fluxes of eight comets as a function of wavelength for heliocentric distances r grouped around 0.3 au (□), 0.7 au(△), 1.0au(0), 2.5au (X) and 3.75au(●). They cover the scattering angles from around 30 to 177°. The continuous curve shows the shape of the observed solar flux and the dashed curve is for Mie calculations with $m = 1.38 - 0.039i$ and $N(a) \propto a^{-3.5}$. Observational data: Bradfield (1974 III) and Kohoutek (1973 XII) (Ney E. P. 1974 *Icarus* **23**, 551); Bradfield (1980 XV) (Ney, E. P. 1982. In *Comets*, ed. L. L. Wilkening, Univ. *Arizona Press,* Tucson, p. 323); P/Crommelin (1984 IV) (Hanner, et al. 1985. *Astr. Ap.* **152**, 177); P/Giacobini-Zinner (1985 XIII) (Hanner, M. S. et al. 1992. *AJ.* **104**, 386); Halley (1986 III) (Bouchet, P. et al. 1987. *Astr. Ap.* **174**, 288; Tokunaga et al. 1988, *AJ.*, **96**, 1971); West (1976 VI) (Ney, E. P. and Merrill, K. M. 1976. *Science* **194**, 1051); Wilson (1987 VII) (Hanner, M. S. and Newburn, R. L. 1989. *AJ.* **97**, 254).

shows that it is rather difficult to extract the properties of dust particles in comets from the study of broad band pass observations. This is borne out from the extensive studies of near infrared colour like J-H and H-K. It also indicates that in general, the results expected from the scattering effects of grains are not drastic.

There have been several refined studies of the continuum measurements of comets in the visible region. However, the results are somewhat confusing. In some comets, the observed wavelength distribution of the scattered radiation has almost the same dependence as that of the solar radiation, while for others, it produces a large reddening. This arises due to several factors. As mentioned earlier, the wavelength dependence of the scattered radiation is derived from observations by comparing the observed continuum with the energy distribution of the Sun. In this procedure, it is very essential to take into account properly the contribution of the molecular emissions, before the dust scattered component can be extracted. This is a difficult and a major problem in scattering studies, as the molecular bands are quite strong and dominate in the visual spectral region. This leaves very few clear windows or spectral regions where the observations could be carried out. However, the situation improves moving towards the red region. It may also be noted that the continuum becomes stronger as the dust to gas ratio increases and the emission bands become weaker as the heliocentric distance of the comet increases. The other difficulty associated with the early observations is connected with poor spectral resolution. A combination of these difficulties and the poor response of the detectors could have been responsible for the differing results of earlier studies. In recent years it has been possible to overcome some of these difficulties and it is now possible to get somewhat reliable and consistent results.

From the observational point of view, it is convenient to describe the scattered radiation or colour from the observed spectra in terms of a quantity called 'reflectivity', which is merely the ratio of the cometary flux to the solar flux at the same wavelength, i.e.,

$$S(\lambda) = \frac{F(\lambda)}{F_\odot(\lambda)}. \tag{9.2}$$

To be consistent with the cometary spectra, the solar spectrum has to be made smooth to the instrumental resolution of the cometary spectra before the ratio is determined. The normalized reflectivity gradient is defined as

$$S'(\lambda_1, \lambda_2) = \left(\frac{ds}{d\lambda}\right)/\overline{S}. \tag{9.3}$$

Here $(ds/d\lambda)$ represents the reflectivity gradient in the wavelength interval between λ_1 and λ_2, and $\overline{S}$ is the mean reflectivity in the same wavelength region defined as

$$\overline{S} = N^{-1} \sum S_i(\lambda). \tag{9.4}$$

The reflectivity gradient $S'(\lambda_1, \lambda_2)$ gives a measure of the colour. If $S'(\lambda_1, \lambda_2) > 0$, it indicates grain reddening.

The reddening curve, in principle, can be extended to the ultraviolet and infrared wavelength regions by combining various observations. However, the main difficulty is that since the observations are performed with different instruments, different wavelength regions etc., it is hard to connect them with one another. An attempt to combine the data obtained from the IUE and ground-based observations, along with infrared observations of comets Bowell, Stephen-Oterma and Cernis, has succeeded in getting the reddening curve for the coma in the wavelength region $\lambda = 0.26$ to 2.5μm. This showed a gradual reddening with an increase in wavelength. In more recent times, the reflectivity and the reflectivity gradient have been determined in a systematic manner from the study of several comets in the spectral region from 0.5 to around $3\mu m$. This broad coverage in wavelength came about as a result of the combination of the scanner observations in the wavelength region of 0.35 to $0.7\mu m$ and the near infrared observations. In these studies, several wavelength regions in the cometary spectra, as far as possible, free of molecular emissions, were selected. This criterion is particularly difficult in the continuum measurements in the ultraviolet region, where they are susceptible to emission lines in the spectral region. The derived average behaviour of S' based on a large number of comets is shown in Fig. 9.2. The results for Comet Halley also show a similar behaviour of the reflectivity gradient with wavelength. These results clearly show that the average S' decreases systematically from ultraviolet into the near infrared wavelength region. In particular, it shows a strong reddening in the ultraviolet region which decreases slowly with an increase in wavelength. The neutral point being around $\lambda \sim 2\mu m$ and the blue beyond $\lambda \sim 3\mu m$. Therefore, comets in general, appear to show an average behaviour for the reddening as shown in Fig. 9.2, although there are some exceptions. In addition, there could be spatial as well as temporal variation in colour in the coma.

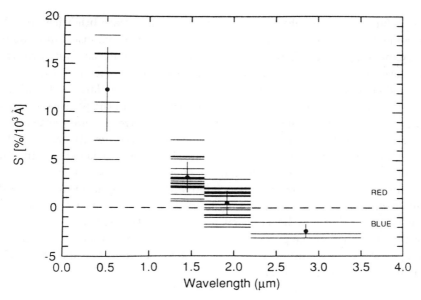

Fig. 9.2. The observed normalized reflectivity gradient $S'(\lambda_1, \lambda_2)$ between wavelengths λ_1 and λ_2 expressed in percent per $10^3 A$ is shown as a function of the wavelength for more than a dozen comets. The horizontal line represents the observational value for each comet over the interval between λ_1 and λ_2. The filled circle represents the mean value of S' with each measured wavelength interval. The dashed line is for neutral scattering with $S'(\lambda_1, \lambda_2) = 0$. Continuum colours due to scattering changes from red ($S' > 0$) to blue ($S' < 0$) as the wavelength changes from optical to near-infrared regions (Jewitt, D. C. and Meech, K. J. 1986. Ap J. 310, 937).

It is also possible to determine the distribution of surface brightness of the continuum radiation. The results based on several comets show that the brightness varies with the nucleocentric distance up to about $10^5 km$ or so, with an average power index of around -1. The particles, generated near the nucleus, as they expand outwards with a constant velocity, will give rise to a surface brightness variation with an index of -1. The close agreement between the observed and the expected variation shows that the fountain model as reasonable. However, with an increase in distance from the nucleus, the effect of radiation pressure on the grains will start becoming important causing distortion in the observed surface brightness curve. This is evident from the observed isophotes of comets which show spherical symmetry close to the nucleus, but distorted at larger distances. Also, the asymmetry in the coma due to sunward ejection of the dust particles becomes prominent. In these situations, the surface brightness in the coma is also a function of the phase angle. Therefore, the surface brightness profile of comets can be described reasonably well with the geometrical dilution of the expanding gas for distances not too far from the nucleus. But, with the increase in distance, the radiation pressure effect becomes important and it dominates at large distances causing deviations from the -1 index law.

The interpretation of reddening results in terms of the physical characteristics of the dust grains is complicated, as it involves several parameters, including the variation in scattering angle, and is also generally based on the Mie theory of scattering for spheres. Also, in general, it is rather difficult to entangle the influence of size and the composition of the dust grains.

In view of the difficulties associated with deriving accurate scattered radiation of dust from continuum observations as well as its interpretation, sometimes it may be sufficient to use a quantity which represents the average property of the grain. This will be particularly useful in the intercomparison of observations of various comets. Therefore, a quantity $Af\rho$ has been defined and it is given by

$$Af\rho = \left[\frac{2\Delta r}{\rho}\right]^2 \left(\frac{F_{com}}{F_\odot}\right) \rho. \tag{9.5}$$

Here A, f and ρ represent the average grain albedo, filling factor of the grains in the field of view and the linear radius of aperture projected on the comet respectively. F_{com} is the observed cometary flux. The filling factor f is given by

$$f = \frac{N(\rho)\sigma}{\pi \rho^2} \tag{9.6}$$

where $N(\rho)$ is the number of grains in the field of view and σ is the average grain cross-section. If the column density $N(\rho) \propto \rho^{-1}$, as in the case of the radial outflow model, the quantity $Af\rho$ becomes independent of the field of view or the geocentric distance. The advantage of the quantity $Af\rho$ is that it can be directly determined from the measurable quantities.

The early reddening observations of the tails of Comets Arend-Roland and Mrkos were found to give a fit with model grains with the refractive index of iron and sizes $\sim 0.3\mu$. The scattered radiation from the head regions of Comets Arend-Roland and Mrkos could also be fitted with particles with the refractive indices between 1.25 and 1.50 (dielectric) and the size distribution proportional to a^{-4} for the sizes of the particles. The interpretation of later observations of Comets Kohoutek, Bradfield, West and others seems to indicate the particle sizes to be submicron or micron and slightly absorbing. The expected intensity distributions have been carried out for various types of silicate materials as well as for various values for the constant refractive index (n) with a small absorption part (k). These comparisons indicate that $n \approx 1.5$ to 1.6, $k \lesssim 0.05$ and the particle size $\sim 0.2\mu$. The observed colour trend of Fig. 9.2 is consistent with dust grain sizes, $a \geq 1\mu m$ and slightly absorbing. It may be mentioned that the expected colour variation of chondritic grains is also consistent with the observed variation

9.1.1. *Phase function*

The average scattering function or phase function, i.e., the scattered intensity as a function of the scattering angle, is another important parameter in defining the nature of the particle responsible for it. The average scattering function can be obtained by observing the coma of the comet as a function of the heliocentric distance or by scanning along the outward direction of the tail. In such observations the quantity that is changing is basically the scattering angle.

It is considerably difficult to derive the phase function, since the physical conditions within the coma change with the scattering angle. The approximate phase function has been derived from the ratio of the visual flux, which represents the scattered radiation, to the infrared brightness that represents the absorbed energy, for the scattering angles between $30°$ and $150°$. Since they both come from the same volume of dust, the ratio is simply proportional to the scattering function of the grains. The derived phase function for several comets is shown in Fig. 9.3. The phase function shows a forward scattering lobe for scattering angle $\lesssim 40°$ with almost a flat shape for an angle between $60°$ and $150°$. It is interesting to know whether the curve increases steeply for $\theta's > 150°$. The observation of Comet Halley carried out upto a scattering angle of $178.63°$ does not indicate a strong backscattering peak. The results for comets Ashbrook-Jackson, Bowell and Stephen-Oterma also show a similar behaviour of Comet Halley. Therefore, there appears to be no evidence for the presence of backscattering peak. The nature of the observations is more in conformity with the expected results from non-spherical particles (Fig. 9.3(b)).

The observed phase function is consistent with the expected distribution of micron size particles with an index of refraction in the ranges $1.3 \lesssim n \leq 2.0$ and $k \geq 0.05$. It is possible to make an estimate of the bolometric bond albedo by integrating the observed ratio over all solid angles.

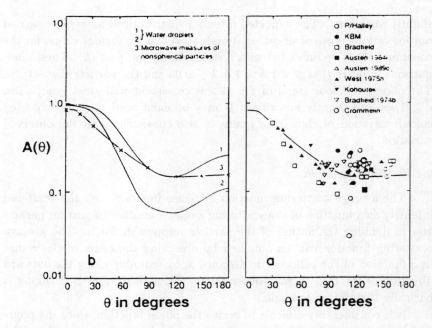

Fig. 9.3. (a) A composite phase diagram of the observed ratio of the reflected to the infrared flux as a function of the scattering angle derived for several comets. For comparison with the data, the curve 3 from (b) is also shown. (b) shows the laboratory data from spherical (curves 1 and 2) and nonspherical particles (curve 3) (Gehrz, R. D. and Ney, E. P. 1992, Icarus **100**, 162).

The value generally lie in the range 0.1 to 0.3.

9.1.2. *Dust production rate from continuum*

The production rate of grains can be calculated from the knowledge of the total number of dust particles in the field of view required to explain the observed scattered flux at the earth, F_{earth}. It is given in gm/sec by the expression

$$Q_D^{opt}(r) = \frac{4\pi \Delta^2 F_{earth}}{\delta} \left[\frac{\overline{M}}{\overline{E}_\lambda}\right] \overline{V} \qquad (9.7)$$

where

$$\overline{E} = \int_{a_{min}}^{a_{max}} \frac{I_0(\lambda)}{2k^2} \frac{R_\odot^2}{r^2} \left[i_1(\lambda, \theta) + i_2(\lambda, \theta)\right] n(a) da \qquad (9.8)$$

and
$$\overline{M} = \frac{\int_{a_{min}}^{a_{max}} \frac{4\pi}{3} a^3 \rho n(a) da}{\int_{a_{min}}^{a_{max}} n(a) da}. \qquad (9.9)$$

Here δ is the linear radius of the projected field of view, ρ is the density of the grain material and Δ is the geocentric distance. The mean velocity of grains is defined by

$$\overline{V} = \frac{\int_{a_{min}}^{a_{max}} a^3 V(a) n(a) da}{\int_{a_{min}}^{a_{max}} a^3 n(a) da}. \qquad (9.10)$$

In the above expression, $n(a)da$ represents the number of grains of size a in the size range between a and $a+da$ and $V(a)$ is the velocity of grains of size a. The summation has to be carried over minimum, a_{min} and maximum, a_{max} size ranges of the size distribution function. The other symbols have their usual meanings.

The calculation of dust production rate from the nucleus requires the knowledge of the density of the grains, the dust particle velocities and the size distribution of grains. However, the density of grains in comets is quite uncertain. Although the instruments on board the spacecrafts to Comet Halley were not designed or calibrated to measure the density accurately, some estimates have been made which are around 0.3 to 1.0 gm/cm^3. The compositional measurements from Giotto and Vega spacecrafts indicate an average value of $\sim 1 gm/cm^3$ for the carbon type of particles and a value $\sim 2.5 gm/cm^3$ for the silicate type particles. The density measurements of the actual interplanetary dust particles indicate a value $\sim 1 gm/cm^3$. In view of the uncertainties, a canonical value of $1 gm/cm^3$ is generally used as the production rate roughly scales with the density.

The velocity of the grains has to be obtained from the study of the dust-gas dynamics in the coma. Either the dust velocities calculated based on a simplified expression for the dust or by the results based on detailed dust-gas dynamics may be used.

The size distribution function for cometary dust is generally taken as

$$n(a)da \propto a^\alpha da \qquad (9.11)$$

where α is a constant. Some variation in the above type of function has also been used. Dust particle detectors were used on Giotto and Vega missions to Comet Halley to record the impact rate and to measure the mass distribution of dust particles. Dust impact detectors (DIDSY) which used the momentum sensors to detect particles were combined with the data obtained from the particle impact analyzer to cover the mass range of around 10^{-12} to 10^{-16} gms. The SP2 instrument on Vega spacecraft comprising of two particle impact sensors (one - an accoustic sensor, and the other — an ionization sensor) was designed to measure the mass distribution over the range 10^{-16} to 10^{-6} gms. The size distribution of grains has therefore been inferred from the *in situ* mass measurements of dust particles. They clearly showed the existence of large quantities of very small particles. The results of Vega 2 can be approximated with power law size distribution function of the type given by Eq. (9.11) with α and size range as follows:

α=-2   for $a < 0.62 \mu m$
α=-2.75   for $0.62 < a < 6.2 \mu m$ and
α=-3.4   for $a > 6.2 \mu m$.

In view of the uncertainties involved in the various quantities, the production rate could be uncertain by factor of two or so. However, this should have a minimal effect on the relative production rate in various comets.

The production of dust can be calculated from Eq. (9.7), provided the scattered continuum is known. As discussed earlier, it is rather difficult to get accurately this quantity due to the presence of strong emission bands in the visual spectral region. This will also introduce uncertainty in the derived production rate of grains. However, the difficulty associated with band emissions improves in the red wavelength region. In addition, the observed fluxes in the near infrared region of comets essentially refer to the scattered radiation by the dust particles. Therefore, it may be advantageous to use the observed fluxes in this spectral region for the calculation of production rate. Some sample calculated results are given in Table 9.1. It is interesting to note that the derived production rate of grains from the observed scattered fluxes and infrared fluxes agree reasonably well.

9.2. Polarization

The polarization of the scattered radiation is an important observation which can give information as to the nature and size of the particle (Chaps. 7 and 8). In fact, it is found that the polarization is quite sensitive to the shape, structure and sizes of the dust particles.

The appearance of bright Comets Arend-Roland and Mrkos gave a good opportunity for making the first good polarization measurements. The measured polarization at $\lambda \sim 5000 \text{Å}$ and for the phase angle of $\sim 100°$ for both the comets is about 20%. Until the recent measurements

Table 9.1. Production rate of grains (Kg/s)

Comet	Date	r(au)	$\Delta(au)$	From IR emission *	From scattered radiation
Halley	25 Aug 85 (Pre)	2.81	3.16	2.4(4)	3.4(4)
	26 Sept 85 (Pre)	2.40	2.18	5.8(4)	5.0(4)
	8 Jan 86 (Pre)	0.90	1.29	2.9(6)	2.2(6)
	2 May 86 (Post)	1.65	0.85	5.3(5)	3.8(5)
	30 May 86 (Post)	2.05	1.75	2.5(5)	2.4(5)
Kohoutek	10 Dec 73 (Pre)	0.65	1.2	2.0(6)	2.5(6)
	16 Dec 73 (Pre)	0.48	1.1	2.0(6)	3.4(6)
	1 Jan 74 (Post)	0.23	1.0	5.0(6)	8.0(6)
	4 Jan 74 (Post)	0.33	0.9	3.3(6)	3.8(6)

* Krishna Swamy, K.S. 1991. *Astr. Ap.* **241**, 260.
Pre: Pre-perihelion observation.
Post: Post-perihelion observation.

on Comet Halley, there were not many systematic continuum polarization measurements on a single comet over an extended range of heliocentric distances and/or for scattering angles from 0 to 180°. The available observations were referred either to the fixed scattering angles or to fixed heliocentric distances. The observations were also referred to mostly in the visual spectral region based on the use of broad-band passes. Therefore, as in the case of continuum measurements, the possible contribution of the molecular emission features has to be taken into account. The observed variations of the polarization with scattering angle for several comets are shown in Fig. 9.4. They show a well defined mean polarization curve as a function of the scattering angle. The polarization shows negative values for scattering angle $\gtrsim 160°$, a gradual increase with a decrease in scattering angle reaching maximum polarization value $\sim$ 20 to 30% around $\theta \sim 90°$ and then decrease for θ's $< 90°$. The extensive polarization measurements carried out in Comet Halley from the visible to the near-infrared wavelength region show a very similar behaviour as shown in Fig 9.4. This general behaviour could be taken as a representative of quite conditions of the coma. There could be a variation of polarization with the level of activity of the nucleus, such as the presence of jets, bursts or in active comets. With

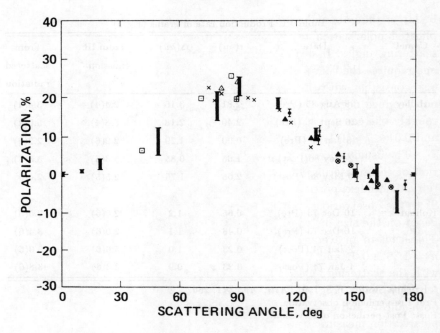

Fig. 9.4. The observed dust polarization in several comets denoted by various symbols is plotted as a function of the scattering angle. The existence of negative polarization for scattering angles around 170° can be seen. The observations clearly delineate a well defined polarization curve as a function of the scattering angles. The dark vertical bars are the calculated values (Krishna Swamy, K. S. and Shah, G. A. 1988, *Monthly Notices*, **233**, 573).

regards to the wavelength dependence of polarization of comets, there were conflicting results. For Comets Bennett and Kohoutek an increase of polarization with wavelength was reported, while for some others like Comets West, Austin and Churyumav-Gerasimenko, neutral polarization in the visible was reported. The multichannel polarization observations of Comet Halley seem to show a slight increase of polarization with the wavelength from the visible to the near infrared.

For interpreting polarization observations, Mie theory has been used extensively. From a comparison of the observed polarization with those of calculations, it is possible to restrict the ranges in the input parameters for the dust particles. The observed wavelength independent polarization can arise if the particle size is such that it lies beyond the peak in the scattering efficiency curve. This observation itself indicates that the particles should be larger than about 1μ. In general, the smaller size particles have a complicated polarization behaviour with respect to the wavelength and the scattering angles. In order to reproduce the observed value of polarization, one requires the complex part of the refractive index to be small. This is also consistent with the result that a high value of the complex part cannot produce negative polarization for large scattering angles. Therefore, roughly one can put a range for the parameters n and k of $1.3 < n < 2.0$ and $0.01 < k < 0.1$ and particle size $\gtrsim 1\mu$.

The usual procedure adopted is to calculate the polarization as a function of the scattering angle for a grid of real and imaginary part of the refractive index, particle size distribution and then to try to get the best fit to the observations. Such an analysis was carried out for Comet Halley based on the size distribution of grains derived from Vega measurements. This leads to an average refractive index of the grain for the visible region, $n = 1.39 \pm 0.01$ and $k = 0.035 \pm 0.004$. The observed linear polarization with the scattering angle for several comets is also consistent with such a refractive index (Fig. 9.4). The calculations for rough particles, including both silicate and graphite grains, also give a good fit to the Comet Halley polarization measurements.

An extensive polarimetric measurements of the tail of Comet Ikeya-Seki carried out in 1965 when the Comet was at $r \approx 0.3$ au showed the surprising result that the polarization value changed from $+0.20$ to -0.42% with scattering angles from 116 to $136°$ respectively, for $\lambda = 0.53$ μm. The neutral point (zero polarization) was found to be around a scattering angle of $125°$. The linear polarization measurements carried out along the tail of Comet Halley in April 1986, at $r = 1.3 au$ showed that they are very similar at 1000 km, 3000 km from the nucleus and for the envelope around the nucleus, for the scattering angles between 125 to $160°$. These observations showed that there was no change in the crossover angle of linear polarization and there was also no drastic modification of the grains as they were transported from the nucleus out to distances of around 1.6×10^6 km.

The expected polarization calculated as a function of the particle size shows that it is possible to produce positive and negative polarizations for grain composition of silicate type and not for that of dirty ice (ice with a small complex part), graphite or iron. Figure 9.5 shows the typical result obtained for grain composition of Olivine. As the polarization is very sensitive to particle sizes, it is possible to get reversal in polarization, if there is a variation of particle size along the tail of the comet.

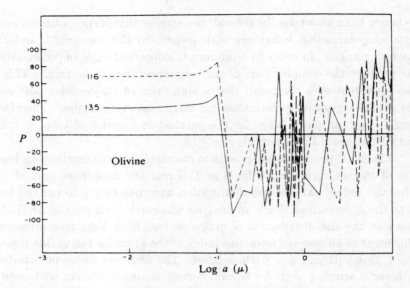

Fig. 9.5. Expected polarization versus particle sizes for $\lambda = 5300 \text{Å}$ and for two scattering angles of 116° and 135°. The oscillations will smoothen out with the use of a size distribution function (Krishna Swamy, K S 1978, *Astrophys Space Sci* **57, 491**).

So far, the discussion pertained to linear polarization. It is also of interest to see whether circular polarization can be detected in comets as it can give additional information about the grains. The expected circular polarization is of the order of 0.5% for certain values of scattering angles which is quite small compared to the linear polarization. The circular polarization has been looked for in Comets Kohoutek, Bradfield and West but with negative results. The average value of circular polarization for Comet Halley is found to be around 1.7×10^{-3}.

9.3. Infrared Measurements

The infrared measurements of comets, particularly in the regions beyond about 1μ are rather recent. This is mainly due to the advances and improvements made in the infrared detectors. There are very few windows that are available for ground based observations, as most of the infrared radiation is absorbed by the Earth's atmosphere (Fig. 9.6). Therefore, most of the observations are carried out with broad band passes. Several such systems exist which have been used by different observers with slight variations in their mean wavelengths and band widths. (Table 9.2).

Table 9.2. Infrared band passes and their wavelengths

Band pass	Arizona System Wavelength (μm)	$\Delta \lambda$ (μm)	IRTF System Wavelength (μm)	$\Delta \lambda$ (μm)
J	1.26	0.20	1.20	0.3
H	1.60	0.36	1.60	0.3
K	2.22	0.52	2.2	0.4
L	3.54	0.97	3.55	1.05
L'	-	-	3.78	0.57
M	4.80	0.60	4.7	0.57
N	10.6	5.	10.50	5.
Q	21.0	11.0	20.6	9.

Adapted from Hanner, M.S. and Tokunaga, A.T., 1991. In *Comets in the Post-Halley Era*, eds. R.L. Newburn, Jr et al. Kluwer Academic Publishers, P. 70.

The infrared observations of comets can provide another independent method for extracting significant information on the physical nature of the cometary grains. This is due to the fact that the observed infrared radiation arises from the re-radiation of the absorbed energy by the dust particles which in turn depends on the nature and composition of the dust. From a detailed comparison of the cometary infrared radiation with the expected infrared fluxes based on grain models, it is possible to infer the physical and the chemical nature of cometary grains. The grains are considered to be in radiative equilibrium with the incident solar energy. The equilibrium temperature of the grain is, therefore, determined by a balance between the absorbed radiation which is mostly in the ultraviolet and visible regions and the emitted radiation which is in the far infrared region. This can be expressed by the condition

$$F_{abs}(a) = F_{em}(a, T_g) \tag{9.12}$$

where

$$F_{abs}(a) = \left(\frac{R_\odot}{r}\right)^2 \int F_\odot(\lambda) \, Q_{abs}(a, \lambda) \pi a^2 d\lambda \tag{9.13}$$

and

$$F_{em}(a, T_g) = \int \pi B(\lambda, T_g) Q_{abs}(a, \lambda) 4\pi a^2 d\lambda \tag{9.14}$$

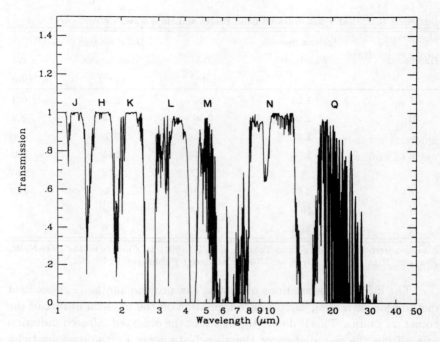

Fig. 9.6. The atmospheric transmission is plotted as a function of wavelength for the region of 1 to $20\mu m$. The spectral band passes corresponding to J to Q are also shown (Courtesy Charlie Lindsey).

where $F_\odot(\lambda)$ representing the incident solar radiation field at wavelength λ, $Q_{abs}(a,\lambda)$ is the absorption efficiency and $B(\lambda, T_g)$ is the Planck function corresponding to the grain temperature T_g. The difference in the geometrical areas between the two sides arises as the grains absorb in one direction and emit in all the directions. The calculation of the grain temperature involves a knowledge of the size and composition of the grains. Knowing the grain temperature, it is possible to calculate the infrared emission from Eq. (9.14).

In general, the emission has to be integrated over the size distribution function to get the total infrared emission from the grains. Therefore, the total infrared emission at the Earth is given by

$$F_{em}(\lambda, r) = \frac{1}{\Delta^2} \int_0^\alpha n(a) \pi a^2 Q_{abs}(a,\lambda) B(\lambda, T_g) da \qquad (9.15)$$

where $n(a)da$ represents the relative number of grains in the size interval between a and $a + da$ and Δ is the geocentric distance of the comet. If there are grains of various types present at the same time, then the total observed infrared radiation is the sum total over all the grain types, j, i.e.,

$$F_{total}(\lambda, r) = \sum_{i=1}^{j} F_{em}(\lambda, r)_i \, x_i \qquad (9.16)$$

where x_i is the fraction of the grain population of type i. The corresponding mass of grains of type i is given by

$$M_i = \int_o^\alpha \frac{4\pi}{3} a^3 \rho_i n(a) da \qquad (9.17)$$

where ρ_i is the density of the ith type of the material.

The first infrared observations made on Comet Ikeya-Seki in 1965 in the wavelength region of 1 to 10μ showed clearly that the comet was very bright in the infrared wavelength region and its colour temperature was higher than that of a black body at the same heliocentric distance. This is due to the fact that for small size grains the efficiency factors at infrared wavelengths are much smaller than unity so that

$$Q_{abs}(a, \lambda) B(\lambda, T_g) \ll B(\lambda, T_g). \qquad (9.18)$$

These results have been confirmed based on the infrared observations of many more comets. Most of the observations on comets before Comet Halley were limited to broad band infrared observations in the spectral region around 2 to $20\mu m$ (Fig. 9.7). However, for Comet Halley it has been possible to get very good broad band and spectroscopic data in the middle infrared and far-infrared wavelength regions based on ground based, airborne and spaceborne instrumentation. So, there exists infrared data in the wavelength region from around 3 to $160\mu m$.

The important observation which gave some clue to the possible nature of the grain was the detection of a broad 10μ emission feature in Comet Bennett. Another feature at 20μ also appears to be present in many comets (see Fig 9.7). However, the whole profile does not show up in these broad band observations as the last band pass is at around $20\mu m$. The observed features at 10 and $20\mu m$ are widely believed to be due to a silicate type of material (Sec. 9.4). The moderate resolution observations of Comet Halley made in the $10\mu m$ region with Kuiper Airborne Observatory have given

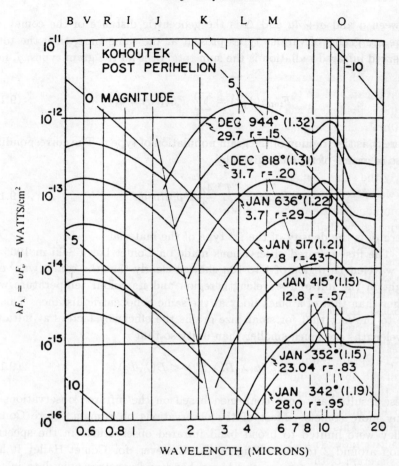

Fig. 9.7. Observed infrared fluxes plotted as a function of wavelength for Comet Kohoutek. The fitted black body temperature for each of the curves along with the factor by which the temperature exceeds the black body temperature is also shown. (Ney, E. P. 1974. *Icarus*, **23** 551.)

important information about the nature of the silicate material.

Observations of the $10\mu m$ feature in the spectra of comets show that the apparent strength of this feature varies from comet to comet and also with the heliocentric distance for the same comet. Early broad-band observations had indicated the apparent absence of the $10\mu m$ feature for heliocentric distances $\gtrsim 1.2 au$. However, moderate resolution spectra of Comet Halley taken at a heliocentric distance of 1.3 au and also observations of other comets show a distinct $10\mu m$ feature suggesting that the silicate feature is also present at larger heliocentric distances. In addition, the strength of the 10 and 20 μm features is a function of the particle size. It becomes weaker with an increase in size of the particles (Sec. 9.3.2). Therefore, silicates can still be the dominant component of the grain even if no $10\mu m$ feature is evident in the spectra. The observations of Comet Halley in the 5 to $10\mu m$ region obtained on 12 December 1985 and 8 April 1986 corresponding to the same heliocentric distance of 1.32 au for pre-and post-perihelion positions agree very well indicating that the dominant grain material was nearly the same for both the dates. Another new emission feature near 3.4 μm was detected in the spectra of Comet Halley. This also appears to be a common feature of most of the observed comets. Comet Halley observations show that the $3.4\mu m$ feature varies with the heliocentric distance of the comet, the feature lying on a background continuum dominated by thermal emission from the cometary dust, scattered incident solar radiation or a combination of the two. The $3.4\mu m$ feature is a characteristic of the C-H stretching vibrations and indicates the presence of some form of hydrocarbons. Taken together, the infrared spectra of comets suggest there are two components to the grains — silicates and some form of hydrogenated carbon.

The grain temperature is generally determined by comparing the profile of the observed fluxes with various blackbody curves. It may be noted that the general shape of the calculated emission curve in the infrared region for larger particle sizes is very similar except for the 10 and $20\mu m$ features. This comes about due to the fact that for the real part of the refractive index (n) not far from unity and for $x = \frac{2\pi a}{\lambda} < 1$, the absorption cross-section of the grain is given by

$$Q_{abs} \approx \frac{8xk}{3} \qquad (9.19)$$

where k is the complex part of the refractive index $m = n - ik$. In these limits the emission given by $B(\lambda, T_g) Q_{abs}$, scales as $B(\lambda, T_g)$, i.e., as a blackbody. However, there is no unique temperature associated with the dust in the cometary coma. The above derived temperature for the grain represents an approximation of an average temperature, since the temperature of the individual grains depends upon the size and composition. From such a procedure it is possible to derive the variation of grain temperature with the heliocentric distance. The grain temperature - distance relation is nearly independent of the grain size for the reasons discussed above.

In the absence of a complete set of data on the refractive indices over the ultraviolet to far infrared region, one can reverse the problem to see what type of materials have absorption properties in the ultraviolet and visible regions that are known to reproduce the observed temperature distribution. Calculations based on various types of grain materials show that $m = 1.6$ or 1.33 with k=0.05 well represents the derived grain temperature variation for Comets Kohoutek, Bradfield and Bennett.

Several attempts have been made to reproduce the observed infrared radiation, from comets using models based on assumed optical properties for the cometary dust particles. These models were essentially based on simple grain population or relatively simple grain mixtures. The observations of various comets including Comet Halley have shown that there are two major components to the grain mixture - silicate rich in olivine and some form of hydrogenated carbon. Such a mixture is consistent with the compositional information available from the Halley fly-bys and from the analysis of interplanetary dust particles. Therefore, a self-consistent cometary emission model may be used taking into account the knowledge derived from the above studies.

It may also be noted that two physical mechanisms have been suggested to explain the $3.4\mu m$ feature. First, the feature could be due to fluorescence process where high energy photons excite large molecules in the coma which subsequently relax and give rise to emissions. Second, the feature could also arise from simple thermal emission from carbonaceous grains. At present there is no preference for one explanation over the other.

The infrared data has been fitted to the results of a cometary coma emission model consisting of grain population of a mixture of olivine rich silicate and amorphous carbon grains. In the absence of refractive index data for cometary silicate and amorphous carbon material, the optical properties of lunar sample 12009.48 for the silicate component and laboratory carbon films for amorphous carbon were used. The infrared emission calculated based on the absorption cross-section derived using Mie theory and for the Halley's size distribution function or for a distribution function of the form $n(a) \propto a^\alpha$ with $\alpha = -3.5$ with a mass ratio of silicate to amorphous carbon particles of around 40 provides a reasonable match to the 3 to $160\mu m$ emission from Comet Halley. The model could also reproduce reasonably well the broad band infrared photometric data of a number of comets from 3 to $20\mu m$. The expected variation of the strength of the $3.4\mu m$ feature as a function of heliocentric distance for Comet Halley is

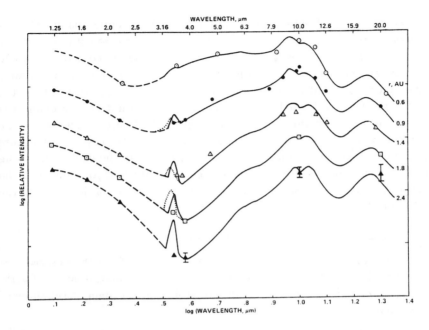

Fig. 9.8. Comparison of the calculated (continuous curves) and observed spectra in the region 1.25 to 20μm of Comet Halley as a function of the heliocentric distance. The calculated curve is for a silicate to amorphous carbon mass ratio of 40. The dashed curves are the shape of the observed scattered solar continuum. The observed profile of the 3.4μm feature is shown by dotted lines in the middle three spectra (Krishna Swamy, K. S., Sandford, S. A., Allamandola, L. J., Witteborn, F. C. and Bregman, J. D. *Ap. J.*, 1989, **340**, 537).

also consistent with the observed variation (Fig. 9.8). At small heliocentric distances, the silicate grains are quite hot and therefore emit a substantial amount of radiation at shorter wavelengths. This radiation raises the continuum level which makes the 3.4 μm feature weaker. However at larger heliocentric distances, the silicate grains are cooler and therefore emit less in the 3.4 μm region, which makes the feature appear stronger. Therefore, it is just a spectral contrast with respect to the emission continuum level and the scattered solar continuum. The thermal emission approach can qualitatively explain the observed behaviour of the 3.4μm feature. The amorphous carbon grains or carbonaceous residue could also in principle account for the 3.4 μm feature observed in Comet Halley. The cometary emission model can reproduce reasonably well the infrared data of various comets as well as of the same comet at different heliocentric distances. This demonstrates that the variation of the grain temperature with heliocentric distance can account for the major changes observed in cometary spectra. In general, the presence of 10 and 20 μm silicate emission can have some effect on the derived blackbody grain temperature. This arises due to the fact that the broad band infrared observations do not clearly define the shape of the continuum around these spectral features. Therefore, the blackbody fit to the observed points gives a somewhat lower temperature compared to the temperature derived by fitting only the continuum using model calculations. Since the spectral region from 5 to 8 μm appears to be devoid of emission lines, it may provide the best region for determining the mean grain temperature.

9.3.1. *Dust production from infrared observations*

The production of dust can also be determined from the observed infrared radiation of comets. The observed infrared radiation at the Earth, F_{earth} is related to the total number of grains in the coma, $N(r)$ through the relation

$$F_{earth} = \frac{E_\lambda}{4\pi\Delta^2} \qquad (9.20)$$

where E_λ represents the infrared radiation of the comet given by

$$E_\lambda = B(\lambda, T_g) Q_{abs}(a, \lambda) 4\pi a^2 N(r). \qquad (9.21)$$

Here Δ is the geocentric distance. A rough estimate for the production rate can be obtained from the relation

$$Q_D(r) \simeq \frac{N(r)}{\tau} \qquad (9.22)$$

where τ is the lifetime of the grain. This gives an average dust mass rate of about 4×10^6 gm/sec for the assumed lifetime of the grain $\tau \simeq 4 \times 10^4 sec$. However, in general, it is necessary to take into account the effect of the size distribution of grains. Therefore, the production rate of grains in gm/sec, as in the case of continuum can be calculated from the relation

$$Q_D^{iR}(r) = \frac{4\pi\Delta^2 F_{earth}}{\delta} \left[\frac{\overline{M}}{\overline{E_\lambda}}\right] \overline{V} \qquad (9.23)$$

where

$$\overline{E_\lambda} = \int_{a_{min}}^{a_{max}} B(\lambda, T_g) 4\pi a^2 Q_{abs}(a, \lambda) n(a) da. \qquad (9.24)$$

The expressions for $\overline{M}$ and $\overline{V}$ are given by Eqs. (9.9) and (9.10). All the symbols which occur in these equations are described in Sec. (9.1.2) or have their usual meanings. The calculation of the production rate of grains from Eq. (9.23) involves the knowledge of density, velocity and size distribution function for the grains. A discussion regarding these quantities is also outlined in Sec. (9.1.2). The production rate of grains has been calculated for various comets. The results for Comet Halley for heliocentric distances between 2.8 and 0.6 au show a close relation with an r^{-4} dependence, while the results for Comet Kohoutek show a close relation for heliocentric distances between 0.15 and 1.5 au with an r^{-2} dependence. It is interesting to see whether there exists a general behaviour of the production rate of dust with the heliocentric distance in comets. This could be carried out by the superposition of the shape of the derived dust production rate with the heliocentric distance of various comets and the results are shown in Fig. 9.9.

The gas (composed of H_2O) and the dust coming out of the nucleus of a comet flow outwards in such a way that the gas carries the dust with it. Therefore, to a first approximation, the heliocentric variation of the dust production rate and the water production rate should be very similar as can be seen from Fig. 9.9. The dependence of r^{-2} for $r \lesssim 1 au$ arises due to the solar flux falling of as r^{-2} and it then deviates from this dependence due to the effect of temperature dependence of the vapour pressure. The ratio of gas to dust for a representative sample of comets is given in Table 9.3. These ratios are to be taken as an indication of the trend of the results due to the uncertainties involved in the water and dust production rates.

9.3.2. Anti-tail

The characteristic feature of Comet Kohoutek was the presence of the anti-tail seen shortly after the perihelion passage. The infrared measurements of this comet made on the same day showed the presence of $10 \mu m$ emission in the coma and in the tail, but not in the anti-tail (Fig. 9.10). The absence of $10 \mu m$ feature in the anti-tail immediately puts a lower limit to the particle size, as can be seen from Fig. 9.11. The figure shows the shape of the emission curves for grains of silicate type for different sizes and for grain temperatures of interest. It shows that the strength of the $10 \mu m$ feature is a function of the particle size and for $a \gtrsim 5 \mu m$, the feature actually disappears. The 10 and $20 \mu m$ features in small size grains show up as they are optically thin. As the size of the particle gets larger and larger, the

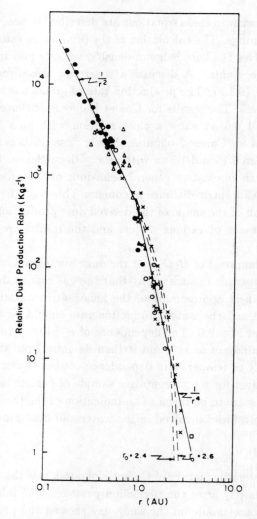

Fig. 9.9. A composite diagram showing the general shape of the variation of the dust production rate with heliocentric distance is derived from several comets, denoted by various symbols. There is a tight relation with heliocentric distance r showing r^{-2} variation up to $r \lesssim 1au$ and r^{-4} for $r \gtrsim 1au$. The dashed curve show the shape of the derived water production for $r_0 = 2.4$ and 2.6 respectively (Krishna Swamy, K. S. 1991. *Astr. Ap.* **241**, 260).

material becomes optically thick and the feature gets washed out. Therefore, the particles present in the anti-tail appear to be of a much larger size

Table 9.3 Gas to dust ratio in comets ($r = 1au$)

Comets	$Q(H_2O)$	$Q(dust)$ $(Kg\,s^{-1})$	$H_2O/dust$ $(Kg\,s^{-1})$
Bennett (1970 II)	1.5(4)	1.5(3)	10
Churyumov-Gerasimenko (1982 VIII)	1.1(2)	5.4(1)	2
P/Crommelin (1984 IV)	1.8(2)	1.2(3)	0.1
P/Encke	2.1(2)	2.8(1)	7
P/Giacobini-Zinner (1985 XIII)	1.2(3)	3.8(2)	3
P/Grigg-Skjellerup (1982 IV)	9.0(1)	2.0(1)	5
P/Halley (1986 III)	1.5(4)	5.4(3)	3
IRAS-Araki-Alcock (1983 VII)	2.2(2)	$2.0(2)^a$	$\lesssim 1.1$
Kobayashi-Berger-Milon (1975 IX)	6.0(2)	1.0(2)	6
Kohoutek (1973 XII)	9.2(3)	4.6(2)	20
Sugano-Saigusa-Fujikawa (1983 V)	8.1(1)	2.0	40
West (1976 VI)	1.5(4)	3.5(3)	4
Wilson (1987 VII)	1.1(4)	1.8(3)	6

(Krishna Swamy, K S 1991. *Astr. Ap.* **241**, 260).

compared to those present in the coma or in the tail region. This conclusion is also consistent with the results based on dynamical considerations of the anti-tail which gives for the particle sizes, $a \gtrsim 15\mu m$. The strength of the silicate feature can change from comet to comet. It can also change with time for the same comet as was the case for Comet Bradfield. This comet had the silicate signature in the observations of March 21.9, 1974 ($r = 0.51au$ and $\Delta = 0.73au$), but was clearly absent in the observations of April 5.8, 1974 ($r = 0.67au$ and $\Delta = 0.69au$). These observations show that the sizes of the emitted grains (Fig. 9.11) could vary from comet to comet, as well as with the heliocentric distance for the same comet. The supporting evidence for the large size particles in the anti-tail also comes from the colour temperature measurements of Comet Kohoutek. It showed that the colour temperature for the anti-tail was much cooler and therefore closer to the black body temperature than that of the coma and the tail temperature.

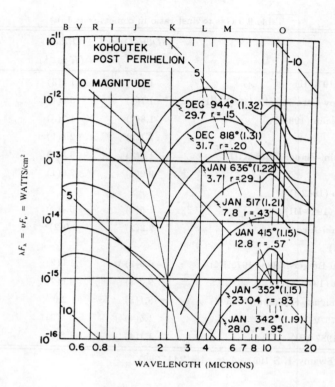

Fig. 9.10. Measured energy distributions in the coma, tail and anti-tail of Comet Kohoutek are plotted as a function of wavelength. Observations refer to January 1.7, 1974. The absence of 10 micron feature in the anti-tail observations can be seen. (Ney, E. P. 1974. *op. cit.*)

9.4. Spectral Feature

The most direct way of getting information about the composition of the grain is through the detection and identification of characteristic spectral features such as the vibration-rotation bands of molecules. It is important to note that such bands retain their identity even when the molecules are in the solid state.

9.4.1. *Silicate signature*

As already remarked, the clue to the possible chemical composition of the cometary grain came from the infrared measurements themselves,

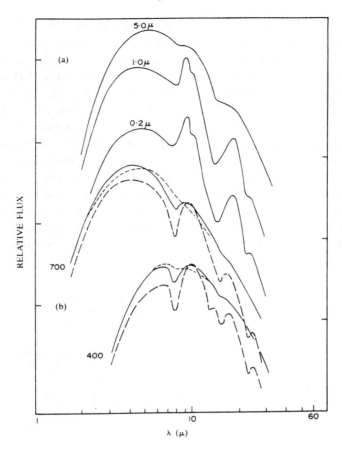

Fig. 9.11. Shape of emission curves for different sizes, grain temperatures and for material of moon samples are shown. (a) for moon sample 12009, $T_g = 550°K$ and sizes of 0.2, 1.0 and 5.0 microns. (b) for moon sample 14321, $T_g = 400$ and $700K$. Long-dashed, continuous and dashed curves refer to particle sizes of 2, 5 and 10 microns respectively. (Krishna Swamy, K. S. and Donn, B. 1979 AJ. 84, 692).

with the detection of a broad emission feature at $10\mu m$ in Comet Bennett. This was confirmed by the observation of many other comets, although the strength may vary from comet to comet with heliocentric distance. In addition to the $10\mu m$ feature, another broad feature at $20\mu m$ appears to be present as well (Fig. 9.7). These two features have also been seen from many of the astronomical sources. This appears to show the common nature of the material present in these objects.

256   *Physics of Comets*

The next logical step is to try to identify the nature of the material responsible for giving rise to these two features. Laboratory measurements of the absorption spectra of various types of rocks and minerals carried out at room temperature or at liquid nitrogen temperature ($89°K$) have a common property in showing strong and broad absorption features around 10 and $20\mu m$ (Fig. 9.12). From the similarity between the laboratory spectra and the cometary spectra, it is generally suggested that the 10 and 20μ features observed in cometary spectra are due to some type of silicate material. The laboratory spectra in addition to showing the main features at 10 and $20\mu m$, also show a few other secondary features which are the characteristics of the particular types of silicate material. However, in the cometary atmosphere, it is quite possible that these secondary peaks could be washed out due to the superposition of a variety of possible types of silicate material that could exist. However, if these features show up

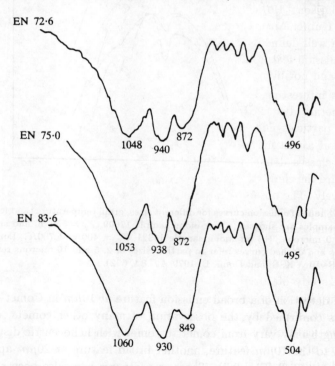

Fig. 9.12. Absorption spectra of Orthopyroxene ionosilicates. Sample curves shown (in cm^{-1}) are for the enstatite-hypersthene series of En_{62}, En_{75} and En_{84}. (Adapted from Lyon. R. J. P. 1963. NASA Technical Note, NASA TN D-1871.)

then it could be used as a diagnostic of the dust composition. The silicate material also shows another feature around 30 to 33μm as was observed in the laboratory absorption spectra of γ-$Ca_2\ SiO_4$, Olivine (Mg_2SiO_4) or other silicates. Unlike the 10 and 20μm features, the feature around 30μm seems to depend more on the chemical composition of the material.

The 10μm feature is attributed to the Si-O stretching vibrations of silicates while the 20μm feature is produced by Si-O-Si bending vibrations. The observed shape of the 10μm feature in comets should be able to provide information about the composition of the grains, since different silicate materials have different band profiles depending upon Si-O stretching frequencies. Comet Kohoutek observed with the moderate spectral resolution in the 10μm region showed a little spectral structure and the entire profile was not obtained. Therefore, the identification of silicate is somewhat uncertain since the entire cometary silicate feature was not available for comparison. However, the observations carried out on Comet Halley with the Kuiper Airborne Observatory and ground-based telescopes clearly showed a well defined spectral feature at 11.3μm located inside the broad 10μm emission feature indicating that the band is produced by crystalline silicates and not by amorphous silicates (Fig. 9.13).

It is interesting to compare the cometary data with the spectra of interplanetary dust particles (IDPs) collected in the stratosphere, as some of these particles may have a cometary origin. Laboratory studies have shown that a vast majority of the collected IDPs fall into one of the three spectral classes defined by their 10μm feature profiles. These observed profiles are the characteristics of olivine, pyroxene and layer lattice silicates respectively. The observed IDP spectra have been used to fit the profile of the 10μm feature as seen in Comet Halley. A good match could be obtained by a mixture containing approximately 55% olivine, 35% pyroxene and 10% layer-lattice silicates (Fig. 9.13). This is in agreement with the results derived from Vega and Giotto mass spectrometer observations.

The study of 10μm feature in a number of comets has shown many similarities but with some disagreements as well. The observed peak seen at 11.25μm is attributed to Mg-rich crystalline olivine. A profile fit to the 10μm feature of some comets indicates the grain composition to be a mixture of olivine, pyroxene and glassy silicates. In comets where the observed 10μm feature is weak or absent, it could be due to the presence of relatively larger size particles which makes this feature weak.

KAO observations of Comet Halley showed the presence of a weak

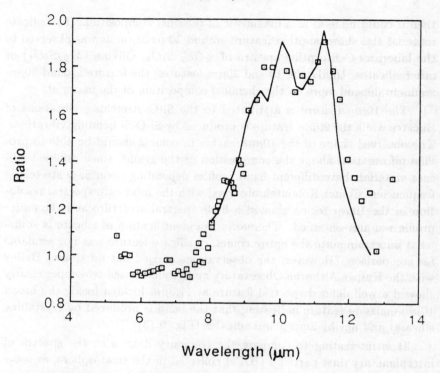

Fig. 9.13. The profile of the 10μm feature derived from the 5 − 13μm spectrum of the central condensation of Comet Halley taken on December 12, 1985 with the Kuiper Airborne Observatory. The solid line represents the fit to the silicate emission by combining spectra from a variety of interplanetary dust particles. The dominant component is olivine with small amounts of pyroxene and layer lattice silicate (Bregman, J. D. et al. 1987. *Astr. Ap.* **187**, 616).

20μm feature. The observations for wavelengths beyond 20μm showed some new weak features around wavelengths 24, 28, 35 and 45μm which have been tentatively identified with olivine particles. This is consistent with the structure seen in the 10μm feature. No spectral features were evident in the observations over the spectral region 40 to 160μm.

9.4.2. *The C-H stretch feature*

A new emission feature near 3.4μm was first detected by Vega 1 spacecraft in the spectra of Comet Halley. This was confirmed by several ground-based observations. The laboratory absorption spectra of organic materials both in gaseous and solid phase generally show a strong feature around

$3.4\mu m$. Most of the C-H stretching vibrations fall in the wavelength regions around 3.1 to 3.7 μm. Therefore, the observed feature in Comet Halley is attributed to C-H stretching of some organic material. The $3.4\mu m$ feature has subsequently been seen from several other comets indicating the common nature of cometary organics. This feature is also seen from some IDPs collected in the stratosphere. The intensity of the $3.4\mu m$ feature as observed by Vega spacecraft seemed to vary inversely with the projected distance from the nucleus in a manner similar to the parent molecule indicating the possible detection of a new parent molecule. It is also not clear whether the carrier is in the gas or solid phase. With the existing observations, it is not possible to specify the nature and composition of the emitting material except to point out some possibilities through the presence of absence of some characteristic spectral features.

The apparent profile of $3.4\mu m$ feature in several comets is around $3.36\mu m$. In addition, there is some evidence for the presence of two weaker bands at 3.28 and $3.52\mu m$. The vibrational frequencies of CH_3 group occur at $3.37\mu m$ and $3.48\mu m$ and that of CH_2 group at $3.42\mu m$ and $3.50\mu m$. Therefore, the observed features in comets appear to indicate the presence of spectral signatures of $-CH_3$ and $-CH_2$ functional groups. The feature at $3.28\mu m$ is the characteristic of aromatic compounds. The structure, width and the intensity of the observed $3.4\mu m$ emission feature suggest that several of these species may be present. Although several suggestions for the possible species such as methanol (CH_3OH), and formaldehyde (H_2CO) have been made, however for a better identification, high resolution observations that can separate the individual fine structures are required. In this connection, the KAO observations of Comet Halley showing the lack of features in the 5.5 to $8\mu m$ region is of interest. The wavelength of the $C-C$ stretching and $C-H$ bending deformations tends to lie in the wavelength region between 5.9 and $25\mu m$. Therefore, the presence or the absence of some of these features can be used as a diagnostic feature of a particular chemical group involved. If the feature at $3.28\mu m$ arises from aromatic hydrocarbons, two other features at 6.2 and $7.7\mu m$ should also be seen which do not seem to be present in the observations. Similarly the absence of strong carbonyl feature in the range 5.5 to 5.9 μm possibly eliminates molecules such as aldehydes and ketones. Therefore, the exact nature of the species responsible for the origin of the $3.4\mu m$ feature is complex and should involve simultaneous and consistent study of all the emissions, their excitations and production rates. The $3.4\mu m$ cometary feature has also been seen from interstellar clouds, molecular clouds, HII regions and so on. But the characteristic feature is different indicating the variation of carbonaceous material present in different environments.

9.4.3. Ice signature

All the available evidence at the present time indicates that H_2O is the major component of the nucleus of a comet (Chaps. 6, 11). Therefore, it is interesting to look for ice grains in the comae of comets. One way to detect such grains is to look for the absorption bands of H_2O or ice in the infrared region of 1 to 4 μm. Laboratory investigations have shown that the 3.1 μm feature is much stronger than the other features occurring around 1.5, 1.65 and 2.0 μm. In a cometary atmosphere these bands can arise by the scattering of small size grains. These bands have been looked for in several comets. The strong 3.1μ ice feature was not detected in Comet West, when the observations were made at $r = 0.53au$. However, one difficulty is that at such small distances the infrared emission is very severe and this could mask the ice feature even if it is present. Therefore, the feature should be looked for when the comet is far away from the Sun. The measurements carried out on the Comet Stephan-Oterma at $r = 1.6au$ did not show other water-ice bands occurring around 1.5, 1.65 and 2.0μ. The observations on Comets Bowell and Cernis gave a slight indirect indication of the presence of 3.1 μm feature. These negative results in general do not imply that water ice may not be present as they could be attributed to various factors. First of all the laboratory measurements have shown that the strength of water-ice features depends upon whether the ice is in the crystalline or amorphous form. A small amount of absorption in the grain can also effectively wash out the features. The strength of the features also depends upon the size of the scattering grains, the temperature of the grains and so on. Therefore, the 3.1 μm feature-search in comets has not given any conclusive results. In spite of all these negative results, the search for the ice bands in comets continues.

9.4.4. Possible new features

The moderate resolution of the spectra of Comets Halley and Wilson obtained in the spectral region 5-13μm showed some features around 6.8, 8.4 and 12.2 μm, which are not easily attributable to silicates. The 6.8 μm feature was seen in December 1985 spectra of Comet Halley. This feature could be attributed to carbonates as seen in IDPs consisting largely of layer-lattice silicates. It could also be due to C-H deformation vibrations in the carbonaceous material responsible for the 3.4 μm feature. The feature at 8.4 μm has been clearly seen in the Comet Halley spectra of December 12, 1985 but not in the spectra of April 8, 1986. The 12.2 μm feature is quite strong in the spectrum of Comet Wilson. This particular feature could also be present in Comet Halley, as broad band observations appear to show an excess emission around 12 μm. These observations indicate that the features are variable. The exact nature of the material responsible for these observed features is not known. The variability of the features could mean that the material has possibly volatile intermediates to the silicates and ices within the comets. Of course, carbonates may constitute a fraction of this material. Several weak features that appear to be present in the $2-13\mu m$ region should be detectable with the high resolution observations of bright comets.

9.5. Properties Derived from *in situ* Measurements

The important information about the chemical composition of dust particles in Comet Halley has been obtained from the dust impact mass analyzer PUMA 1 and 2 on Vega and PIA on Giotto spacecrafts. Dust particles striking a silver target placed in front of the mass spectrometer generate a cloud of ions and the positive charge are mass analyzed, which indicates its chemical composition. Several thousand mass spectra of dust particles were recorded by the instruments. These studies indicated broadly three classes of particles: (1) mostly made up of light elements such as H, C, N and O indicative of organic composition of grain called 'CHON' particles (2) similar to CI Chondrites but enriched in carbon and (3) primarily O, Mg, Si and Fe suggestive of silicate grains. Therefore, the Comet Halley grains were found to be essentially composed of two end member particle types - a silicate and a refractory organic material (CHON) in accordance with the infrared observations.

The possibility of the presence of core-mantle structure of dust particles comes from the fact that on the average CHON ions appear to have a higher initial energy than the silicate ions. The observed ion abundances have been transformed to atoms based on laboratory calibration from the knowledge of the ion yields with different projectiles. The resulting composition of Comet Halley's dust is given in Table 11.2. The abundances of rock forming elements in Halley's dust are within a factor of two relative to the solar system abundances. The abundances of H, C and O are more than that of CI-chondrite believed to be the unaltered meteorites from the early solar system which are available for laboratory investigation. This can be interpreted to mean that the Halley dust is more primitive than CI-chondrites. The puzzle of carbon depletion in comets also seems to have been resolved as they are tied up in the refractory organics. Due to uncertainties in converting the ion intensities to atomic abundances, there could be a factor of two uncertainties in the final derived abundances. The derived gas to dust ratio is $\sim$ 1.1 to 1.7. An estimate for the density of silicate dominated grain has been made and indicates a value $\sim 2.5 gm/cm^3$, while CHON dominated grain indicates a value $\sim 1 gm/cm^3$. Several refractory grains show $^{12}C/^{13}C \approx 5000$ which is drastically different from the normal value $\sim$ 89. If this is genuine then it reflects that the carbon coming out of different nucleosynthesis sites has been incorporated in the Halley dust particles.

The possible presence of polymerized formaldehyde $(H_2CO)_n$ (also called polyoxymethylene, POM) in the dust grains of Halley has been proposed based on the results of the positive ion cluster analyzer (PICCA) experiment conducted aboard the Giotto spacecraft (Fig. 6.15) although other possibilities exist. The dust grains contain a large number of organic compounds (Table 13.4), e.g., unsaturated hydrocarbons like pentyne, hexyne, etc; Nitrogen derivatives like hydrocyanic acid etc; Aldehydes and acids like formaldehyde, formic and acetic acid etc. and so on. Many more molecules are expected to be present in the dust. It is quite possible that a large number of molecules might have also been destroyed due to the high velocity of impact $\sim 78 km/sec$.

9.6. Albedo of the Particles

The average value for the albedo of cometary particles can be obtained from a combination of very simple physical arguments and observations. The observation required for such a study is the scattered radiation at the visible wavelength and the total integrated infrared flux from a comet. These are readily available for many comets. The basic physical idea behind such a simple method is that the continuum in the visual region arises due to the scattering process which depends on the scattering efficiency of the grain. On the other hand the observed infrared radiation is dependent on the amount of impinging energy absorbed by the grain, which depends on the absorption efficiency of the grain. For small optical depths and also neglecting phase dependence effects, the optical surface brightness can be written as

$$S_{opt}(\lambda) = \frac{F_\odot(\lambda)\tau}{4\pi} = \frac{F_\odot(\lambda)N_d l \pi a^2 Q_{sca}(a,\lambda)}{4\pi}. \quad (9.25)$$

Here N_d is the column density of the dust particles and l is the representative path length. The infrared surface brightness can be approximated as

$$S_{ir} = \frac{\langle Q_{abs}\rangle \pi a^2 F_\odot N_d l}{4\pi}. \quad (9.26)$$

Here $\langle Q_{abs}\rangle$ represents a mean absorption efficiency for the grain and $F_\odot = \int F_\odot(\lambda)d\lambda$, the integrated solar radiation. It is possible to derive an expression for the average albedo γ of the particles from the above equations. This is given by

$$\frac{\gamma}{1-\gamma} = \frac{F_\odot S_{opt}(\lambda)}{S_{ir} F_\odot(\lambda)}. \quad (9.27)$$

The estimated values of albedo for several comets lie typically in the range of about 0.1 to 0.3. In the above method, the phase dependent scattering is neglected (Chap. 8). Allowing for the phase dependent scattering, the value of albedo for Comet West lies in the range of about 0.3 to 0.5.

This method has been used to map the spatial variation of an average albedo of dust grains in Comet Halley at $r = 1.7 au$ using the simultaneous two dimensional images made in the infrared region at $10.8\mu m$ and in the visible region with the continuum filter at 6840A. The albedo map for Comet Halley shows that the value lies in the range of around 0.25 to 0.45. These are systematically higher than the values of 0.07 to 0.15 derived for Comet Giacobini-Zinner based on the same method, but is consistent with the values derived for other comets from their scattering function.

9.7. Continuum Emission in the Radio Region

As discussed in Chap. 7, the laboratory experiments carried out with clathrate-hydrate snows for cometary conditions indicated that an icy-grain halo around the coma should exist with particles of millimeter or centimeter sizes. Because of the large particle sizes, the thermal emission from such grains should mainly be in the radio spectral region. The continuum emission at 3.71 cm was in fact detected for the first time in Comet Kohoutek. Later on such emissions have also been detected from a few other comets.

9.8. Radiation Pressure Effects

The dust particles released from the nucleus are subjected to radiation pressure forces which push them to different distances from the nucleus and this ultimately gives rise to the observed dust tail as discussed in Chap.7. The dust tail is composed of various Syndynes and each curve is defined in terms of a parameter $\beta \equiv (1 - \mu)$, which is the ratio of solar radiation pressure to gravity. These Syndynes when projected on to the photographs of the tail give a range of β values which encompass the observed tail. The maximum value of β fixes the minimum size of the particles that can exist in the tail. Others will essentially be pushed away from the system. From such comparisons, it is possible to get the maximum value of β denoted as β_{max} for various cometary tails. Table 9.4 gives a list of β_{max} values obtained for various comets. They show that there are no particles in the tail which are subjected to radiation pressure forces beyond about $\beta_{max} \simeq 2.5$. This apparent cut-off in the value of β can be used to infer the properties of the dust particles. However, the interpretation is not straightforward as it involves various parameters of the grains, which means one has to invoke various grain models. As an example, Fig. 9.14 shows a plot of the variation of β as a function of the radius of the particle for different types of particles. It can be seen that for small size particles most of the curves are flat except for the silicate type (Basalt) of materials. The observed maximum value of $\beta_{max} \simeq 2.5$ seems to lie in between these two general shapes of the curves.

9.9. Summary

We may now briefly summarize the main characteristic properties of cometary grains based on the discussion presented so far. In order to reduce the ambiguity, it is important as well as necessary to consider various types of observations which have to be satisfied simultaneously by any grain model. However, it may not be practicable in all the cases. In the absence of such detailed studies, it is even worthwhile to restrict the parameters which define the grain characteristics within certain possible ranges.

Table 9.4. Values of β_{max} obtained for various comets* in comets ($r = 1 au$)

Comet	β_{max}	Comet	β_{max}
1957 III	0.55	1970 II	1.9
1957 V	2	1970 II	3.8[a]
1962 III	2.5	1973 XII	0.8[b]
1965 VIII	1.1-1.4	1976 VI	2
1965 VIII	0.8 (farthest part)	1976 VI	2.5
1965 VIII	2.5 (near the head)	Several Comets	2.5

[a] from IR photometry; [b] from Colorimetry (*Saito, K., Isobe, S., Nishioka, K. and Ishii, T. 1981. *Icarus* **47**, 351.)

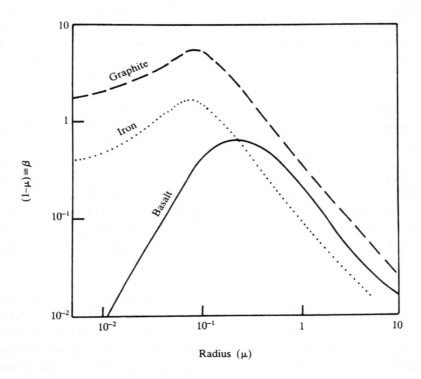

Fig. 9.14. The variation of radiation pressure force $(1 - \mu) \equiv \beta$ with the radius of the particle for three types of material is shown. (Adapted from Saito, K., Isobe, S., Nishioka, K. and Ishii, T. 1981. *Icarus* **47** 351.)

The optical continuum measurements and the variation of grain temperature with heliocentric distance indicate the refractive index of the grain material to be similar to that of silicate or ice. The polarization measurements as well as scattering phase function in the visual region give a limit for the refractive index as $1.3 < n < 2.0$ and $k \lesssim 0.1$. The silicate nature of the grains also comes from the interpretation of 10 and $20\mu m$ emission features.

Most of these results are interpreted based on single particle models. If one introduces multi-component models based on different compositions and the size distribution functions, the situation becomes complicated with the introduction of many more variable parameters into the problem. It should also be noted that all the above discussions are based largely on the application of the Mie Theory to spherical and homogeneous particles. More complications will enter into the problem if the grains are irregular in shape, fluffy, porous, inhomogeneous in composition or the aggregate of very small grains, etc. It is hard to handle all these complications purely from theoretical means.

The crystalline nature of silicate grains is revealed by the structure seen in the $10\mu m$ feature from Comet Halley and also in other comets. Further support for the olivine nature of grains comes from the detection of several features in the wavelength region 20 to 40 μm in Comet Halley. The evidence for the second component to the cometary grain came from the discovery of CHON particles in Comet Halley as well as the detection of 3.4 μm feature in the infrared observation. The two major components of the refractory grain mixture - silicates and some form of carbon are also consistent with the analysis of interplanetary dust particles. The unsuspected presence of a large number of very small particles was also detected in Comet Halley. The emission from a two component model based on a mixture of silicate and carbon particles matches well with the observed heliocentric variation of 3-20 μm emission from comets. Therefore, the variation of grain temperature with the heliocentric distance can account for the major change observed in the cometary spectra. The general shape of the heliocentric variation of dust and H_2O production rates in comets are very similar in nature supporting the hypothesis that gas and dust flow together as they are released from the nucleus. The chemical composition of dust has been derived for the first time based on Comet Halley observations. The puzzle which had existed before Comet Halley that there is a depletion of carbon in the coma gas appears to have been resolved as it is tied up in the grains. These CHON particles are made up of highly complex molecules and there is evidence to show that it can also be a source of cometary gas. Therefore, efforts are being made to study these problems through semi-empirical, numerical as well as through laboratory investigations.

Problems

1. Derive an expression for the variation of the temperature of a black body as a function of distance from the Sun. Assume that the energy distribution of the Sun can be represented by a black body of temperature $6000°K$. What will happen if the body is a planet with no atmosphere or if there is an atmosphere?

2. Calculate the grain temperature for graphite grains of $a = 0.2\mu m$ at $1 au$ from the Sun. How much does this value differ from that of black body temperature at the same distance?

3. Suppose in the above problem, the grain has some impurity which gives $Q_{abs} = 1$ at $1 mm$. What will happen to the temperature of the grain?

4. Calculate the number of electrons and atoms or molecules required to explain the observed scattered radiation in comets at $r = 1$ au and for $\lambda = 5000 Å$. Take the value of the scattered intensity to be 50% of the solar intensity at that wavelength. What is the expected wavelength dependence of these?

5. Suppose the heavy elements like O, C, N hit the grain and stick to it with a probability α. If the temperature of the gas is T and assuming the mean velocity to be given by the Maxwellian distribution, deduce the expression for the rate of growth of the grain. What is the time required for an ice grain of initial radius of 0.01μ to grow to 1μ, if $\alpha = 1$, $T = 100°K$ and $n_H = 10/cm^3$?

6. A grain releases a molecule and reduces in size due to protons hitting the surface of the grain. Deduce an expression for the rate of decrease of grain size with time, if β is the probability for the release of a lattice molecule when the proton hits the surface. What is the time required to destroy completely an ice grain of size 1μ with $N_H = n_e = 10/cm^3$, $T = 10^4 K$ and $\beta = 0.1$?

7. Assuming solar constant of $1.4 \times 10^6 ergs/cm^2$ sec and spherical completely absorbing particles of density 3.0, calculate the limiting radii of

particles to be retained in the solar system released from Comet Encke at aphelion distance of 4.10 au and perihelion distance of 0.34au.

8. Compare the importance of radiation pressure on the motion of the dust compared with the force of gravity at the orbit of Venus. Calculate the time required for the dust to be driven out to the distance of the Earth by radiation pressure alone.

References

The early work on continuum studies can be found in these two references:
1. Liller, W. 1960. *Ap. J.* **132** 867.
2. Remy-Battiau, L. 1964. *Acad. r. Belg. Bull. el. Sci. 5eme Ser.* **50** 74.

The following papers may be referred for later work:
3. Jewitt, D. C. and Meech, K. J. 1986, *Ap. J.* **310**, 937
4. A'Hearn, M. F. et al. 1984, *AJ* **89**, 579.

The early measurement of polarization was done by
5. Bappu, M K V and Sinvhal, S D 1960. *M N* **120** 152.

The existence of negative polarization in comets was shown in the following two papers:
6. Kiselev, N N and Chernova, G P 1978 *Soviet Astron.* **22** 607.
7. Weinberg, J L and Beeson. D E 1976 *Astr Ap* **48** 151

For later work the following may be consulted:
8. Dollfus, A. and Suchail, J.-L. 1987. *Astr. Ap.* **187**, 669.
9. Dollfus, A. 1989, *Astr. Ap.* **213**, 469.
10. Mukai, T., Mukai, S. and Kikuchi, S. 1987, *Astr. Ap.* **187**, 650.

The first infrared measurements made on Comet Ikeya-Seki is reported by
11. Becklin, E E and Westphal, J A. 1966. *Ap J* **145** 445.

The following paper reports the first detection of 10μ emission feature in Comet Bennett:
12. Maas, R., Ney, E P and Woolf, N J 1970. *Ap J* **160** L101

The interpretation of infrared observations can be found in the following papers:
13. Krishna Swamy, K S and Donn, B 1968. *Ap J* **153**, 291
14. Oishi, M., Okuda, H. and Wickramasinghe, N C 1978. *Publ. Astro Soc Japan* **30** 161
15. Krishna Swamy, K. S. et al. 1988. Icarus **75**, 351.
16. Hanner, M. S., (ed.) 1988 *Infrared observations of Comets Halley and Wilson and properties of grains*, NASA Conference Publication 3004.

The determination of grain albedo can be found in the following paper:
17. O'Dell, C R 1971 *Ap J* **166** 675.

The first detection of continuum emission at 3.7 cm in Comet Kohoutek is reported in
18. Hobbs, R W., Maran, S. P., Brandt, J. C., Webster, W. J. and Krishna Swamy, K. S. 1975. *Ap. J* **201** 749.
19. Hobbs, R. W. Brandt, J. C. and Maran, S. P. 1976, *Ap. J.* **218**, 573.

Infrared studies of IDPs can be found in.
20. Sandford, S. A. and Walker, R. M. 1985, *Ap. J.* **291**, 838.

In situ size distribution function is given in the paper
21. Mazets, E. P. et al. 1986, Nature **321**, 276.

The following papers may be referred for the study of 10 and 3.4 μ features.
22. Hanner, M. S., Lynch, D .K. and Russell, R. W. 1994. *Ap. J.* **425**, 274.
23. Disanti, M. A., Mumma, M. J., Geballe, T. R. and Davis, J. K. 1995, Icarus 116, 1.

CHAPTER 10

ION TAILS

The ion tails of comets provide a unique and natural place for the study of various plasma processes. Many of the features seen in the ion tails of comets change on a short-time scale as well as on a long-time scale. The formation of ion tails is basically due to the interaction of the solar wind plasma with the cometary plasma. The detailed dynamical model calculations for such an interaction process had indicated the presence of Large Scale Structures like bow shock, ionosphere and so on. The *in situ* measurements of Comets Giacobini-Zinner and Halley showed not only the presence of these features but also indicated the plasma to be highly complex, containing instabilities, waves, turbulence and so on. Some of these aspects will be discussed here. In addition, some of the large scale features that have been seen in the ion tail of comets will also be discussed.

10.1. Evidence for the Solar Wind

The existence of the solar wind was first postulated based on the study of the ion tail of comets. The model was put on a firm footing with the work of Parker based on the hydrodynamic expansion model of the solar corona. The model predicted in detail the expected nature of the solar wind as well as the resulting shape of the interplanetary magnetic field. These were confirmed later on with observations made through satellites.

The idea of the solar wind came from the photographic observations of comets which showed that certain features like knots in the ion tails were moving away along the tail. The velocity and the acceleration of the features obtained from successive photographs showed clearly that these features were moving in the anti-solar direction with high velocities. The typical velocity is about 100 km/sec. A measure of the deceleration is generally expressed in terms of the effective gravity parameter $(1-\mu)$ as defined in Chap. 7. The typical values of $(1-\mu) \approx 10^2$ were quite commonly observed in the ion tails of many comets. This indicates an outward force which is 10^2 times larger than the solar gravity. One force which could operate and give rise to this motion on an atom or a molecule is the radiation pressure. The expression for the radiation pressure is given by

$$F_{rad} = \left(\frac{F_\odot}{c}\right)\left(\frac{\pi e^2}{mc}f\right) \tag{10.1}$$

where $F_\odot$ is the solar flux at the comet and all other quantities have their usual meanings. It was found that the Eq. (10.1) fails to explain the observed accelerations of features in the ion tail by several orders of magnitude. This led Biermann to hypothesize the existence of the solar wind, i.e., the high velocity corpuscular radiation coming from the Sun, which accelerates the ions in the tail through the momentum transfer. In fact he developed a very simple expression for this interaction and obtained the equation

$$\frac{dv_i}{dt} \approx \frac{e^2 N_e v_e m_e}{\sigma m_i}. \tag{10.2}$$

Here the quantities with the subscript e refer to electrons and i to ions, N_e is the electron density and v_e is the velocity of electrons, σ is the electrical conductivity. In order to explain the observed accelerations, the above equation requires the solar wind velocity to be of the order of a few hundred kilometers per second and the density of about 600 electrons/cm^3. These values may be compared with the present day velocity of about 400 to 500 km/sec and the density of about 5/cm^3.

10.2. Dynamical Aberration

More support for the solar wind hypothesis came from the study of the orientation of ion tails of comets. The tail axes of ion tails are found to lag behind by a few degrees with respect to the radius vector. The lag arises due to the resultant effect of the solar wind velocity and the velocity of the comet. This is known as the dynamical aberration. Figure 10.1 shows the geometry of the comet tail on the plane of the sky. Here r, t and $-v$ denote the radius vector, the tail vector and the negative velocity vector of the comet. For the interpretation of ion tail orientation, it is convenient to use

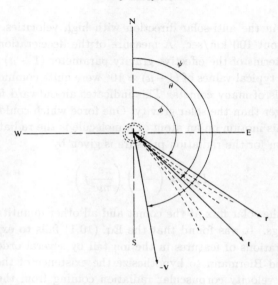

Fig. 10.1. The geometry of the comet tail (dashed lines) projected on to the plane of the sky. The position angles ϕ, θ and ψ denote the prolonged radius vector, the comet tail and the direction of comet's velocity back along the orbit respectively. (Belton, M. J. S. and Brandt, J. C. 1966, *Ap. J. Suppl.* **13**, 125).

the plane of the comet's orbit as the reference frame. The observational data is the position angle θ of the tail axis on the plane of the sky. From a knowledge of the various quantities as shown in Fig. 10.1, the orientation of the tail in the comet's orbital plane can be calculated (Fig. 10.2) with the assumption that it also lies in the same plane. The aberration angle ϵ is then the angle of the tail with respect to the radius vector. The basic equation used for the interpretation of dynamical aberration observation is given by

$$\mathbf{t} = \mathbf{w} - \mathbf{v} \qquad (10.3)$$

where $\mathbf{t}$ is the tail vector which lies in the orbital plane of the comet, $\mathbf{w}$ is the solar wind velocity and $\mathbf{v}$ is the comet's orbital velocity. Physically the above equation means that the direction of the tail is the direction of the solar wind as seen by an observer riding on the comet. If the solar wind velocity is resolved into radial (w_r) and azimuthal (w_ϕ) components, the aberration angle for a comet near the solar equator is given by

$$\tan \epsilon = \frac{v \sin \gamma - w_\phi \cos i}{w_r - v \cos \gamma} \qquad (10.4)$$

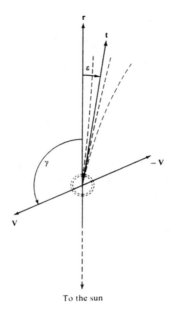

Fig. 10.2. The geometry of the comet tail projected on to the plane of the comet's orbit. The aberration angle ϵ and the angle γ are shown. (Belton, M. J. S. and Brandt. J. C. 1996, *op. cit.*).

where i is the inclination of the comet's orbit to the plane of the solar equator and γ is the angle between the radius vector and the direction of v. All the quantities in the above equation are known except for w_r and w_ϕ. Since the value of w_ϕ is small, the value of w_r can easily be calculated from Eq. (10.4). Since the observations on comets cover a wide variety of situations, it is possible in principle to get not only w_r and w_ϕ but also their variations. If w_ϕ has an appreciable value, it should show up in the aberration angle between the direct (D) and retrograde (R) comets. The above method has been applied to about 60 comets covering approximately 1600 observations. The resulting averages for the aberration angles for direct and retrograde comets are

$$<\epsilon>_D = 3.7°$$

and

$$<\epsilon>_R = 5.5°.$$

Assuming w_r and w_ϕ are the same for the two cases and using the average values for other parameters, the resulting mean values for w_r and w_ϕ are

$$<w_r> = 450 \pm 11 km/sec$$

and

$$<w_\phi> = 8.4 \pm 1.3 km/sec.$$

In the above discussion, it was assumed that the tail of the comet lies in the orbital plane of the comet. This assumption can be relaxed and the velocity field of the solar wind can also be taken into account. The following variations for w_ϕ and w_θ which are found to be consistent with the observational and theoretical evidences have been used in the comet tail orientation problem

$$w_r = \text{constant}, w_\phi = w_{\phi,o}\frac{(\cos b)^{2.315}}{r} \tag{10.5}$$

and

$$W_\theta = w_m \sin 2b \tag{10.6}$$

Table 10.1. Parameters of solar wind derived from ion tails.

Parameter	Published sample	Total sample (includes unpublished data)
ω_r (km/sec)	402 ± 12	400 ± 11
$\omega_{\phi,o}$ (km/sec)	7.0 ± 1.8	6.7 ± 1.7
ω_m (km/sec)	2.6 ± 1.2	2.3 ± 1.1
Number of Observations	678	809

(Brandt, J. C. and Mendis, D. A. 1979. In *Solar System Plasma Physics*, Vol. II, eds. C. F. Kennel, L. J. Lanzerotti and E. N. Parker, Amsterdam: North Holland Publishing Company.)

where b is the solar latitude, $w_{\phi,o}$ and w_m denote the azimuthal speed in the plane of the solar equator at 1 au and the maximum value of w_θ. The results of analysis are given in Table 10.1. It is also possible to get the minimum solar wind speed which actually refers to the larger values of ϵ. The minimum value obtained for the solar wind speed is around $225 \pm 50 km/sec$. All these results are in good agreement with the direct space probe measurements. The presence of the azimuthal component of the solar wind was also first deduced based on the study of ion tails which was later confirmed through the direct space probe measurements, vindicating the simple aberration picture of the ion tails of comets. It is also remarkable that it can give so much of information without a detailed knowledge of the interaction between the solar wind and the tail plasma.

From the above discussion, it is apparent that the derived results based on the use of Eq. (10.3) are in good agreement with the gross observed properties of plasma tails. Having established the validity of Eq. (10.3), it is now possible to reverse the problem and calculate approximately the expected general shape of the plasma tail from the same equation. This formalism is known as the Wind Sock Theory.

10.3. Theoretical Considerations

The comet acts as an obstacle to the free flow of the solar wind. Therefore one can approach the problem of the interaction of the solar wind with the cometary atmosphere purely from the theoretical considerations. However, this is a complicated problem as the interaction is coupled through

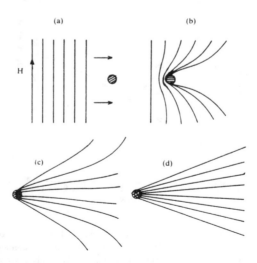

Fig. 10.3. A schematic representation of the "piling up" and "curling up" of interplanetary magnetic field by the coma of the comet. The gradual formation of the ion tail is shown in Figs. a, b, c and d. (Adapted from Alfven, H. 1957. *Tellus* **9**, 92).

the interplanetary magnetic fields. The basic idea of the comet-solar wind interaction was provided by Alfven in 1957 and this was that the cometary ions follow the frozen-in magnetic field lines of the solar wind. These field lines wrap around the comet's ionosphere and it is finally dragged into the tail as shown in Fig. 10.3. This is the basis of all the plasma models.

In a simple model, the plasma can be assumed to act as a fluid in the sense of fluid dynamics, allowing the use of simple hydrodynamic methods to predict the flow pattern of the plasma. Even without detailed modeling, it was suggested around 1964 that the expansion of the ionized coma gas should act as an obstacle to the supersonic solar wind flow which should result in a bow shock at a distance of around 10^4 to 10^5 km from the nucleus. The particle distribution functions can usually be determined from the conservation laws, namely, continuity, momentum and energy equations

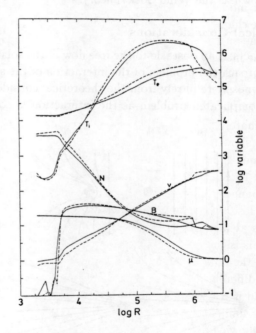

Fig. 10.4. Variation of physical quantities as a function of distance from the nucleus (R in km) for axisymmetric model (dashed lines) and the three dimensional model (solid lines). Ti=ion temperature (K), Te=electron temperature (K), v=velocity (km/sec), N=number density (/cm^3), B=magnetic field strength (nT) and μ=mean molecular weight (Wegmann, R., Schmidt, H. U., Huebner, W. F. and Boice, D. C. 1987, *Astr. Ap.* **187**, 339).

including the effect of electric and magnetic fields. These are to be supplemented with the equations for the magnetic field and the electric field. These equations constitute the magneto hydrodynamic equations (MHD) for the plasma and provide a good description of most of the plasma processes in the coma. The results of accurate one-and-two-dimensional models with spherical gas flow showed that a weak bow shock is expected with a Mach number 2. In recent years, more realistic and sophisticated models have been constructed, which takes into account various physical processes including chemistry in the neutral and ionized coma and its interaction with the solar wind. The interaction of the ionized coma with the solar wind lead to deviations from spherical symmetry. The assumption of axial symmetry is also destroyed due to the embedded magnetic field. Therefore a detailed description of the region would require a 3-dimensional magnetohydrodynamic model. These MHD models describe the macroscopic flow patterns and the characteristic boundaries expected from such interactions. The result of one such detailed 3-dimensional MHD calculation is shown in Fig. 10.4. The model takes into account momentum exchange from ion-neutral collisions, photoionization, mass-loading through ion pick up and the Lorentz forces of the transverse magnetic field. The simulation also takes into account the detailed physics and chemistry of the coma together, as they are intimately connected.

The results show that a bow shock occurs around 5×10^5 km from the nucleus for Comet Halley parameters. The ions are heated by the shock to around 2.2×10^6K. However they cool down very fast due to collisions and charge exchange. The temperature of electrons is not affected by the shock and they lose their energy due to electron impact reaction. A sharp boundary exists at a distance of around 4800km where the flow velocity falls steeply. This is the contact surface which separates the contaminated solar wind from the pure cometary plasma inside. Therefore at this surface two plasmas of different origin converge. The magnetic field is compressed by the shock and it reaches a value of about 53 nT in the stagnation zone. The value drops suddenly at the contact surface. The variation of physical parameters as shown in Fig. 10.4 arise as a result of balance of different effects.

The overall morphology of the flow of plasma near a comet is schematically shown in Fig. 10.5. They can be understood as follows. At large distances from the comet the interaction of the solar wind is collisionless as the cometary material is neutral. Once the ions are created, they are

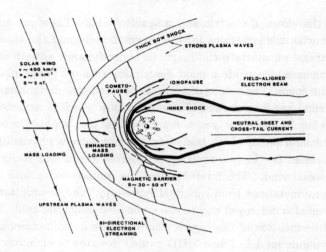

Fig. 10.5. Schematic representation of the global morphology produced by the interaction of the solar wind with the cometary atmosphere shows the flow pattern and the presence of various discontinuities arising from such an interaction (Mendis, D. A. 1988, Ann. Rev. Astr. Ap. **26**, 11).

accelerated due to the interplanetary magnetic field (**B**) and the motional electric field of the solar wind, $\mathbf{E} = -\mathbf{V}_{SW} \times \mathbf{B}$ where $V_{\mathbf{SW}}$ is the solar wind velocity. The most important source of ionization of the gas is the photoionization by the extreme ultraviolet radiation. The charge exchange transfer of solar wind protons with cometary neutral also contribute to the ionization process. The ions formed move with a typical speed of around 1 km/sec in the cometary frame of reference, while the solar wind velocity is 400 km/sec. The newly formed ions which are initially at rest are accelerated by the motional electric field of the solar wind leading to a cycloidal motion with both gyration and $\mathbf{E} \times \mathbf{B}$ drift motion. The gyration speed depends upon the angle between the solar wind velocity and the magnetic field. This assimilation of the cometary ions into the magnetized solar wind is called 'mass loading' of the inflowing solar wind. The dynamics of the solar wind containing cometary ions depend upon the overall pressure and density associated with these ions. The thermal speed of ions, picked up by the solar wind, is of the order of solar wind speed. The picked up energy of the ions in the upstream in the solar wind reference frame is typically about 10-20 KeV for O^+, whereas the solar wind protons have thermal energies $\sim 10 eV$. Therefore the picked up ions are quite hot and the pressure due to these ions dominate the total pressure even at large cometocentric distances. With the decrease in cometocentric distance, more and more ions are added to the solar wind and hence the total pressure increases. As solar wind progressively gets 'mass loaded', due to momentum conservation, the solar wind slows down. As the solution of the fluid equation cannot go continuously from supersonic to subsonic state, this results in a weak bow shock at a critical level of mass-loading. This depends crucially on the mean molecular weight. Therefore the cometary bow shock arises purely as a result of mass-loading process and is unique among the solar system objects. Many of the energetic ions like O^+, C^+ and H^+ are generated upstream of the comet. Moving inwards from the shock, the solar wind continues to interact with the cometary material of increasing neutral density. These outflowing neutrals play an important role in slowing down the incoming solar wind through collisions. Therefore cometary ions which are produced down-stream of the shock are less energetic ($E \sim 1 keV$) compared to those produced upstream of the shock. In a real situation there is a slow gradation in the energy of the picked up ions and hence in the solar wind reference frame, energy of the pick-up ions decreases with decreasing cometocentric distance. Therefore the plasma found in the inner regions of coma contains ions of different populations made up of energetic ions picked up far upstream, and relatively cold ions picked up locally. Finally, the velocity decreases rapidly, giving rise to a sharp boundary called 'cometopause', In this transition region, collisions dominate. Near 'cometopause', the hot and warm cometary ions disappear from the plasma through charge exchange collisions with the neutrals. Inside the 'cometopause', the speed is very low, the ions are abundant and are modified by collisions and chemistry. This also leads to compression of the magnetic field giving rise to the magnetic barrier. The boundary of the field free cavity is called 'diamagnetic cavity boundary surface'. It is also called contact surface, the ionopause or the tangential discontinuity. This boundary separates the purely cometary plasma and the mass-loaded solar wind plasma. The basic physical mechanism of its formation is due to the balance between the outward ion-neutral frictional force (due to the flow of neutrals past stagnated ions) and an inward directed electromagnetic (**J** x **B**) force. Here the mass loading term is less important. The presence of another shock inside the cometary ionopause has been suggested in order to decelerate the supersonic outward flowing cometary ions and divert them into the tail, but it has not yet been seen.

10.3.1. Comparison with observations

The *in-situ* measurements of Comet Giacobini-Zinner and more so of Comet Halley, which provided detailed data, confirmed the large scale picture of the solar wind interaction with comets at close heliocentric distances. The presence of a bow shock at a distance $\sim 1 \times 10^6$ km from the nucleus in Comet Halley was clearly identified by instruments aboard spacecrafts, measuring plasma, magnetic field and plasma wave. The detection of bow shock in Comet Halley has resolved the controversy about its existence. The spacecraft measurements, has also given detailed information on the structure of the bow-shock. For example, the water group ions O^+, OH^+, H_2O^+ and H_3O^+ showed a sudden enhancement following the bow shock. Inside the bow-shock, in the cometo-sheath region the flow speed was found to decrease continuously due to ion pick up and the plasma becomes more dominated by cometary ions. The phase space density distribution of cometary H^+ ions showed a shell-like structure at a distance

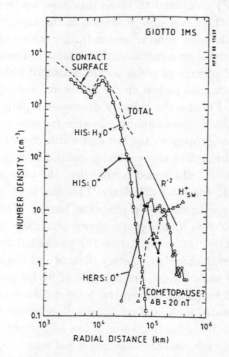

Fig. 10.6. Number density profiles of water group ions and solar wind protons as measured by Giotto ion mass spectrometer (see Ip, W.-H. 1989, *Ap. J.* **343**, 946).

of around 8×10^6 km upstream of the bow-shock in Comet Halley. This pattern arises due to the fact that the velocity distribution of pick up ions which should be in the form of a ring with velocity component $v_\perp = V_{SW} \sin \alpha$ perpendicular to the interplanetary magnetic field and a drift velocity $v_{11} = V_{SW} \cos \alpha$ along **B** and relative to the solar wind is unstable and leads to various plasma instabilities. As a result of wave-particle scattering, the H^+ ions diffuse along the sphere transforming the ring-like structure to shell-like structure.

The existence of cometopause at a distance of around 10^5 km came from the observation that the density of the solar wind protons was found to decrease rapidly approaching this distance, while the comet ion density increased rapidly with $1/R^2$ radial dependence for $R < 10^5$km (Fig. 10.6). At a cometodistance $\sim 1.4 \times 10^5$ km there was also a sudden decrease of electrons with energies $\geq 10eV$ while the magnetic field strength increases from 6nT to 26nT. For radial distances $< 2 \times 10^5$ km the dominant ions change progressively from O^+, OH^+ and H_2O^+ to H_3O^+ (Fig. 10.7). The dominant ions inside 10^4 km are H_2O^+ and H_3O^+ which continuously

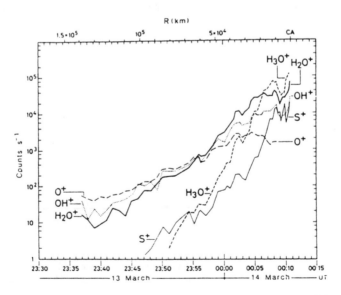

Fig. 10.7. The variation of abundances of several species as measured by ion mass spectrometer on Giotto from Comet Halley. R is the distance of Giotto from the nucleus and the time refers to UT at the ground station (Balsiger, H. et al. 1986, *Nature* **321**, 330).

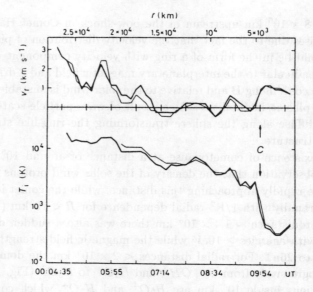

Fig. 10.8. Radial velocity and temperature of ions with mass/charge ratio of 18 (dark lines) and 19 (light line) measured inbound by Giotto of Comet Halley (Balsiger, H. et al. 1986, *Nature* **321**, 330).

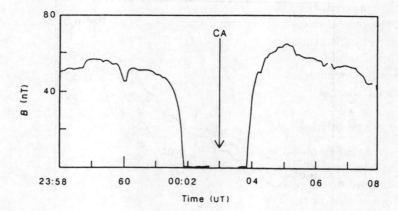

Fig. 10.9. The variation of magnetic field versus time as measured by the magnetometer on board Giotto of Comet Halley (Neubauer, F. M. ESA Sp-250, Vol. 1, p. 35).

increase towards the nucleus as indicated by the models. The ions C^+ and S^+ were also found to be very abundant. The ion temperature dropped from 2600K to 450K at a distance of around 4700km (Fig. 10.8). These results confirmed the presence of tangential discontinuity. The interesting result that Giotto magnetometer measurements showed is that there was a dramatic drop in the magnetic field $\sim$ 40nT to zero at cometocentric distance $\sim$ 4600 km from the nucleus inbound and $\sim$ 3900 km outbound (Fig. 10.9). The region bound by these two distances is devoid of magnetic field. The location at which the inner edge of the ionopause occurs can be estimated from the balance between the magnetic force and the ion-neutral drag force. This gives a distance $\sim$ 4335 km for inbound and $\sim$ 3470 km for the outbound for the Giotto encounter. These are in reasonable accord with the observed values mentioned above.

The electron temperature derived from the abundance ratios of $CH_3OH_2^+/H_3O^+$ and H_3O^+/H_2O^+ measured by Giotto over the distance of 2000 to 20,000 km, show the expected radial dependence of increase of T_e with increase in distance, except for a sharp peak around 11500 km. This peak coincides with the unexpected dense plasma seen in Comet Halley observations just ahead of the magnetic free cavity. The observed and the model based electron temperatures are in reasonable agreement inside the magnetic field cavity where the physical conditions are simpler, as they are not directly influenced by the highly variable solar wind conditions. But outside the contact surface there is some discrepancy between the calculated and the general shape of the observed variation. The sharp peak present in electron temperature around 11500 km, corresponding to the dense plasma feature seen in Comet Halley, is not a general feature of cometary coma as shown by the model. Mostly, it should be specific to conditions existing in Comet Halley, at the time of Giotto flyby.

10.4. Instabilities and Waves

It was suggested that there could be some instabilities present in the plasma due to the mass-loading of the solar wind by the newly created ions. The nature of the pick up cometary ions by the solar wind depend upon the orientation of the solar wind flow to the interplanetary magnetic field (IMF). When the IMF is orthogonal to the solar wind V_{SW}, the $\overline{V}_{SW} \times \overline{B}$ motional electric field makes the newly created ions gyrate around the magnetic field. In this case the anisotropy in their gyro velocities (as they can have both the zero velocity and twice the solar wind velocity), can give rise to three types of low frequency instabilities, namely the ion-cyclotron instability, a parallel propagating nonoscillatory mode and a fluid mirror instability. In the other extreme case, when the IMF is parallel to the solar wind velocity vector, there is no $\overline{V}_{SW} \times \overline{B}$ solar wind force acting on the cometary ions. For such a situation the ions form a beam in the solar wind plasma frame moving at a velocity $-\overline{V}_{SW}$ relative to the ambient plasma. This can lead to two types of instabilities, a right handed resonant helical beam instability and a nonresonant instability. However the coupling between the solar wind and the cometary ions becomes complicated if the solar wind flows obliquely to the magnetic field at an angle θ. Therefore, in principle, various kinds of plasma instabilities and waves could be generated due to the ion pick up by the solar wind, depending upon the complex local conditions of the plasma. The *in situ* plasma and magnetic field measurements of Comets Giacobini-Zinner and Halley by the spacecrafts have shown the presence of a large number of waves of various kinds. They are highly complex and difficult to interpret. Only a few of these results will be mentioned here.

One of the main results that came out of the study of the Comet Giacobini-Zinner by the ICE spacecraft is the detection of the water group (16 to 19 amu) ion cyclotron frequency from spectral power analysis. This has also been detected in Comet Halley observations. Several peaks with cyclotron fundamental at 7 mHz and its harmonics at 14, 21, 29 and 39 mHz has been seen in the cross-spectral densities of the solar wind proton velocity and magnetic field. The fundamental mode at 7mHz is linearly polarized and the higher harmonics are either linearly or highly elliptically polarized. The cometary influence extends to a distance of around 2×10^6 km.

Waves have been detected upstream of the bow shock for Comets Giacobini-Zinner and Halley. A variety of polarizations have been detected from circularly polarized waves propagating at large angles relating to the ambient field to highly elliptical polarized waves. The detection of short duration magnetic pulses during $\theta = 90^0$ intervals was rather an unexpected result. The field variations have a duration of around 6 to 7 secs and is comparable to the proton cyclotron frequency in the spacecraft frame. These pulses are mostly transverse oscillations.

At distances $\sim 3 \times 10^4$ to 18×10^4 km from the comet the magnetic field experiment detected mirror mode waves which are characterized by irregular dips in the magnetic field. Similar structures have been seen from Giotto Spacecraft in Comet Halley and ICE on Comet Giacobini-Zinner. Higher frequency whistlers have been seen near the bow shock of Comet Giacobini-Zinner and it is an integral part of the magnetosonic wave.

Several mechanisms for the formation of whistlers have been proposed such as through generation of dispersive whistlers derived from hybrid simulation results, generation of pick up of heavy ions and protons at the distorted steepened fronts of the magnetosonic waves and so on.

The waves of extremely low frequency (ELF) in the region 10 to 1500 Hz and of very low frequency (VLF) in the region 10^3 to 10^6 Hz generated by cometary pick up ions and photoelectrons are present in Comets Giacobini-Zinner and Halley upto distances of about 2×10^6 km from the nucleus. The Halley emission were about an order of magnitude more intense than as seen from Comet Giacobini-Zinner. Some of the wave modes that has been seen in Comets are shown in Fig. 10.10 They show the electron plasma oscillations (EPO), the ion acoustic waves, electromagnetic whistlers and lower hybrid resonance (LHR) waves.

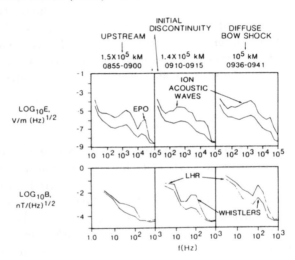

Fig. 10.10. Several wave modes seen in Comet Giacobini-Zinner in the upstream and down stream are shown. This includes electron plasma oscillations (EPO), ion acoustic waves, electromagnetic whistler modes and Lower hybird resonance (LHR) waves (Scarf F. L. et al. 1987, *Astr. Ap.* **187**, 109).

An indication of the presence of strong turbulence even upto distances $\sim 2 \times 10^6$ km from the nucleus came from the detection of strong wave activity. In addition to high level plasma activity, large-amplitude magnetic field variation were observed. The turbulence measured by Vega and Giotto Spacecrafts of Comet Halley showed that the magnetic field fluctuations were smaller compared to that observed in Comet Giacobini-Zinner. There is a clear correlation between the magnetic field variations and solar wind plasma measurements which suggest the waves are fast-mode magnetosonic waves. The observed magnetic field turbulence spectrum follows a power law spectrum with an index of about 2, which is sufficiently different from the fully developed Kolmogorov spectrum with a spectral index of 5/3. This indicates that the cascade has not progressed far enough.

The detection of strong turbulence has led to several theoretical investigations. Both analytical treatments and numerical simulations have been applied to interpret these results. They could arise due to the presence of cyclotron harmonic emission, which would add power at the higher frequencies; or if ions have a finite temperature, it will shift the wave frequencies to higher values or due to non-linear effect of the waves and so on. It is quite possible that several processes could lead to a smooth power law spectrum. It is found that resonant wave-particle interactions and ion-shell formation can lead to a wave power law spectrum with an index of around 2.0.

Though a large number of substructures have been identified in the *in situ* measurements, the mechanism of formation of such structures is not clear.

10.5. Acceleration of Cometary Ions

The possibility that cometary ions could be accelerated to high energies in the coma due to the turbulent plasma environment came from the analogy with the diffuse shock acceleration at the Earth's bow shock. The existence of such accelerated ions in Comet Giacobini-Zinner came from the observations carried out with the ICE spacecraft. The measurement showed large flux of cometary ions of about 100 keV even upto distances $\sim 10^6$ km from the nucleus. Large anisotropy in their distribution was also detected. The *in situ* measurements of Comet Halley confirmed the presence of appreciable fluxes of energetic cometary ions in the coma, but it extended even beyond 100 keV and upto about 0.5 MeV.

The maximum energy that a newly formed ion can have in the rest frame of the spacecraft is given by

$$E_{max} = \frac{1}{2} m_i (2V_\perp)^2 = 2 m_i V_{SW}^2 \sin^2 \alpha$$

where $2V_\perp$ is the total velocity comprising of the drift speed $V_\perp$ and roughly an equal amount of gyration speed. V_{SW} is the solar wind speed and α is the angle between the solar wind velocity and the interplanetary magnetic field. For $V_{SW} \simeq 400 km/sec$, the maximum energy E_{max} that water ions can have ~ 60 keV corresponding to $\alpha = 90°$. There should be a sharp cutoff in the energy distribution beyond this value. However the observations indicating that the energy spectra extended upto ~ 0.5 MeV clearly showed that significant ion acceleration must have occurred. The value of $\alpha = 90°$ or $0°$ correspond to the extreme cases of velocity vector of the ions in the direction of the solar wind (E_{max}) or in the opposite direction (E_{min}) and should therefore lead to anisotropies in the energy distribution.

The possible mechanisms for the acceleration of ions has led to several studies of wave-particle interactions. The cometary ions could be accelerated through diffuse shock acceleration process near the bow shock. This process is basically related to random multiple scattering of charged particles across the shock front, separating the supersonic and subsonic sides, a form of first order Fermi acceleration. Another mechanism is the stochastic acceleration process, a form of second order Fermi acceleration process. Here the mechanism is the scattering of the particles by the waves in the turbulent cometary plasma. In the diffuse acceleration process, the accelerated ion population builds up ahead of the bow shock with a characteristic length scale given by K/V_{SW} where K is the diffusion coefficient of the medium normal to the shock. For the case of ion acceleration at the bow shock of Comet Halley, the estimated distance is $\sim 10^6$ km. The diffuse shock acceleration process is more important in this region and for energies ≤ 100 keV. For energies $> 100 keV$ and for distance $> 5 \times 10^6$ km far from the bow shock, the acceleration through stochastic process or the second-order Fermi acceleration process is of major importance. Support for the stochastic acceleration process also seems to come from the observations that can be fitted well with an exponential velocity distribution of cometary ions. Other mechanisms that have been considered for ion acceleration process include the lower hybrid turbulence that could be generated by the comet ion pick up process as well as by the bow shock formation and ion cyclotron wave absorption.

The presence of intense electron plasma oscillations and ion-acoustic waves in the Comets Giacobini-Zinner and Halley indicate that electron heating and acceleration should take place. Some of the processes considered for electron heating are ion acoustic wave instability, cross-field streaming instabilities, lower hybrid wave acceleration and so on.

10.6. Large Scale Structures

Different kinds of plasma processes must be taking place in the ion tail of comets. This is evident from the photographs of bright comets which show a wide variety of unusual large scale dynamical phenomena. A wide variety of plasma tails can be seen from the Atlas of Cometary Forms. Many of these interesting phenomena appear to be controlled to a large extent by the microscopic plasma physical processes arising out of the comet-solar wind interactions. There are many other features which arise due to the dominance of the solar wind. Therefore any change in the solar wind property in a small or on a large scale is reflected directly in the behaviour of the cometary plasma. Some of the characteristic features arising out of these processes are given in Table 10.2. It should be noted that most of the observed features are transient in nature. Since the ion tail is basically a plasma column of large extent and is also bright, it gives a splendid opportunity for the study of spatial and temporal variations. It can also help in distinguishing the phenomena connected with the processes taking place inside the tail like instabilities etc. from those introduced by the fluctuations in the solar wind.

Table 10.2. Plasma process and the possible associated events.

Physical process	Lead to observed features
Change in the solar wind conditions	Changes the orientation of the ion tail, bends the tail
Development of instabilities	Enhancement of ionization, flares, condensations, filaments, rays, helices, waves, etc.
Magnetic field lines disconnected	Disconnected tails

10.6.1. *Tail rays or streamers*

The filamentary structure in the ion tails of comets is quite common and is also very conspicuous (Fig. 1.19). The shape of these streamers appears to show the importance of magnetic fields. Since the CO^+ ions present in the ion tail follow the direction of the magnetic field, the streamers are made visible through CO^+ emission. These rays going in the anti-solar direction grow with time and can extend from smaller lengths of about 10^3 km to longer lengths of about 10^6 km. The rays are distributed almost symmetrically around the comet tail axis. They also seem to turn towards the main tail axis. The time scale for coalescing process is around 10 to 15 hours. Since the time of formation of a new streamer is about one hour, usually streamers as many as 20 to 25 can be seen at any given time. These streamers do not interact with each other. They appear to originate from a small volume very close to the nucleus, at a distance of the order of 10^3 km from the nucleus. In addition to seeing the streamers originating from the coma, streamers starting from the tail have also been noticed. Since these rays do not originate from the ionized region, the mechanism of formation of these rays has to be different. Two possibilities have been suggested either they are produced by nonlinear compressional waves or the cometary ions trace the magnetic field lines. Observations appear to show the first explanation to be unlikely although there are problems with the second explanation. The ion stream could also arise due to the effect of magnetic sector boundaries in the IMF (Sec. 10.6.6) The IMF following the sector boundary reconnects at the front side of the ionosphere. This can lead to ion streamers. In the coma, in addition to magnetic lines of force, neutral sheets or surfaces also curl around and their behaviour will be similar to the observed reversal of directions in the interplanetary medium. It has been suggested that the streamers represent the enhanced plasma which is confined to these neutral tubes or surfaces. These ions are concentrated in the neutral zones as a result of the field gradients on either side of the neutral regions. Also these neutral regions prevent the field lines of opposite polarity merging with each other leading to annihilation. These neutral tubes are nearly cylindrical, centred on the tail axis and when seen edge-on, they appear as rays. They appear symmetrical about the tail axis and this is consistent with the observations. It is quite possible that strong shock waves in the sunward direction can generate keV electrons and energetic protons. These high energy electrons streaming along the magnetic field lines can result in high ionization in a short time scale which can give rise to streamers.

10.6.2 *Knots or condensations*

The plasma tails are filled with structures such as bright knots or condensations which are basically ion concentrations. With the use of time sequence photographs, it is possible to follow these knots as a function of time for a period of a few hours to a day. From such measurements, the velocity and acceleration can be determined. Their velocities lie in the range of about 20 to 250 km/sec. Even though the collision mean free path for the particles is large, it is still a reasonable approximation to treat the interaction between the solar wind and the cometary plasma as a fluid-like interaction. Based on such a hypothesis, several other mechanisms for the observed acceleration have been suggested. One suggested process is that the acceleration will result when the plasma is squeezed out of the flux tubes by the magnetic pressure gradients. Another point of view is that the enhancement of the momentum transfer takes place due to some form of instability setting in the plasma. It is not clear at the present time whether the observed motions are the mass motions or some form of disturbance propagating down the tail.

10.6.3. Oscillatory structure

The photograph of Comet Kohoutek taken on 13 January 1974 when it was at $r = 0.58$ au and $\Delta = 0.81$ au, clearly showed the presence of a wavy structure in the ion tail, moving approximately at a speed of about 235 km/sec. (Fig. 1.17). The feature appeared to be a helical structure. The radius and the wavelength of the structures are about 2.3×10^5 km and 1.4×10^6 km respectively. The wavy structure had an extension of about 3.6×10^6 km and was about 1.6×10^7 km away from the nucleus. These oscillations have been interpreted in terms of "kink instability" based on the studies of the Earth's geomagnetic tail. The theoretical work developed for the Earth's geomagnetic tail has been successfully applied to the comet tails. In brief, an azimuthal component of the magnetic field is produced, in a cylinder of plasma when a current flows through the longitudinal magnetic field. This finally gives rise to a helical field. The phase speed of the helical kink turns out to be simply the Alfven speed of the tail plasma, which is given by

$$C \approx \frac{B}{(4\pi\rho)^{1/2}} \qquad (10.10)$$

where B is the magnetic field and ρ is the mass density. For a number density of CO^+ ions of 10 to $10^3/cm^3$, the magnetic field lies in the range of

$$100\gamma \leq B \leq 1000\gamma.$$

10.6.4. "Swan-like" feature

Another large scale dynamical structure in the tail of Comet Kohoutek can be seen in the photograph of 11 January, 1974 (Fig. 1.18). This is generally termed as "Swan-like" feature because of its appearance. The size of the feature is estimated to be about 5×10^6 km and is situated at a distance of about 15×10^6 km from the nucleus. The feature was found to move away from the nucleus with an apparent speed of about 250 km/sec. The Swan feature is not a mass flow but rather a propagating hydromagnetic wave phenomenon. In fact it could be the kink instability in an advanced stage of growth.

10.6.5. Bend in the tail

The changes in the solar wind conditions can reflect directly in the observations of ion tails. This was established for the first time, based on the photographic observations of Comet Bennett. It was shown that the kinks observed in the tail of this comet were correlated with the solar wind events as observed from the satellite measurements.

A striking example of this type can be seen from the photograph of Comet Kohoutek taken on 20 January 1974, where a large bend can be seen. This is shown in Fig. 10.11. It is interesting to investigate whether

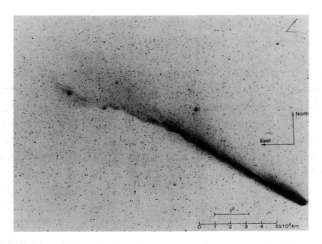

Fig. 10.11. Photograph of Comet Kohoutek taken on January 20, 1974 which shows the large scale bending of the tail. (Brandt, J. C. and Chapman, R. C. 1981. *Introduction to Comets*. Cambridge: Cambridge University Press.) Illustration credited to Joint Observatory for Cometary Research, NASA.

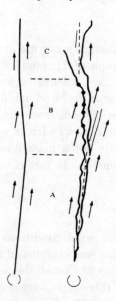

Fig. 10.12. Schematic representation of Fig. 10.11. The left portion shows overall kink and the right portion shows some details. The inferred solar wind direction is shown by arrows. (Adapted from Jockers, K. 1981. *Icarus* **47** 397).

this bend is correlated with the solar wind properties for that date. The properties of the solar wind are obtained by the observations made by the Earth orbiting Satellite IMP-8. These observations have been used in conjunction with the theory to predict the nature and structure of the ion tail projected on to the sky for the day of the observation. There was a good agreement between the expected and the observed comet tail configurations. These results clearly showed that the observed large-scale tail curvature coincided with the compression region of a high speed solar wind stream. Generally a high speed solar wind stream causes enhanced geomagnetic activity which usually shows up in the geomagnetic records. A geomagnetic storm was actually detected on January 24 to 27, corresponding to the Comet Kohoutek observations. Figure 10.12 shows schematically the tail configuration and the bend. The direction of the solar wind is shown by the arrows. It can be seen that in region A, the tail has the same direction as the direction of the solar wind. This means that the tail has taken the equilibrium position. In region B, the flow of the solar wind is across the tail and has not yet adjusted to the direction of the solar wind. On the other hand the segment C is well aligned with the head of the comet, as the tail has not yet been affected by the solar wind. This is a striking example which clearly shows that the comet tails are very effective for monitoring the changing conditions of the interplanetary plasma.

10.6.6. Disconnection events

Another interesting and characteristic feature quite often seen in the plasma tail of comets is the disconnection events which have been noted since early times. The photograph of Comet Morehouse shows the abrupt disconnection of the tail as shown in Fig. 10.13. The disconnection events appear to be cyclic in the sense that after a tail gets disconnected a new

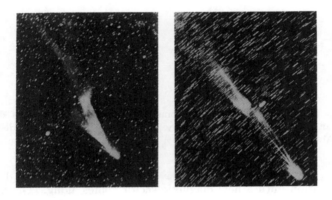

Fig. 10.13. Photograph of Comet Morehouse showing a tail disconnection event. Left and right photograph refer to September 30, 1908 and October 1, 1908 respectively. (Niedner, M. B. and Brandt, J. C. 1979. *Ap. J.* **223** 671).

tail appears to form and this process seems to repeat itself. A possible mechanism for such phenomena can be understood from Fig. 10.14. The explanation is essentially related to the presence of magnetic sector boundaries in the interplanetary magnetic field (IMF). A sector boundary is defined as the boundary between a region in which the field lines are directed primarily towards the Sun (in the ecliptic plane) and a region in which the field lines are directed primarily away from the Sun (in the ecliptic plane). When the comet ionosphere intercepts the magnetic sector boundary of the IMF at which the magnetic field direction reverse by $180°$, it disconnects the existing tail and it will move in the anti-solar direction. A new tail will then form with the reconnection of the field lines with the opposite

294   Physics of Comets

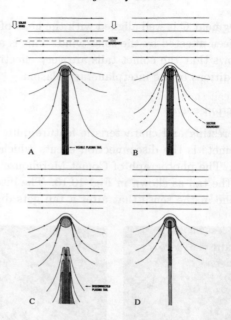

Fig. 10.14. Process of formation of disconnected plasma tail of a comet due to the passage of an interplanetary sector boundary. Tail rays are shown by the shaded portions. (Niedner, M. B. and Brandt, J. C. 1979. op. cit.).

field orientation until a new magnetic sector comes along. The process is repeated. Therefore if a bright comet is observed for a long time, it is possible to observe many events as described above. In the reconnection phase, the magnetic energy may be dissipated in the form of an increase in brightness or ionization or flares etc., and it may show up in the observations. There is some observational evidence to suggest that these things do happen. The study based on many other comets including Comet Halley seems to be consistent with this interpretation. Another possibility for the disconnection events is that the high velocity stream of the solar wind hitting the comet can disrupt the cross tail currents which disconnect the tail.

It is clear from the discussion presented so far that the whole problem of comet-solar wind interaction is a highly complex one. The exact nature and the location of the different discontinuities depend upon the various physical processes that can take place during the interaction. What happens in a particular situation depends upon the complex interaction of the plasma, turbulence, solar activity and so on. However the large scale structure as predicted by the MHD calculations like bow shock, ionopause, magnetic cavity and so on have been confirmed from the *in situ* measurements of Comets Giacobini-Zinner and Halley. This shows that the mechanism of interaction between the solar wind and cometary plasma giving rise to various structures is fairly understood at the present time. The measurements also showed the plasma to be highly complex in showing the presence of strong turbulence, various instabilities, waves, high energy particles and so on. Various mechanisms have been proposed to explain the origin of these observed features.

All the properties of the ion tail that have been discussed so far can be classified as belonging to the "typical" comets. In these comets the ion-tails are more or less similar and behave in a certain manner except for the presence of certain characteristic features in the tail in some cases. The major aim has been to understand the gross behaviour of such "typical" comets.

Problems

1. Take for the velocity of the solar wind $v = 450 km/sec$ and the proton density $n = 5/cm^3$. Calculate the magnetic field required to counteract the solar wind pressure.
2. Does the magnetic field exist in comets? Can you suggest some method to probe the magnetic field associated with a comet?
3. Calculate the Alfven speed in the tail of comets.
4. Discuss the relative importance between cometary plasma and laboratory plasma.

References

The evidence for the existence of solar wind is discussed in this paper
1. Biermann, L. 1951, Zs. f. *Astrophysik* **29**, 274.

For a complete theoretical discussion of the solar wind, refer to
2. Parker, E. N. 1963, *Interplanetary Dynamical Processes*. New York: Interscience Publishers.

The phenomena of dynamical aberration is discussed in the following papers.
3. Belton, M. J. S. and Brandt, J. C. 1966 *Ap. J. Suppl.* **13**, 125.
4. Brandt, J. C. and Mendis, D. A. 1979, In *Solar System Plasma Physics*, Vol. II. eds. C. F. Kennel, L. J. Lanzerotti and E. N. Parker. Amsterdam: North-Holland Publishing Company p. 253.
5. Hoffmeister, C. 1943, *Zs. f. Astrophysik* **23**, 265.

The Wind-Sock theory for the ion tail is discussed here.
6. Brandt, J. C. and Rothe, E. D. 1976, In *The Study of Comets*, eds. B. Donn. M. Mumma, W. Jackson, M. F. A'Hearn and R. Harrington, NASA SP 393, Washington D. C. p. 878.

The basic idea of the solar wind interaction is given in this paper.
7. Alfven, H. 1957, *Tellus* **9**, 92.

The theoretical model of the Comet-Solar Wind interaction is from
8. Wegmann, R. et al. 1987 *Astr. Ap* **187**, 339.

The confinement of ions very close to the nucleus was discussed by
9. Wurm, K. 1963 In *The Moon, Meteorites and Comets* eds. B. M. Middlehurst and G. P. Kuiper, Chicago: The University of Chicago Press p. 575.

The following atlas gives cometary plasma tails of various forms.
10. Rahe, J. Donn, B and Wurm, K. 1969 *Atlas of Cometary Forms*, NASA SP-198, Washington D. C. GPO.

A good discussion of the *in situ* plasma measurements can be found in the following papers:
11. Flammer, K. R. 1991 *In Comets in the Post-Halley* Era. eds. R. L. Newburn, Jr. et al. Kluwer Academic Publishers. P. 1125.
12. Tsurutani, B. T. 1991, *In Comets in the Post-Halley* Era. eds. R. L. Newburn, Jr. et al. Kluwer Academic Publishers. P. 1171.
13. Ip, -W. H. and Axford, W. I. 1990, In *Physics and Chemistry of Comets*, Springer Verlag, p. 172.

The electron temperature is discussed in the paper
14. Eberhardt, P. and Krankowsky, D. 1995. *Astr. Ap.* **295**, 795.

CHAPTER 11

NUCLEUS

Though the nucleus is responsible for most of the observed phenomena, its nature is the least understood. As the nucleus cannot be observed directly from the Earth because of its small size, the information with regard to their possible nature, structure and composition has to come from indirect means. The fly-bys to Comet Halley in 1986 has given new insights into this area.

11.1. Theory of Vaporization

When the solar radiation impinges on the surface of the nucleus, part of the energy is absorbed and part of it is reflected. The reflected part depends upon the albedo of the nucleus. The absorbed energy gives rise to the surface temperature which sublimates the material and also gives rise to the thermal emission. The absorbed energy also depends upon the angle between the incident radiation and the normal to the surface.

The amount of energy radiated from the surface is dependent upon the temperature, which in turn is related to the chemical composition and the structural properties of the nucleus. The sublimation of the gases depends upon the latent heat and its variation with temperature. The penetration of the heat flow inside the nucleus also depends upon the conducting properties of the material. The presence of dust in the coma can also have an effect on the vaporization of the volatile components. Therefore the amount of volatile constituents that come out of the surface of the nucleus is a very complicated function involving many unknown factors. Hence, until we have better ideas with regard to the above processes, one has to resort to simplifying assumptions in writing down the equation for the energy flow.

The temperature of the nucleus is determined by the balance between the amount of absorbed and the emitted energy. The energy balance equation for the simple model of a spherical nucleus which takes into account the thermal re-radiation and the heat used to transform the vaporizing ice into gas for the steady state condition can therefore be written as

$$\frac{F_\odot(1 - A_v)}{r^2} \cos z = (1 - A_{IR})\sigma T^4 + Z(T)L(T). \tag{11.1}$$

Here $F_\odot$ is the solar radiation at 1 au. A_v and A_{IR} are the bond albedo of the nucleus in the visible and infrared region and r is the heliocentric distance. $Z(T)$ and $L(T)$ represent the vaporization rate of the gas in molecule/cm^2/sec and the latent heat for sublimation in ergs/molecule. T is the surface temperature and z is the solar zenith distance. The vaporization rate $Z(T)$ can be deduced by relating it to the equilibrium vapour pressure of ice. Under the equilibrium conditions the number of molecules coming out of the nuclear surface will be balanced by the number hitting the surface. The number of molecules hitting the surface for such a situation is equal to $\frac{1}{4}N\bar{v}$. Therefore

$$Z(T) = \frac{1}{4}N\bar{v} \tag{11.2}$$

where N is the gas density and $\bar{v}$ is the mean speed for the Maxwellian distribution of velocities which is given by

$$\bar{v} = \left(\frac{8kT}{\pi m}\right)^{1/2} \tag{11.3}$$

Here m is the average molecular weight of the gas. Using the above relation for $\bar{v}$ and with the use of the relation $P = NkT$, $Z(T)$ can be written as

$$Z(T \text{ or } r) = \frac{P}{(2\pi mkT)^{1/2}}. \tag{11.4}$$

The total production rate of the molecule is given by

$$Q(r) = S < Z(r) >$$

where $< Z(r) >$ represents the production rate averaged over the solar zenith distance z and S is the surface area of emission. The integration over

z can be avoided under some simplifying assumption. For a non-rotating nucleus of radius r_n, the cross section for the absorption of solar energy is πr_n^2, while for a fast rotating nucleus the incident energy is uniformly distributed over the entire surface $4\pi r_n^2$. The corresponding value of $<\cos z>$ is 1 and $1/4$ for the slow rotator and fast rotator models respectively. The equations show that even the very simple energy balance equation requires the specification of the six parameters, namely A_v, A_{IR}, L, P, η and S. Here η represents the geometrical factor related to $\cos z$. Since it is a multivariable problem, it is necessary to make some assumptions to limit the number of free parameters. This depends upon the problem under investigation. It then fixes certain parameters and others are derived from comparing with observations. This procedure is just a method of simplifying the problem and therefore the results have to be weighted in terms of the assumptions made in the problem.

Since H_2O is the dominant component of the nucleus of a comet, the values for L and P corresponding to H_2O are generally used. The vapour pressure can also be calculated from Clasius-Clapeyron equation

$$P = P_r \, exp\left[\frac{L(T)}{RT}\left(\frac{1}{T_r} - \frac{1}{T}\right)\right]$$

where T_r and P_r are the reference point. The vapour pressure in general is very sensitive to temperature. Therefore, the general variation of the production rate of H_2O and OH as a function of the heliocentric distance should reveal something about the vaporization of ice from the nucleus of a comet and should in turn give information about the overall nature of the nucleus of a comet. A simultaneous solution of Eqs. (11.1) and (11.4) iteratively will give the unknown $Z(T)$ and T, for assumed values of the parameters that enter in the above two equations. The results for the variation of Z as a function of the heliocentric distance for different cases are shown in Fig. 11.1. The curves depend upon the values of A_v and A_{IR}. The curves are almost the same if $A_v = A_{IR}$. For unequal values of A_v and A_{IR} the maximum difference in the results arises in the shift of the curves in log r by about ± 0.2. Figure 11.1 shows r^{-2} dependence in the beginning and then shows a rapid fall in the sublimation rate of water for distances between 2 and 3 au. The calculated behaviour of the curve is basically related to the steep dependence of vapour pressure on temperature. The sublimation rate is nearly proportional to the impinging energy if the temperature is greater than a certain critical value. This is the case for small values of r

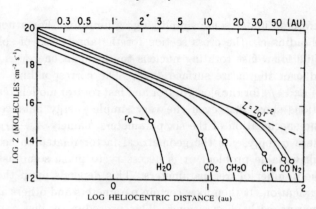

Fig. 11.1. The vaporization rate Z, for various snows as a function of the heliocentric distance. The distance after which the vaporization becomes negligible is denoted as r_0. (Delsemme, A. H. 1982. In *Comets*, ed. Wilkening, L. L. Tucson: University of Arizona Press, p. 85).

and hence Z varies as r^{-2}. For distances greater than a certain value, the sublimation rate falls down steeply with temperature and this gives rise to the observed effects. The cometary observations also show a similar cut-off around these distances, after which they are not generally seen. This effect is particularly so in the case of short period comets where the cut-off is around $r \sim 3 au$. These observations are in striking agreement with the expected curve for the sublimation of H_2O. Therefore the observations by itself seem to indicate that H_2O may be the dominant constituent of the nucleus. In addition, the vaporization theory gives the total production rate of rough order with the observed rates of several comets where the water controls the sublimation process.

Of course there are some comets which have been seen much beyond $r \sim 3au$, where it is unlikely that the water will vaporize. Therefore for such comets the volatile gases may not be controlled by pure H_2O or pure CO_2 (Fig. 11.1). This led to the suggestion of clathrates or hydrates. In these, the frozen gases like CH_4, CO_2, etc. should be in a hydrated form in the nucleus, such as $CH_4.6H_2O$ and $CO_2.6H_2O$ etc. These are stable only at very low temperatures. They are much less volatile than CH_4 or CO_2 themselves. The clathrates are basically loose crystal lattices of H_2O in which the molecules like CH_4, CO_2, etc are loosely packed. These molecules are not chemically bound but they are bound by the weak Van der Waals forces. The cut-off distances for such types of clathrates lie in the range of about 5 au or so. Therefore the comets which have been observed beyond $r \gtrsim 3au$ are likely to arise out of this structured ice material. This was a significant step in the understanding of the cometary nuclei.

The early results of the vaporization of H_2O from the nucleus based on the steady state properties of the nucleus had met with reasonable success in explaining the gross observed nature of the light curves of various comets. But, the observations of several comets have shown that the nature of the light curve for Pre-and Post-perihelion passages is not symmetrical in shape, but rather there is a difference between the two portions. Observations also indicate jets of gas coming out from discrete active regions of the nucleus. The photographs of the nucleus of Comet Halley taken from Giotto spacecraft as well as the isotopes of these photographs and also of other comets show a general emission of the gas from the nucleus in addition to the jets of gas. The activity level of the nucleus has been estimated to be anywhere between a few percent to about 50% or so. However, due to various kinds of uncertainties involved at the present time, observations does not appear to put constraints on the relative amounts of the outgassing from the active and inactive parts of the nucleus. Therefore, the total production rate of H_2O as derived from the ground based or satellite observations should in general reflect the average production rate due to the general emission of the gas and jets. In some of the studies A_{VIS}, A_{IR}, L and P are fixed and the resulting sublimation rate is then averaged with the geometrical factor involving $\cos z$ to determine the emission area to explain the observed production rates. The asymmetric nature of the observed light curve for pre-and post perihelion passages is attributed to the geometrical factor arising due to the different exposed areas from the non-spherical nucleus. The assumption of using constant values of A_{VIS}, A_{IR}, L and P means that the material property of the nucleus is the same for all the comets. A *priori* it is not possible to say how good is this assumption.

Several attempts have been made to refine the models that take into account various other factors like, thermal properties of the surface layers, the rotation period of the nucleus, the effect of diurnal heating and cooling etc. To take care of the temperature distribution with depth, an additional term $-\kappa_S \frac{dT}{dZ}|_s$ has to be added to the right hand side of the Eq. 11.1, which represents the heat flux conducted downward into the nucleus. Here κ_S is the thermal conductivity of the surface material, $\frac{dT}{dZ}|_s$ is the surface temperature gradient and z is the depth below the surface. To evaluate the conduction term in Eq. 11.1 the heat diffusion equation

$$\frac{\partial T}{\partial t} = -K\nabla^2 T = K\frac{\partial^2 T}{\partial x^2}$$

has also to be considered. However the new term depends upon thermal conductivity of the material of the nucleus which introduces uncertainty in the temperature calculation. The temperature with depth varies due to the rotation of the nucleus and the amplitude decreases exponentially with depth. The scale length associated with it is called the thermal skin depth and depends on the conductivity of the medium. For a H_2O material and for a rotation period of about 20 hours the skin depth is $\sim$ 3 to 30 cms depending upon the compact or fluffy nature of the nucleus. These refined models are useful but they have their own limitations due to the uncertainty in the thermal conductivity which depend strongly on the state of the nuclear material, whether porous or compact in nature. Gas phase conductivity may also contribute to the heat flow. The resultant effect may be that H_2O production may also take place from the sub-surface layers and then flow out through the surface. These are all functions of the position on the surface due to the rotation of the nucleus. Hence, the theoretical modeling of the sublimation of H_2O has serious limitations due to lack of knowledge of porosity and its size, the actual material property of the nucleus and so on. Therefore simple models are quite useful in trying to understand the general trend as well as qualitative study of comets.

It will be very useful to find an analytical expression for the curve of Fig. 11.1 which is based on the vaporization of water from an icy nucleus. It can be written as

$$Z = Z_o \alpha \left(\frac{r}{r_o}\right)^{-m} \left[1 + \left(\frac{r}{r_o}\right)^n\right]^{-k}$$
$$= Z_o g(r) \qquad g(r) = \alpha \left(\frac{r}{r_o}\right)^{-m} \left[1 + \left(\frac{r}{r_o}\right)^n\right]^{-k}$$
(11.5)

where $r_o = 2.808 au$, $m = 2.15$, $n = 5.093$ and $k = 4.6142$. The value of α is chosen such that $z = z_o$ for $r = 1 au$, which gives $\alpha = 0.111262$. Equation (11.5) fits the data within $\pm 5\%$. As will be discussed later, the vaporization from an icy nucleus represented by Eq. (11.5) can explain the observed nongravitational forces in comets (Sec. 11.9).

11.2. Outbursts

Outbursts have been seen quite frequently in comets. It is basically related to a sudden increase in the production of volatiles from the nucleus. Comet Schwassmann-Wachmann is well known for its large outbursts and flares up frequently. These outbursts can brighten the comet by a factor of 100 or more compared to the quiescent brightness. Initially, the comet looks star-like in appearance. After the outbursts, a halo forms and eventually it fades away leaving behind the original nucleus. The whole process might take around 3 to 4 weeks. The outbursts from Comet Schwassmann-Wachmann have been seen regularly from heliocentric distances of around 5 to 7 au. Comet Halley showed an outburst when it was at $r = 14 au$. It is rather difficult to understand such a phenomena based on the sublimation of H_2O from the nucleus, as the production rate of H_2O falls down steeply for $r \gtrsim 3au$. A number of suggestions and ideas have been proposed to explain the origin of these outbursts. It has been suggested that exothermic chemical reactions involving free radicals or pockets of volatile gas stored beneath the surface takes place until the pressure would build up which leads finally to explosions. The phase change from amorphous below about 140K to crystalline ice at higher temperatures is an exothermic process giving about 24 cal/gm, which is a substantial amount of energy and could trigger the outbursts. Since the phase change is critically dependent upon the temperature, the outburst should depend on the heliocentric distance. The vaporization of pockets of CO_2 or CO from water ice surroundings could produce the outburst. In addition, the outbursts from Comet Schwassmann-Wachmann seem to occur from the same location. Therefore, there appears to be no single explanation for the presence of outbursts in comets. It is quite possible that different processes operate in different comets or for the same comet at different times. The inferred results from the observations of Comet Halley can provide some restrictions on the number of possibilities. For example, the low density and low tensile strength for the nucleus inferred for Comet Halley, can be affected by the mechanisms of pressure build up around pockets of gas. Since the observations seem to indicate that the whole nucleus is not active at any given time but only partially and if due to some mechanism this could be increased, then automatically the comet will brighten up enormously. Therefore it appears that outbursts should be a common phenomena of all comets. Only those which are bright enough to be detected against the background brightness will be detected.

11.3. Albedo and Radius

The dimension of the nucleus of a comet cannot be obtained directly as it cannot be resolved. The general method which is commonly used is from the photometry of the comet at far off distances from the Sun. At these distances the comet is more or less stellar in appearance. This comes about due to the fact that at these distances, the temperature of the nucleus will be so low that the sublimation of the gases from the nucleus cannot take place. Therefore one is essentially seeing the nucleus of the comet. Hence the observed radiation at these distances essentially comes from the reflection of the incident solar radiation by the nucleus of comet. For such a situation, the brightness should vary essentially as $r^{-2}\Delta^{-2}$.

From such photometric observations, the radius of the nucleus R can be obtained from the relation

$$R^2 = r^2 A^{-1} \phi(\alpha)^{-1} 10^{0.4[V_\odot - (V_c - 5\log\Delta)]} \tag{11.9}$$

where $\phi(\alpha)$ is the phase function according to Lambert's Law, A is the albedo for the nucleus, $V_\odot$ and V_c are the absolute magnitudes of the Sun and the comet respectively. Equation (11.9) has been applied to several comets using photometric magnitudes and a reasonable range of values for the albedo of the nucleus. However, it is possible to get simultaneously the values of radius and the albedo of the nucleus from the following simple physical considerations.

The photometric observations of comets made at great distances from the Sun give only the total brightness which is AS where S is the effective cross section of the nucleus for the reflection of the solar radiation. It can be estimated from the observed magnitudes. It is also possible to get the value of the quantity $(1 - A)S$ from the theory of vaporization of ice-water from the nucleus when the comet is near the Sun wherein the vaporization is maximum. The integration of Eq. (11.1) over the nucleus gives the relation

$$\frac{F_\odot(1 - A_v)S}{r^2} = 4\sigma(1 - A_{IR})ST^4 + 4SZL. \tag{11.10}$$

For reasonable values of the albedo $A_v (0 \leq A \leq 0.7)$ and for the heliocentric distance less than about 0.8 au, the radiative contribution term in the above equation can be neglected. Therefore Eq. (11.10) reduces to

$$(1 - A_v)S = \frac{4SZLr^2}{F_\odot} = \frac{QLr^2}{F_\odot} \tag{11.11}$$

where $Q = 4SZ$ represents the production rate of water which can be obtained from the observed production rates of OH or H in comets (Chap. 6). The value of L, the latent heat for vaporization of water is 11500 calorie/molecule for $T = 200°K$. From a knowledge of AS and $(1 - A_v)S$ it is then possible to get the albedo and radius from the following equations:

$$S = (1 - A_v)S + AS \qquad (11.12)$$

and

$$A_v = \frac{A_v S}{A_v S + (1 - A_v)S}.$$

These methods when applied to various comets give for the radius of the nucleus the values in the range of about 1 to 5 km (Table 11.1). The method is valid if the effective area is roughly the same for the reflected light and for the vaporization equilibrium. However, there could be a difference in the areas in a real situation. In addition, there is some difficulty in separating the contributions to brightness from the nucleus and coma. The quantity (1-A)S can also in principle be determined from the infrared observations of the nuclei of comets. To a first approximation, (1-A)S is directly proportional to the thermal emission in the 10-20μm region provided the vaporization process from the nucleus does not take a significant fraction of the absorbed energy. It is also possible to make an estimate of the size of cometary nucleus from the band width of the back-scattered radar measurements provided the rotation and the orientation of the spin axis is known from some other means. The derived radius for the nuclei of comets based on various methods roughly lies in the range of around 2 to 8 km or so.

Table 11.1. Albedo and nuclear radius.

Comet	Albedo (A)	Nuclear radius (R)
Tago Sato Kosaka	0.63 ± 0.13	2.20 ± 0.27 km
Bennett	0.66 ± 0.13	3.76 ± 0.46 km

Delsemme, A. H. and Rudd, D. A. 1973 *Astr. Ap.* **28** 1.

The direct determination of the size of the nucleus of a comet was made possible by the study of Comet Halley. The images taken by the spacecrafts

Fig. 11.2. Composite image of the nucleus of Comet Halley made from several images obtained from Halley Multicolour Camera on board Giotto Spacecraft (Courtesy Keller, H. U., Copyright Max-Planck - Institut für Aeronomie, Lindau, Germany.)

directly give the projected dimension. Based on different aspects seen by the three spacecrafts, Vega 1, 2 and Giotto, it is then possible to reconstruct the actual three dimensional shape of the nucleus. Figure 11.2 shows one best image obtained from Giotto. It clearly shows the irregular shape of the nucleus, more characterized as a 'peanut' or 'potato-shaped' nucleus. From reconstructed images, the dimension of the nucleus is estimated to be $16 \times 8 \times 7.5$ km. The estimated total surface of the nucleus is about 400 ± 80 km^2 and its volume about 550 ± 165 km^3.

An estimate of the geometrical albedo has also been made for Comet Halley from the Vega images and the inferred value is 0.04. The geometrical albedo obtained for several comets could vary from as low a value as 0.02 to perhaps as high as 0.10 to 0.15. The surface temperature of the nucleus of Comet Halley estimated from Vega measurements is of the order of 300-400 K, which is somewhat higher than the expected temperature $\sim 190K$ from a vaporizing dirty ice nucleus corresponding to that distance. This has been interpreted to mean that the surface may be covered with a mantle of a darker substance.

11.4. Rotation

The rotational state of the nucleus of a comet is important for an understanding of the physical properties of the nucleus of a comet and its surface. The evolution of the spin state and the nucleus gives information about the changing pattern of the observed activities of a comet. It also plays a key role in a better understanding of the activity of the nucleus of a comet resulting due to gas evaporation from the nucleus, generally called 'non-gravitational force'. These studies are being carried out observationally to deduce the rotation period of the nucleus and the orientation of the spin axis of comets and its interpretation in terms of physical and dynamical models.

The time period of nuclear rotation can be estimated from some time-dependent property associated with the comet. It could be the observation of a certain repetitive feature, sequence of images of the nucleus, from an analysis of modeling of light curves or the non-gravitational force and so on.

The successive expanding halos with velocities of about 0.5 km/sec seen in the photographic observations of several comets have been interpreted on the basis of the rotation model for the nucleus. Parabolic envelopes have also been seen in the coma of comets. With the assumption that these arise out of the repetitive ejection of material from a single active area, it is possible to make an estimate for the rotation period of the nucleus. The procedure is to get the time at which the material was ejected, called the zero time. If the zero time can be found from concentric halos, then the time interval between the two zero times gives the rotation period. From the measured diameters of the halos or from the latus recta of the parabolic envelopes, the rotational velocities of a large number of comets have been determined. This method is generally known as 'zero age' method. The basic assumption involved in the halo method is that the halo structure arises mainly due to the rotation of the nucleus. Some doubts have been expressed about the accuracy of these time periods.

Many comets seem to show broad, fan-shaped coma coming out of the central condensation and in the general direction of the Sun. The width appears to change from comet to comet and with time. These have been

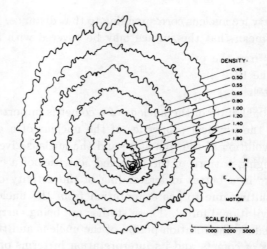

Fig. 11.3. The desitometer tracings of the photograph of June 25, 1927 in the inner part of the fan of Comet Pons-Winnecke is shown here. The jet near the nucleus in the sunward direction and the curvature effect as one goes outwards can be seen. The circled dot denotes the direction of the Sun. (Sekanina, Z. 1981, *Ann. Rev. Earth Planet Sci.*, **9**, 113). Reproduced with permission from Annual Reviews Inc.

interpreted in terms of anisotropy in the out-gassing. This can clearly be seen from the densitometer tracings of the inner part of the Comet Pons-Winnecke which are shown in Fig. 11.3. These have been interpreted in terms of the orientation of the spin axis of the nucleus.

The continuous monitoring of continuum or molecular emission lines over an extended period of time can give information about the periodicities connected with the rotation of the nucleus. The rotational state of the nucleus can also in principle be determined from the physical modeling of the observed cometary activities, even though it involves many assumptions. Based on various methods, the time periods of rotation of many comets have been determined. Unfortunately there is lot of variation in the derived periods for the same comet. The diversity in the derived values are illustrated with two examples.

For Comet Encke, the zero age method gave a value of 6.5 hrs. The periodic variation in the photometric magnitude gave a value of 22.4 hrs. The photometric data obtained near aphelion could be satisfied with periods of 22.6 hrs, 15.1 hrs or 7.5 hrs, but the most probable value being 15.1 hrs.

For Comet Halley several values have been derived. The 'zero age' method gave a period of about 10 hrs. The most direct evidence came from the morphology of sun ward jets. A rotational period of around 52 hours was estimated from the processing of spiral jets seen in the photographs taken during the 1910 apparition. The comparison of Vega 1 and Vega 2 camera observations gave a rotational period of 52-54 hours. However the images of features like jets near-nucleus show a period $\approx$ 2.2 days. A time period of about 7.4 days was obtained from the time series studies of the narrow-band photometric observations of the coma. The same time period was also indicated by the morphology of jets of free radicals, Lyman α emission and the 18 cm OH line, but nothing around a period of 2.2 days. Therefore there appears to be two rotation periods corresponding to 2.2 and 7.4 days.

For a non-spherical nucleus there could be several possible rotation states. It can rotate around the long or short axis. It can also have nutation motion. The rotation could be in its lowest rotational state energy or in an excited state. Depending on the state of the nucleus, it can have one or two time periods and their harmonics. Several attempts have been made to understand the observed periodicities of Comet Halley in terms of rotational, precessional or nutational motion of the nucleus. These studies show that a non-spherical nucleus and in particular, one ellipsoidal in shape can have two modes around its center of mass. The 7.4 days period could be attributed to rotation along its long axis, while the rotation around the short axis could give 2.2 day period.

Based on the above brief discussion, it is evident that the periodicities depend upon the state of the nuclei. In particular, on the shape, activity and their location, jets, how the gas comes out and so on. This leads to the complex nature of the periodicities in the observed spectra than indicated by the assumed simple models. Hence the interpretation of the observed periodicities is a difficult matter. Therefore, it is not surprising that there is a lack of consensus on the observed periodicities for Comet Halley even though vast amount of data exist.

11.5. Density:

The determination of bulk density of the nucleus of a comet is an important parameter as it can give information about the structure of the nucleus. It can be estimated provided the mass and the volume of the nucleus are known. However they are difficult to determine. Therefore the derived values are uncertain and they vary from comet to comet. The method that is commonly used is to make an estimate of the mass from the study of the non-gravitational perturbations of the orbital motion of periodic comets. With a reasonable value for the assumed radius of the nucleus, it is then possible to make an estimate for the average density of the nucleus. The derived values for Comet Kopff, Temple 2 and Giacobini-Zinner is around 0.1, 0.54 and 0.19 gm/cm^3 respectively. For Comet Halley it has been possible to get a reasonably good estimate for the volume of the nucleus from the images taken by the spacecrafts. Hence the bulk density can be determined. Unfortunately even here there is a wide spread in the derived values of the density as the determined mass crucially depends somewhat upon the assumptions made with regard to the out gassing mechanism. Therefore depending upon the emphasis put on the various sources of information on the water production rate, the values can have a wide variation between 0.2 to 1.5 gm/cm^3 or fitting the data points with two curves gives a density around 0.28±0.10 and 0.65±0.19 gm/cm^3 and so on. Therefore there is still uncertainty in the derived bulk density for Comet Halley. Therefore, the general conclusion is that the results appear to indicate a somewhat lower value for the bulk density of the nucleus of comets. This indicates that the nucleus could be fluffy and a loosely packed structure.

11.6. Chemical Composition

The information about the chemical composition of the nucleus has to come from the observed chemical composition of the gaseous coma. From the observed species in the coma of many comets, it is clear that the volatiles coming out of the nucleus must basically be made up of the abundant elements H, C, N and O. (Table 4.2) It is of interest to know quantitatively the abundances of elements H, C, N, O and others as well as the volatiles present in comets. It is not possible to derive the elemental abundances based on the observation on a single comet as the data on production rate of all possible molecules, radicals, neutrals and ions are not available. Therefore this has to be determined in a rough way by combining all the available information on the gaseous and dust components of various comets. In fact, even before Comet Halley observations, a *heuristic* model for the nucleus had been constructed based on the observations of several comets, which is as follows.

The elements H, C, N and O present in volatiles can in principle be studied through the atomic resonance lines or from simple molecules. Hence it is possible to deduce an average value of H/O, C/O, S/O and N/O for

Comets. Using a mean dust to gas ratio of 0.8 and with the reasonable assumption that the cometary dust is represented by the mean abundances of the C1 Chondrites, it was then possible to construct a heuristic model for the elemental abundances in a comet, based on the observed elemental ratios. The significant result that emerged out of this simple approach is that hydrogen is depleted by a large factor with respect to cosmic abundance.

More striking was the depletion of carbon by a factor of 4 or so. This is generally known as 'carbon depletion problem' in comets. This could mean

Table 11.2. Elemental abundances in comets (an average heuristic model).*

Element	Atomic	Sun(a)	Cometary abundances		Total
			Dust(b)	Gas	
H	1.0	12.0	8.89 ± 0.08	9.22 ± 0.15	9.39 ± 0.12
C	12.0	8.56 ± 0.04	8.49 ± 0.08	8.25 ± 0.20	8.69 ± 0.14
N	14.0	8.05 ± 0.04	7.20 ± 0.12	7.95 ± 0.22	8.02 ± 0.22
O	16.0	8.93 ± 0.04	8.53 ± 0.05	8.95 ± 0.15	9.09 ± 0.10
Na	23.0	6.33 ± 0.33	6.58 ± 0.20	-	6.58 ± 0.20
Mg	24.3	7.58 ± 0.05	7.58	-	7.58
Al	27.0	6.47 ± 0.07	6.41 ± 0.10	-	6.41 ± 0.10
Si	28.1	7.55 ± 0.05	7.85 ± 0.04	-	7.58 ± 0.04
S	31.1	7.21 ± 0.06	7.44 ± 0.12	6.48 ± 0.26	7.48 ± 0.24
K	39.1	5.12 ± 0.13	4.88 ± 0.18	-	4.88 ± 0.18
Ca	40.1	6.26 ± 0.02	6.38 ± 0.11	-	6.38 ± 0.11
Ti	47.9	4.99 ± 0.02	5.18 ± 0.18	-	5.18 ± 0.18
Cr	52.0	5.67 ± 0.03	5.53 ± 0.09	-	5.53 ± 0.09
Mn	54.9	5.39 ± 0.03	5.28 ± 0.15	-	5.28 ± 0.15
Fe	55.8	7.67 ± 0.03	7.30 ± 0.07	-	7.30 ± 0.07
Co	58.9	4.92 ± 0.04	5.06 ± 0.22	-	5.06 ± 0.22
Ni	58.7	6.25 ± 0.04	6.19 ± 0.18	-	6.19 ± 0.18

* Normalized to Magnesium (Adapted from Delsemme, A. H. 1991. In *Comets in the Post-Halley Era*, eds. R. L. Newburn Jr. et al., Kluwer Academic Publishers, p. 377.
(a) Anders, E and Grevesse, N. 1989, *Geochim. Cosmochim. Acta*. **53**, 197.
(b) Jessberger, E. K. and Kissel, J. 1991 In Comets in the Post-Halley Era, eds. R. L. Newburn et al., Kluwer Academic Publishers, p. 1075.)

that either that carbon could be tied up in the grains or it could really be less in comets. Rest of the elements were more or less similar to cosmic abundances. The observations of Comet Halley helped in resolving this carbon depletion problem as carbon is mostly tied up in organics, called the CHON particles.

The *in situ* measurements of Comet Halley, made with impact ionization mass spectrometer on board the Vega spacecraft, have given the elemental composition of dust. This is given in Table 11.2. The results are based on 79 high quality mass spectra. The contribution from the gaseous component has to be added to the dust component, to get an estimate of the total elemental abundances. A knowledge of the dust to gas ratio is required, for estimating the contribution from the gas component. Using a mean value for the dust to gas ratio of 0.8, the derived elemental abundances for the gas phase as well as the total abundance of gas and

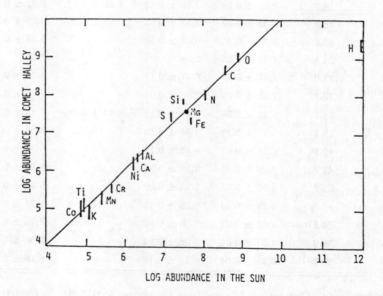

Fig. 11.4. A comparison of the elemental abundances in Comet Halley and in the Sun showing that they are very similar. This result can probably be generalized to other comets. The abundances refer to $\log N(H) = 12.0$ (Delsemme, A. H. 1991. In *Comets in the Post-Halley Era*, Eds. R. L. Newburn, Jr. et al., Kluwer Academic Publishers, p. 377).

dust is given in Table 11.2. These results are very similar to the results derived before Comet Halley as discussed earlier, which was based on the study of various comets. A direct comparison of the abundances of Comet Halley and in the Sun is shown in Fig. 11.4. It shows that the elemental abundances in Comet Halley are very similar to the solar values except for hydrogen.

Table 11.3. Volatile fraction of molecules in the nucleus of Comet Halley. (An average heuristic model)

92.0% with O	5.6% with N	2.2% Hydrocarbons	0.2% with S
78.5 H_2O	2.6 N_2	1.5 C_2H_2	0.1 H_2S
4.5 HCO.OH	1.0 HCN	0.5 CH_4	0.05 CS_2
4.0 H_2CO	0.8 NH_3	0.2 C_3H_2	0.05 S_2
3.5 CO_2	0.8 N_2H_4		
1.5 CO	0.4 $C_4H_4N_2$		

(Adapted from Delsemme, A. H. 1991. In *Comets in the Post-Halley Era*, eds. R. N. Newburn Jr., et al. Kluwer Academic Publishers, p. 377.)

The detailed results derived for Comet Halley also make it possible to build a consistent model for the volatiles escaping from the nucleus. A heuristic model for Comet Halley is shown in Table 11.3. This is derived by adjusting the elemental abundances in the molecules so that it balances with the total elemental ratios as given in Table 11.2. Some of the chosen molecules as given in Table 11.3 are based on some of the complex molecules seen from the dust grains in Comet Halley as well as from other considerations. As can be seen, about 80% of the volatiles make up the ice are water. The remaining 20% accounts for the other molecules listed in Table 11.2. Due to various uncertainties and errors involved in different quantities, the results of Table 11.3 should be taken as an indication of the trend of the results rather than the actual values.

There appears to be a good determination of the chemical composition of Comet Halley in particular and the comets in general. The similarity of the elemental abundances in Comet Halley and that of the solar system value suggest that Comet Halley is unlikely to be differentiated radially. Also if all comets including Comet Halley originate from the same reservoir, then they should all have a similar pristine chemical composition.

11.7. Mass Loss

It is interesting to make an estimate of the order of mass loss from the nucleus of a comet per revolution. The average production rate at 1 au obtained from the observations of OH and H in comets is around 10^{29} to 10^{30} molecule/sec. (Chap. 6) which corresponds to about 10^7 to 10^8 gm/sec. Therefore in an apparition, the total amount of material lost is around 10^{14} to 10^{15} gm. This corresponds roughly to about 1% of the total cometary material. The observed mass loss rates indicate that the lifetime of short-period comets is limited to about 10^4 to 10^5 years.

11.8. Structure

There is no direct method of knowing the internal structure of the nucleus. Therefore one has to infer the possible nature of the structure of the nucleus from the observed characteristics of comets. The observational fact that the splitting of the nuclei is quite commonly seen in comets implies that the nuclei should be quite brittle. Otherwise it will not break that easily. The important point is that the splitting process does not totally destroy the nucleus. The classic example is that of Comet Shoemaker-Levy 9 which disintegrated into 21 pieces due to tidal forces of Jupiter (Figs. 1.9, 1.10). The splitting phenomenon also gives a chance to compare the different split components with the original nucleus and through this one may be able to get some information about the changes with depth. The splitting of Comet West into four components (Fig. 1.8) has been analyzed in some detail. The components are observed to separate out with extremely small velocities of the order of one meter per second. After their initial velocity they are all subjected to non-gravitational separation forces produced as a result of the jet action resulting out of vaporization of ice (Sec. 11.9). The observations can be fitted very well with such a physical model. The observations of several other split comets, which have different time-period can also be explained in a similar manner. These results show that the different pieces coming out of comets of extreme ages show the same behaviour with regard to the vaporization process and this suggests that the stratification of the nucleus may not be that important. The outburst seen in many comets indicate the nucleus could be fragile as well as porous. In general the new comets appear to exhibit more activity than short-period comets at large heliocentric distances. This may indicate that the outer layers of new comets may be extremely loose with more volatile material contained in it or there could be a halo of volatile material around the nucleus.

The erosion of the nuclear surface can take place during its long exposure at far off distances from the Sun due to the bombardment of cosmic rays, the solar wind etc.

The gross material strength of comets can be obtained from the estimates of the tidal forces acting on the Sun grazing comets and from other considerations. The gravitational compression force ($\lesssim 10^5$ dynes/cm^2) is quite small and cannot compress the icy material of the nucleus. The tensile strength against the tidal disruption of Sun grazing comets indicates a value in the range of 10^3 to 10^5 dynes/cm^2. Similar order of strength of the material in the nucleus is estimated from the period of rotation of the nucleus. These values are small and they indicate the material to be weak. The strength of the material can also be estimated based on the study of meteor trajectories as they plunge into the Earth's atmosphere since they are believed to be cometary debris (Chap. 13). They also give value of about 10^3 to 10^5 dynes/cm^2. In addition to this, the particles collected at high altitudes appear to be fluffy. The estimated low bulk density for the nuclei of comets indicate the material to be fulffy and loosely packed in the nucleus.

There are also other types of observations which can give some clue to the structure of the nucleus. It is well known that comets go around in their orbits many times. These observations show that only a thin layer of the material comes out of the nucleus keeping the core of the nucleus intact. More support for this comes from the Sun grazing comets which survive even after such an encounter. Also the jet force arising out of the vaporization of the gases modifies only the trajectory of the comet and it does not destroy the nucleus as a whole. All these observations appear to show that the nucleus cannot be composed of a cluster of small particles termed as the sand-bank model. In the sand-bank model the particles are independent of each other but they follow similar orbits around the Sun. In this case there is no real body at the centre although there is a high concentration of particles at the centre. The images of the nucleus of Comet Halley taken from Vega and Giotto spacecrafts have clearly shown the nucleus to be a one solid chunk. The images have also given enormous information about the surface morphology and topology of the nucleus of Comet Halley. They showed the nucleus to be irregular in shape with active areas where from emerge jets of gas. There are dark areas as well as craters, valleys, mountains, depression of the region etc., have also been seen. It also showed the roughness of the nucleus.

There are other spectroscopic observations which also give information about the structure of the nucleus. The study based on the gas production rates of various molecules shows that they are unrelated to the dynamical age of the comet or to the line emission to continuum ratio i.e. gas/dust ratio (Chap. 4). Also comets with wide varying time periods $\sim 10^2$ to 10^5 years, show a similar appearance in their spectra in the ultraviolet region (Chap. 6). These results show that to a first approximation, there is little variation in the gross chemical composition of the material of the nucleus. The noncorrelation with the dynamical age of the comet indicates the nucleus to be homogeneous in composition at least up to a certain depth. Even if the stratification is present it may not be severe enough to produce any noticeable effect in the gross chemical composition of comets. This could be interpreted as the absence of any differentiation process in the nucleus unlike in the case of planets where this process is important. The anomalous abundance of ^{26}Mg found in some chondrules in Allende meteorite suggests that ^{26}Al should have been present in the abundance of the original material out of which comets were formed. Since the half life of ^{26}Al is only 0.7 million years, it would have decayed during the early stages of the solar system, within a few million years or so. It is possible that the heat generated in the decay process of ^{26}Al could have an effect on the internal structure of the nucleus.

11.9. Non-Gravitational Forces

The orbit of a comet can be computed by well-known methods based on Newton's law of gravitation. The accuracy of the determined orbit depends upon the number of observations available for the comet. The orbit of short-period comets can be determined accurately as they can be observed many more times than that of long-period comets. The orbit of a comet is generally perturbed strongly by the planets as they enter the inner part of the solar system. The perturbation induced due to such an encounter can be incorporated in the orbit calculation and these complications have become easier to handle with the availability of high speed computers. However in spite of such accurate orbit calculations, it has been noticed that they seem to differ from the observed position of comets at each revolution. The well-known comet for which this is striking is the Comet Encke, discovered in 1786 and having a period of 3.3. years. It persists in returning at each revolution about $2\tfrac{1}{2}$ hours too soon. This was noticed by Encke himself as far back as 1820. Observationally it is found that out of several comets studied some came earlier than the predicted time and others later than the predicted time. This means that some are accelerated and the others decelerated, which demonstrated conclusively that there was an additional nongravitational force acting on the comet which was not included in the orbit calculations. It was noticed from observations that this nongravitational force decreased substantially with an increase in the distance between the comet and the Sun.

The equation of motion of a comet in the rectangular coordinate system used in the orbit calculation procedure can be written as

$$\frac{d^2\mathbf{r}}{dt^2} = -\mu\frac{\mathbf{r}}{r^3} + \frac{\partial R}{\partial \mathbf{r}} \qquad (11.14)$$

Here μ is the product of the gravitational constant and the mass of the Sun. The term on the right-hand side of Eq. (11.14) takes care of the planetary perturbations i.e., R is a planetary disturbing function. The above equation was generalized to include nongravitational forces as

$$\frac{d^2\mathbf{r}}{dt^2} = -\mu\frac{\mathbf{r}}{r^3} + \frac{\partial R}{\partial \mathbf{r}} + F_1\mathbf{r} + F_2\mathbf{T} + F_3\mathbf{N}. \qquad (11.15)$$

Here F_1, F_2 and F_3 represent the additional acceleration components, F_1 is along the radius vector defined outward along the Sun-Comet-line, F_2 is perpendicular to r in the orbit plane and towards the comet's direction of motion and F_3 normal to the orbit plane. r, T and N are the three unit vectors along the three direction of forces. Since the nature of F_1, F_2 and F_3 was not known, one can write

$$F_i = A_i f(r) \qquad (11.16)$$

and to see what form of $f(r)$ can remove the observed discrepancy. The component of the force F_3 is generally present for active comets, but it is difficult to determine a meaningful solution due to its periodic nature and also the average non-gravitational acceleration is determined from the solution over three or more apparitions. Therefore for most of the comets the value of A_3 is put equal to zero. The change in the semi major axis introduced due to the radial and transverse perturbing acceleration (R_P, T_P) is given by the equation

$$\frac{da}{dt} = \frac{2}{n(1-e^2)^{1/2}}\left[(e\sin\nu)R_P + \frac{P}{r}T_P\right]$$

where n, e, ν and r represent the orbital mean motion, eccentricity, true anomaly and the heliocentric distance respectively. P presents the orbital semi-latus rectum, $a(1-e^2)$. The derived empirical function $f(r)$ was found to agree well with the form $g(r)$ given by the Eq. (11.5). Knowing the function $f(r)$, the constants A_i can be calculated. These nongravitational parameters are actually obtained from the least square fit to the definitive orbit of the comet. Therefore the empirical work on the nongravitational force has been put on a firm basis based on a physical model. In recent years the expression (11.5) has been used in all the orbit calculations.

These ideas fit in well with the matter ejected from a rotating icy model of the nucleus which can be understood from Fig. 11.5. When the solar radiation impinges on the nucleus, the material evaporates from the surface and moves out in the direction of the incident radiation. This results in a jet action, a force pushing the comet away from the Sun. If the nucleus is rotating then there will be a time delay between the heating and the ejection of the gas from any point on the surface. If the rotation is in the same direction as its motion around the Sun, the delayed jet action will have a forward component which will increase the orbital period of the comet (Fig. 11.5b). As a result it will go into a higher orbit and so it will show up later than the predicted time. On the other hand, if the nucleus rotates in the opposite direction to the motion around the Sun (Fig. 11.5c) the jet action will have a backward component. This will reduce the period of the comet and hence it will show up earlier than the predicted time. Therefore, depending on the direction of rotation of the nucleus, some of

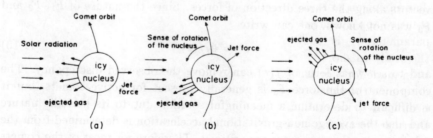

Fig. 11.5. The result of outguessing on the cometary nucleus: (a) the nucleus does not rotate; (b) the nucleus rotates along the direction of motion around the Sun. The jet force is also along the direction of motion increasing thereby the period of the comet; (c) the nucleus rotates in the opposite direction to the direction of motion around the Sun. The jet force is now in the opposite direction to its motion resulting in a decrease in the period of the comet.

them will be accelerated and some will be retarded, which is consistent with the observation.

It is now possible to make a meaningful comparison of the nongravitational parameters A_1 and A_2 for different comets as they represent the relative mass loss rates. However, there was still a major problem requiring the outgassing to be much higher than what was known at that time based on the line emission of C_2 and CN in the visual region of the spectrum (Chap. 6). Otherwise the jet force was not enough to exert an adequate force on the comet. This problem was resolved with the discovery of the presence of a huge halo of hydrogen gas around Comet Bennett in the 1970's through the detection of the Lyman α line radiation (Chap. 6). The observed decrease of nongravitational effect with an increasing distance from the Sun can also be explained based on the vaporization model. The nongravitational parameter indicates that about 1% of the material is lost from each comet in one revolution which is consistent with other estimates (Sec. 11.7). To take care of the active emission areas from the surface of the nucleus, an effective sublimation rate $Z(z,r)$ (molecules/cm^2/s) at a heliocentric distance r and the Sun's local zenith angle z is defined as

$$Z(z,r) = Z_0(r)\mathcal{G}(z,r)$$

where Z_0 is the sublimation rate at the subsolar point and $\mathcal{G}(z,r)$ (≤ 1) is the relative sublimation rate at the Sun's Zenith angle z. If there are several sources then it has to be averaged to get the rotation-averaged acceleration.

The transverse component given by the parameter A_2 is quite sensitive to the effects of episodial events in the orbital motions of comets. The determination of non-gravitational parameters require at least three consecutive apparition and in addition at least three independent values of the parameters A_2 is required for the study of temporal variations. This limits the study to short period comets for which astrometric observations have been made for 5 or more apparitions. Such observations have been used to determine the values of A_1 and A_2 for a large number of comets. These have been used to understand the diurnal effects and seasonal effects in comets. Diurnal effect refers to the short term variations arising due to outgassing from uneven exposure of individual sources to Sunlight caused by the rotation of the nuclei. Seasonal effects are the long term variations associated with producing the asymmetric curve in the production rate with respect to perihelion.

In the discussion so far, it has been generally assumed that the parameter representing the component of the force normal to the orbital plane, A_3

is equal to zero. This is due to fact that the introduction of the parameter A_3 did not seem to improve the orbital solutions implying that it is hard to extract the value of A_3 i.e. essentially indeterminate. Therefore A_3 is the least understood parameter and it requires further study. With this in view, several attempts have been carried out, in recent years, to detect the contribution from the normal component of the transferred momentum. Using various mathematical techniques, the three non-gravitational parameters A_1, A_2 and A_3 have been derived for a number of comets. The results are rather consistent in showing that the value of A_3 is not large. Therefore with the availability of accurate values for the parameters A_1, A_2 and A_3 representing the component of non gravitational perturbation in the radial, transverse and normal directions it may be possible to study the time evolution of the orbits, precession of the spin axis, wobbling and so on.

All the studies carried out so far for the last two decades or so, were based on the symmetric models for the non-gravitational effects in the equation of cometary motion. The implicit assumption in the symmetric model is that the non-gravitational effects are the same on either side of the perihelion position. However, in a real situation the non-gravitational effects are more of perihelion asymmetrical in nature. The evidence for such an effect comes from the non-random distribution of the active regions on the surface of the nucleus and the observed asymmetrical nature of the light curve of comets, which basically reflect their outgassing histories. Therefore the results based on symmetrical non-gravitational acceleration model is being questioned. However, it should be noted that the symmetric models have been very successful in providing accurate ephemerides but the derived physical properties of comets should be taken with caution. Therefore attempts are being made to take into account the asymmetrical nature of the non-gravitational force in comets by considering the out gassing as accurately as possible.

An approach that is being attempted is a slight modification of the symmetrical non-gravitational acceleration model. In this procedure the possible asymmetry in the out-gassing is taken into account by shifting the time (DT) for a few days before or after perihelion passage so that it coincides with the maximum value of the water vaporization curve. DT=0 corresponds to the symmetric case. Therefore the expression for the asymmetric non-gravitational acceleration model at any time t is given by $g(r')$, where $r' = r(t') = r(t - DT)$. The new calculations are similar to the symmetric model case except that the function $g(r')$ is evaluated for the heliocentric distance corresponding to the time $t - DT$, rather than for t itself. The time shift DT is varied till a best fit to the anisotropic observations is found and this in general agrees with the shift seen in the maximum of the light curve from the perihelion. It is found that the asymmetric model appears to provide better orbital solution compared to symmetric models. The new approach is being applied to the study of several comets. The results based on a few comets have shown that the radial and transverse non-gravitational parameters derived from non-symmetric models are different from symmetric models. Therefore the inferred physical properties like rotation and precession of cometary nuclei as derived from symmetric models have to be re-examined.

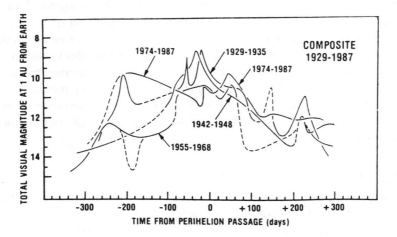

Fig. 11.6. A composite figure showing the light curves of Comet Schwassmann-Wachmann 2 for the period from 1929-1987. The negative times correspond to the inbound orbit and positive times to outgoing orbit. The long-term variations as well as erratic behaviour can be seen (Sekanina, Z., 1993. *Astro. Ap.* **271**, 630).

Since the non-gravitational force is the resultant effect of the redistribution of the momentum that is transferred to the nucleus by the sublimating gas, it is a complicated function of several factors like shape, structure, nature and areas of active regions, jets, direction of the ejection of gas and so on of the nucleus. The resultant effect is to complicate the rotational as well as the orbital motion of the nucleus of a comet, which shows up as peculiarities in the observations (Fig. 11.6). It also depends upon sudden activation or deactivation of discrete emission centres in the nuclear surface.

In conclusion, the vaporization theory of water from the nucleus can explain reasonably well the visibility of comets at around the heliocentric distance of 3 au. It can also explain the nongravitational forces seen in comets. However, the mechanism of formation of outbursts seen at large heliocentric distances is not yet understood. The dimension of the nucleus of Comet Halley deduced from the images taken from spacecrafts is about $16 \times 8 \times 7.5$ km. The nucleus is also found to be irregular in shape with several active areas from which jets of gas emanate. The geometrical albedo is about 0.04. The rotational period of several comets indicates a value of around a few hours. There appears to be two rotational time period for Comet Halley corresponding to 7.4 and 2.2 days. The two time periods could arise from an elliptical shape of the nucleus with rotations along the two axes. A heuristic model for the chemical composition of comets has been derived based on ground based observations of various comets and *in situ* measurements of Comet Halley. Around 80% of volatiles in the nucleus are made up of H_2O, around 10% of CO and the rest accounting for all the other molecules. The nucleus should also contain complex molecules as revealed by the dust composition in Comet Halley. The observed characteristics of comets can also be interpreted in terms of the nucleus being mostly fragile and undifferentiated.

Problems

1. Do you expect the vaporization rate of gases from the nucleus to be constant at every revolution around the Sun?
2. Consider a nucleus made up of ice 1 km in radius, with mass loss rate of 10^8 gm/sec and a period of 5 years. If the comet is assumed to be active for about a year at every revolution, how much does the nucleus shrink per revolution? What is the lifetime of the nucleus?
3. A comet of radius 1 km is moving with an orbital velocity of 60 km/sec at 1 au. Calculate the amount of out-gassing required if the gas comes out at 1 km/sec from the sun-lit hemisphere of icy nucleus so that the non-gravitational force is about 1% of the orbital force.
4. How do you reconcile the fact that the nucleus contain mostly water but the observed geometrical albedo for Comet Halley is 0.04.
5. If the nucleus of a comet was made up of 80% CO_2 and rest is made up of other molecules like H_2O, CO and others, what will be the scenario of such a comet?

6. What are the consequences if the nucleus is a hard solid body instead of fragile and fluffy in nature?

References

The following paper gives a good account of the vaporization theory
1. Delsemme, A. H. 1982, In *Comets* ed. Wilkening, L. L. Tucson: University of Arizona Press, p. 85.

The following papers may be referred for later work.

2. Dobrovlśkij, O. V. and Markovich, M. Z. 1972. In *The Motion, Evolution of orbits and origin of Comets* (Proc. IAU Symp. 45), ed. G. A. Chebotarer et al., Reidel Publication, p. 287.
3. Fanale, F. P. and Salvail, J. R. 1987, *Icarus* **22**, 535.
4. Prialnik, D. and Bar-Nun, A. 1990. *Ap.J.* **363**, 274.
5. Rickman, H. 1991. In *Comets in the Post-Halley* Era. eds. R. L. Newburn Jr. et al. *Kluwer Academic Publishers*, p. 733.

The existence of non-gravitational force in comets was proposed for the first time in a classic paper by

6. Whipple, F. L. 1950. Ap. J. **111**, 375.

For later work the following papers may be consulted:

7. Marsden, B. G., Sekanina, Z. and Yeomans, D. K. 1973, *A J.* **78**, 211.
8. Sekanina, Z. 1993, *Astr. Ap.*, **277**, 265.
9. Belton, M. J. S. 1991, Ref. 5, p. 691.

The following papers give a good account of the chemical composition and cometary activity of the nucleus.

10. Delsemme, A. H. 1991, Ref. 5, p. 377.
11. Sekanina, P. 1991, Ref. 5, p. 769.

CHAPTER 12

ORIGIN

There are many ideas and hypotheses with regard to the origin of comets. Some of these date back to Laplace and Lagrange. Dynamical simulations in conjunction with available observations, although scarce, have given new insights into this area of study. Some of these aspects will briefly be discussed in this chapter.

12.1. Evidence for the Oort Cloud

During early times, the emphasis in the study of comets was mainly on the determination of their orbits. This resulted in the accumulation of a lot of data with regard to the orbital characteristics of comets, which gave an insight into some of the physical problems. One such fundamental aspect, which came out of these studies - namely, nongravitational force has already been discussed in the previous chapter. Another significant result refers to the problem of the origin of comets themselves. Several general characteristic properties can be noted just from the stability of comets. The long-period comets, mostly in parabolic orbits, appear to come from all directions in the sky. The short period-comets, have low inclination to the ecliptic plane and have a strong association with the planetary system. Any reasonable theory of the origin of comets has to explain these features. Various concepts, ideas, hypotheses and theories, have been put forward over the years discussing the merits and demerits of interstellar vs solar system origin of comets. Therefore the whole subject was in a confused state. In a classic paper, Oort in 1950 proposed the unification of the origin of comets within a reasonable and consistent framework. This has given rise to a rapid development of the subject and a better understanding of the whole phenomena.

Even with limited observational material available on about 19 long-period comets at that time, Oort showed that a simple plot of the number of comets versus the inverse of semi major axis, $1/a$ (equivalent to orbital energies) of the original orbit gave a conspicuous peak near value zero (i.e. nearly parabolic orbits) (Table 12.1). The original orbit refers to the orbit of the comet before it enters the planetary system. The peak observed in the $1/a$ distribution cannot be due to chance but represents the real characteristic property. This is based on the fact that the observed dispersion in $1/a$ is much smaller than it would have been if it had passed through perihelion passage once. Even one passage can bring in a dispersion in $1/a$ which is much larger than the observed values. Therefore most of the comets must have come into the solar system for the first time, generally called 'new' comets. Most of these comets appear to come from the region of say 30,000 to 50,000 au. This led Oort to recognize the existence of a spherical cloud of comets around the Sun at this distance but still gravitationally bound to it. This is generally called the "Oort cloud". It should be pointed out that as far back as 1932, Opik had suggested the possibility of the presence of such a cloud surrounding the solar system. With more

Table 12.1. Distribution of original values of semimajor axes, a(au).

	$1/a$		Number of comets
	$\leq$	0.00005	10
0.00005	-	0.00010	4
0.00015	-	0.00015	1
0.000020	-	0.00020	1
0.00025	-	0.00050	1
0.00050	-	0.00075	1
	>	0.00075	0

Oort, J. 1950, *Bull. Astr. Inst.* Netherlands **11**, 91.

accurate and high precision data available in recent years on about 200 long-period comets the idea proposed by Oort has been confirmed and all the results deduced earlier based on the limited data remain more or less the same. Figure 12.1 shows the histogram of the number of comets versus

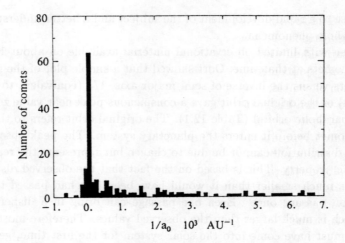

Fig. 12.1. A histrogram of the number of comets plotted as a function of the original inverse semi major axes for the observed long-period comets. The sharp peak of comets near zero value of original 1/a represent the new comets from the Oort cloud. A few hyperbolic comets are likely to arise from the error in their orbit calculations (Marsden, B. G., Sekanina, Z. and Everhart, E. 1978, *AJ.* **83**, 64).

the original 1/a values based on about 200 comets with well-determined orbits. The period corresponding to the peak value of (1/a) is about 4×10^6 years and the mean aphelion distance $\sim 50,000$ au. These distances are almost comparable to the distances of nearby stars.

The comets in the Oort cloud are constantly being perturbed gravitationally by the passing stars and clouds so the orbit of the comets is modified continuously. The effect of these perturbations is to induce a velocity at that distance which is comparable to the velocity of escape from the solar system. It follows therefore that the effect of these perturbations is to completely randomize the velocity distribution of comets. This explains the observed spherical symmetry of the cloud.

Due to the influence of stellar perturbations in the Oort cloud, the comets can escape and occasionally enter into the planetary system, some of which finally end up as long-period comets. Therefore the observed peak in the 1/a distribution function is maintained with a time scale of around 4×10^6 years by a balance between constant supply of comets from the Oort cloud compensated with the losses as they enter into the planetary system. On subsequent returns through random walk in orbital energy due to planets, several of these comets can be captured as short period comets. The total number of comets in the spherical cloud was estimated by integrating the number density which was derived on the assumption of hydrostatic equilibrium and with the use of free fall velocity. The estimated total number of comets is around 2×10^{11}. With regard to the question of formation of comets, it was argued that they probably originated in the first instance within the planetary system itself and ended up in the spherical cloud. In spite of several difficulties, Oort's investigation is remarkably successful in reproducing the main characteristic features of the observed orbits of comets and also in keeping the attractive feature of the formation of comets along with the solar system. While this model is generally accepted, the details pertaining to physics, dynamics, structure and stability of the Oort cloud as well as the formation of comets are the subject of study at the present time.

12.2. Evolution and Properties of Oort Cloud

The transformation of comets from the Oort cloud to the observable comets can be understood in a quantitative fashion in the following manner. Due to stars perturbation, many comets leave the cloud for ever and some others enter the planetary system. Among these some may happen to come close and are observable as "new" comets. When a fraction of these comets encounter the planets, particularly Jupiter, they are perturbed and leave the system altogether after their first encounter. Some of these get caught in the solar system, and are seen as long-period comets. By repeated encounters with the giant planets, enough of them are progressively transferred from long-period orbits to intermediate-period orbits and finally into short-period orbits. A large number of investigations have been carried out for an understanding of some of these problems through analytic, semi-analytic and numerical calculations. With the availability of high speed computers, the Monto Carlo approach has been used to simulate the dynamical evolution of comets in the Oort cloud, wherein it is possible to investigate the effect of varying initial conditions. The idea is to consider a large number of comets in the Oort cloud and study their time evolution by taking into account the perturbations due to various sources. The orbits of comets could be followed until they escape from the system in a hyperbolic orbit. It is the cumulative effect of the changes in the comets orbital energy during each passage through the planetary system which is important in deciding the fate of the final orbit of the comet. Due to large computer time involved, some approximations have to be made to obtain results within a reasonable length of time. This makes it rather difficult to compare the results of different investigators. Nonetheless it is still useful in arriving at some of the general conclusions.

The results of Monto Carlo simulation for 10^5 hypothetical comets gives a good match to the observed distribution in $1/a$. (Fig. 12.1). On the average a long-period comet with perihelion distance $< 4au$ makes around five passages through the planetary region before taking on some final state with a mean life time of about 6×10^5 years. In addition to the perturbing forces normally considered, arising due to planetary encounters and passing stars that continually change the orbital elements of comets, other type of perturbing forces have been shown to be important. One is due to the existence of Giant molecular clouds with a typical mass $\sim 3 \times 10^5 \ M_\odot$ and radius $\sim 20pc$ respectively. The number of encounters of the solar system with Giant molecular clouds is around 1 to 10 during its lifetime. This has been shown to be a major perturber of the orbits of comets in the Oort cloud. Just as stars passing the Oort cloud perturb the orbit of comets, in a similar way the distribution of stars in the Galaxy can also have a major perturbing effect on the orbits arising due to 'tidal' distortions. This arises

Table 12.2. Influx rate of comets from the Oort clouds.

Perturber	Relative number
Random stars	1.0
Intermediate size molecular cloud	124
Vertical galactic tidal force	1.72
Close stellar passage ($D_\odot = 10^4 au$)	62.1

Adapted from Fernandez, J. A. and IP. W. -H. 1991. In *Comets in the Post-Halley Era.* eds. C. L. Newburn et al. Kluwer Academic Publishers. p. 487.

due to the gradient of the gravitational force in the solar neighborhood which is slightly different at the position of the Sun and the comet. As a result the galactic acceleration is different at the two locations which leads to a net tidal force acting on the comet with respect to that of the Sun. This effect has been taken into account in a more realistic manner in recent dynamical calculations. The relative importance of various perturbers for the net influx of comets can be seen from Table 12.2. The results are for a radial distribution of comets in the Oort cloud of the form $n(r) \propto r^{-3}$ and normalized to unity for the influx rate caused by random passing stars. For ranges around $7 \times 10^3 \lesssim a_{\text{original}} \lesssim 4 \times 10^4$ au of the Oort cloud the effects of forces other than planetary perturbations are important.

Physically, if the angular momentum per unit mass ($h = (2\mu q)^{1/2}$, $\mu = GM_\odot$, q = perihelion distance) is greater than a certain amount, it is removed from the Oort cloud. However due to perturbation of the angular momentum caused by the perturbers, the loss in comets will be filled in at any time. From these, the fraction of comets filled in due to the cumulative effect of all the perturbers can be calculated. This involves the angular dependence factor in the calculations which bring in the anisotropies in the aphelion directions. These are relevant when comparing with the observed distribution of comets. The permanent perturbers like passing stars and the galactic tides will try to fill the respective cones of comets with semi major axis roughly in the range 1.2×10^4 to 3.5×10^4 au and almost nil for semi major axis $\lesssim 1.2 \times 10^4$ au. For $r \gtrsim 3.5 \times 10^4$ au, the loss cones are filled in from the outer portions of the Oort cloud so that there is a constant steady supply of comets. The fact that certain clustering in position and angle in the observed comet distribution still exists point to the fact that the effects of the latest encounters with some perturbers has not been washed out and so it must be of recent in origin. Of particular interest is the observational data of comets which show deficiency at the galactic equator and poles which is in accord with the expected distribution due to galactic tides acting on the Oort cloud.

The total number of comets in the Oort cloud has to be derived from a combination of the observed flux of comets and the dynamical simulation of comets. This has to take into account the comets which cannot be seen due to their larger perihelion distances and also the limiting absolute magnitude for the observed flux of comets. Therefore, the values derived by various investigators differ. The population of comets brighter than absolute magnitude 11 in the Oort cloud could roughly be in the range of (0.4 to 1.3) $\times 10^{12}$. The population of comets in the inner region of the cloud could be larger by a factor of 5 to 10 than of the above values. These estimates are based on the average flux of comets observed at the present time.

The total mass of the Oort cloud depends upon the uncertainties in the sizes of the nucleus of comets and their bulk density. It has been estimated, based on the bulk density of material which can range between 0.2 to 1.2 gm/cm^3, and the derived cometary mass distribution using mass-brightness relation. The estimated total mass for the Oort cloud, according to various investigators, range from 14 to 1000 Earth masses, with a reasonable value nearer to the lower limit. The original mass of the Oort cloud must have been larger by a factor of $\sim$ (2-5) due to the loss of the comets, (during the life time of the solar system), which may be in the range $\sim$ 40 to 80%.

The structure and stability of the Oort cloud has also been studied based on the N-body simulation with the inclusion of various perturbations. This approach tries to simulate the actual physical system, wherein the interaction of all the bodies is taken into account. Therefore this method overcomes several limitations of the Monte Carlo simulations method, but could be restrictive because of excessive computer time involved. It is possible to study physical processes through this approach, such as the exchange of energy and angular momentum between the cloud and the stars and so on. A comparison of the mean square velocity of the cloud with the escape velocity at any particular layer would show whether it can escape from the system. The results show that the inner regions are stable, while the outer layers can escape from the system within times of the order of the age of solar system. These outer layers will be replenished by comets from the inner region of the Oort cloud, which acts as the reservoir of comets.

The studies based on Monte Carlo calculations have indicated that influx of cometary flux from the outer regions of the Oort cloud could vary roughly by a factor of 2 to 3. These are a sort of common phenomena. Sometimes it is possible that the passing stars happen to penetrate the Oort cloud, or the stars could be of relatively high mass, or pass the cloud with small impact parameter. In these situations the orbit of comets could be changed by a larger amount. The net effect will be to scatter more long-period comets into the solar system, which will increase drastically the observable comets giving rise to cometary shower (Fig. 12.2). Monte Carlo simulations have been carried out for such situations to get an idea of the expected results. The impact rate increases by a factor of around 300 for the case of star passing at a distance of $\sim$ to 3×10^4 au from the Sun; and seems to last for around 2 to 3×10^6 years. The increase in flux rate arises due to the fact that comets coming from the inner part of the Oort

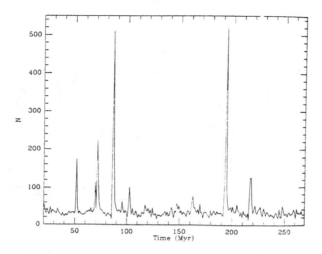

Fig. 12.2. The results of Monte Carlo simulation showing the number of new long-period comets entering the terrestrial planets region $q < 2au$ from the Oort cloud. The spikes represent the comet showers arising due to random stars panetrating the Oort cloud (Courtesy Heisler, J).

cloud have relatively shorter time periods compared to those coming from outer regions and hence perturbed more due to planetary perturbations and in turn make more revolutions. However such close stellar encounters are quite rare and the frequency of occurrence could be perhaps once in 3 to 5×10^8 years or so. In a similar way, a sudden increase in the influx of comets from the Oort cloud can increase by a factor as high as 10^3 for a close encounter with a Giant molecular cloud ($M \sim 5 \times 10^5 M_\odot$) or even higher rate of influx for the penetrating Giant molecular cloud. But such occurrences may occur at intervals of several times 10^7 years. The observed distribution of comets is a complicated function of the various physical mechanisms which could operate on the Oort cloud. The enhancement of ejection of comets could vary in a periodic fashion, random or in some complicated way, depending on the manner in which the disturbances act on the Oort cloud. This could have catastrophic or moderate effects on the geophysical, biological or climatic changes on the earth. They could possibly appear in terrestrial records. In view of these considerations, it has been suggested that the random cometary showers could contribute substantially for the formation of craters on planets and the Moon and also the biological extinction events on the Earth. Also the statistical study

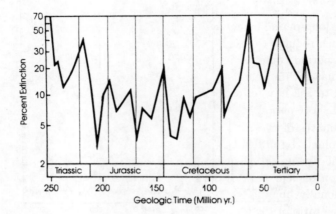

Fig. 12.3. A statistically significant mean periodicity of around 26 million years is indicated by the Extinction records for the past 250 million years. (Raup, D. M. and Sepkoshi, J. J. 1984, *Proc. Nat. Acad. Sci.* **81**, 801).

of the fossil records has been interpreted in terms of biological extinction events that may be taking place with a time scale of approximately 26×10^6 years (Fig. 12.3). An interesting measurement was carried out at the sea floor looking for iridium concentration for the period from 33 to 67 Myr ago. The results showed an increase by a factor of around 13 near the Cretaceous-Tertiary boundary (referring to 66 Myr). This can be explained in terms of a strong comet shower which enhanced the iridium concentration. The expected value from such a process agrees within a factor of two of the observed value seen at the Cretaceous-Tertiary boundary. At present these may be considered as possibilities rather than evidence in support of them, as more work needs to be done from both observational and dynamical points of view.

The Oort cloud has been considered to be a cold storage place for long-period comets without much happening until it is perturbed by a perturbing force. In recent years, it has been suggested that during the life time of the Oort cloud the outer layers of the cometary nuclei could be modified due to various physical processes such as, bombardment by galactic cosmic rays, solar wind, heating due to nearby supernova explosions, luminous stars passing by, impact due to various debris and so on.

12.2.1. Short period comets

The long-period comets, after random-walk through the planetary system, which reduces the semi-major axis of the orbit, could finally give rise to short period comets. The important work of Everhart in 1972 showed clearly that such a dynamical mechanism can be an efficient process. The question to ask is whether such a mechanism can produce the number of observed short-period comets.

Estimates of the expected number of short period comets from long period comets by diffusion process has been made and it varies considerably. The dynamical calculations have shown that the ratio of the initial population of near parabolic orbits to the number of long period comets gravitationally bound to the system after N perihelion passages varies as $N^{1/2}$. The ratio of the observed long period comets to short period comets is ~ 5 and hence $N \sim 25$. This indicates that after around 25 passages or so, around 90% of the original comets are lost from the system. It is quite possible that some of the long period comets can be transformed into short period comets after an unusually long number of revolutions. It is estimated that around 2×10^3 revolutions are required for transforming a long period comet to short period comet with $P < 20$ years. The transformation can also take place only over a restricted region, $4 < q < 6$ au and $0 < i < 9°$, with an efficiency factor of one captured comet for about 10^2 long period comets. In addition, other physical processes such as sublimation, breaking of the nucleus and so on could also prevent the long-period comet turning into a short-period one. Therefore the existing dynamical calculations show that roughly one comet out of a total of 10^3 comets can end up as a short-period comet. Hence the major difficulty in the process of producing long-period comets to short-period comets through diffusion process in the solar system is that the resulting number of short-period comets are too small compared to the observed number. Also the long-period comets tend to preserve their inclination as they evolve into short-period comets, in contrast to those of observed short-period comets, whose orbits are confined to inclinations $30°$ or so. In addition, the original population of long period comets with random inclinations would lead to a certain fraction of short period comets with retrograde orbits, which is not seen. These are some of the major difficulties and therefore it is necessary to consider other mechanisms for the production of short period comets.

In order to produce low inclination, short period comets, the source has also to be confined to a flattened disk of comets. This has led to the hypothesis that the trans-Neptunian population of comets could be the source of short period comets. The possibility of the existence of such a source of comets had been suggested by Kuiper in 1951. He postulated a remnant of the accretion disk of planetesimals in the solar nebula, which never managed to accrete into planets. This is now termed as Kuiper belt. This idea has been elaborated later on by several investigators and showed that such a residual population of planetesimals could have remained bound to the solar system. More recent detailed simulation studies have shown that the short period comets can be produced from the low-inclination hypothetical comets starting from the Kuiper belt. The short period comets produced from such a comet belt are estimated to be about 300 times more efficient than produced through long period comets from the Oort cloud. There is the possibility that the Kuiper belt continues and merges with the inner Oort cloud, but the dynamical calculations seem to favour two separate comet populations. It will be interesting to look for the difference in the two populations of comets such as in composition and so on. So far no such difference has yet been seen.

The estimated number of comets in the Kuiper belt, to maintain the current population of short period comets, is around 10^8 to 10^{10} in the ecliptic plane beyond the orbit of Neptune between 30 and 50 au or so. The estimated total mass in the Kuiper belt can have values in the range of 0.0025 to 0.02 Earth masses for a range of average mass of the nucleus $(3.2 \times 10^{17} - 3.8 \times 10^{16}$ gms) and for population of 4×10^8 comets. Several limited observational searches have been carried out to look for slow moving objects in the outer solar system. Some of them are: (i) photographic survey of 6400 sq. deg. carried out in the R magnitudes upto $m_R \approx 19.5$, (ii) CCD survey of 4.9 sq. deg. in V magnitude upto $m_V \simeq 22.5$. These gave negative results. However a recent survey of 1.2 sq. deg. of the ecliptic plane imaged to apparent magnitude 25, has detected 7 trans-Neptunian objects, at heliocentric distance of $\sim$ 30 to 50 au. There exists some unpublished less deep surveys, which had detected 10 more objects. More recently about 30 objects have been detected based on Hubble Space Telescope observations.

12.3. Origin of the Oort Cloud

The reservoir of comets generally called the Oort cloud is quite well established. The observable long-period comets can be understood reasonably well, based on the time evolution of new comets coming out of the Oort cloud. Therefore the origin of long-period comets essentially reduces to the problem of the origin of the Oort cloud. Various hypotheses have been proposed to explain the origin of the Oort cloud. They can all be divided roughly into two groups-interstellar and the solar system. The other possibility of comets forming at the present location of the Oort cloud itself is not feasible due to the extremely low densities, which make it difficult for the growth of the particles.

If interstellar comets do exist and enter the solar system at times then they should be seen with hyperbolic orbits. However, no such comet has been observed for the last 200 years or so. This observation can be used to put a limit on the number of interstellar comets in the solar neighborhood to be $\leq 10^{-4}$ au^{-3}. This shows that the number of interstellar comets entering the solar system at the present time is not large. In addition, capturing hyperbolic comets from interstellar medium is a highly unlikely process as a third body is required to dissipate the excess kinetic energy involved. It is quite possible that passage of the solar system through a Giant molecular cloud or dense region might result in the capture of comets leading to bound orbits around the Sun. One can also think of other scenarios which could have given rise to the Oort cloud. Although these are possible in principle, however there appears to be no concrete viable models based on such mechanisms.

The other possibility that comets is of solar system in origin implying that they were formed at the same time and as a part of the formation of Sun and planets has received considerable interest and attention. On this hypothesis, cometary material should be similar to that of the solar system material. This is borne out by the measurement of the isotopic ratio of $^{12}C/^{13}C$ in comets ~ 100, comparable to the terrestrial value of 89.

However, the formation of comets in the inner solar system is highly unlikely due to the fact that the presently known chemical composition of comets requires a temperature at the time of formation to be quite low $100K$ to keep the volatiles like H_2O, CO_2, CO, NH_3 and CH_4 from evaporating. This led to the other possibility that the comets were formed in the outer parts of the nebula that formed the planets. The chemical composition of the solar system bodies can roughly be divided into three classes depending upon the characteristics of the elements present in them. For example, gases such as hydrogen, helium and noble gases stay as gas even at low temperatures, ice melts at moderate temperature and lastly the terrestrial materials like silicon, magnesium, and iron melt at higher temperatures. Table 12.3 shows the relative abundances of various elements in the solar

Table 12.3. Relative abundances of atoms by mass in the solar system.

Elemants		Sun	Terrestrial planets and meteorites	Jupiter	Saturn	Uranus, Neptune and comets
Gaseous	H He	1.0	Trace	0.9	0.7	Trace
Icy	C N O	0.015	Trace	0.1	0.3	0.85
Earthy	Mg Si Fe etc.	0.0025	1.0	Trace	Trace	0.15

Whipple, F. L. 1972. In *Motion, Evolution of Orbits and Origin of Comets* eds. Chebotarev, G. A. Kazimirchaka-Polonskaya, E. I. and Marsden, B. G. dordrecht: D. Reidel Publishing Company, p. 401.

system bodies. It is clear that Jupiter and Saturn were formed mostly of the original solar material like the Sun, while Uranus, Neptune and Comets were formed in the colder regions which account for the icy material. Therefore, probably comets were formed beyond Saturn.

Additional evidence for the low formation temperature of $T \leq 25K$, comes from the observation of S_2 in Comet IRAS-Araki-Alcock. Additional support comes from the study of ortho and para forms of molecular hydrogen. The energy difference between the lowest level of the two forms is around 24 cm^{-1} and hence the ratio of the two will depend on the temperature. The equilibrium value is 3.0 corresponding to a statistical equilibrium temperature of around 60K. The measured ratio of ortho to para of 2.2 to 2.3 from the high resolution 2.7 μ m band of H_2O in Comet Halley indicate that the temperature of formation of ice $\sim$ 25K. Thus there are several observational results which seem to show that comets are formed far from the Sun and in the cooler regions of the solar nebula, placing them beyond

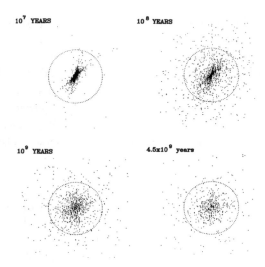

Fig. 12.4. The results of dynamical evolution of a cloud of comets ejected from Uranus-Neptune zone is shown projected onto a plane perpendicular to the galactic plane for several time intervals. The dotted circle with a radius of 2×10^4 au separates the inner and outer Oort cloud (Duncan, M., Quinn, T. and Tremaine, S. 1987, *AJ.*, **94**, 1330).

Uranus-Neptune zone, possibly the Kuiper belt.

The ejection of minor bodies appears to be a natural consequence of the accumulation process of giant planets. Due to gravitational perturbations of various planets, the orbit of some of these bodies can detach themselves from the bound system. Monte Carlo simulations have demonstrated clearly that such mechanisms can be present in the system. The results of such calculations have shown that the planets tend to eject icy planetesimals from their present zones to eccentric long period orbits with semi-major axis $\leq 10^3$ au. At these distances the perturbations due to stellar and Galactic sources tend to be important to push the orbits further into the region of the Oort cloud and thus detaching them away from the zone of influence of planetary perturbations. This can finally lead to a cloud with more comets at its center compared to its outer region, similar to the Oort cloud. A typical result of the Monte Carlo calculations for the assumed initial hypothetical comet distribution confined to the ecliptic plane is shown in Fig. 12.4. They show the expected distribution in the plane perpendicular to the plane of the Galaxy for different times. The comets are more or less randomized both in eccentricity and inclination in a time period of around

4.5×10^9 years and for distances beyond around 10^4 au. It is also found that relatively more icy planetesimals are ejected out of Uranus-Neptune region compared to Jupiter-Saturn region. Out of the total number of comets ejected and finally form a Oort cloud, roughly around 40% of those comets are lost from the system in a time scale of the solar system. They are lost due to various perturbation forces as discussed earlier. The dynamical calculations, thus indicate that the Oort cloud can be produced as a natural by-product of the various processes that could take place in the solar system during its lifetime.

Although the presence of the Oort cloud is well-founded, questions pertaining to its origin, physics, dynamics, structure and stability are still uncertain. One method to get some idea of the age of the cloud is through the study of the isotopic ratios of elements. The $^{12}C/^{13}C$ ratio available for comets indicates a ratio of about 100. This ratio is more like the solar system value (89) rather than the average value of about 30 to 50 of the interstellar material (Chap 5). This seems to show that comets were formed out of the solar nebula material about 5 billion years ago. In addition, the presence of low temperature volatiles in comets seems to imply that they are formed in cool regions of space which exist in the outskirts of the solar system.

Dynamical studies which take into account the various perturbations indicate that the Oort cloud can account for the observed long period comets. The perturbation of the Oort cloud due to the encounter with Giant molecular cloud and galactic tidal force is found to be important, in addition to planetary encounters and passing stars. The passage of a perturber close to the Oort cloud can give rise to shower of comets. It can have severe effect on the Earth's environment, as revealed by the interpretation of geophysical, biological and terrestrial records. The formation of short period comets from long period comets through diffusion process in the solar system has some difficulties, including accounting for the observed number of short period comets. This has led to the hypothesis that the source of short period comets may lie beyond the orbit of Neptune between 30 and 50 au, probably the remnant of the non-accumulated material. This region is generally known as Kuiper belt. As ejection of minor bodies appears to be a natural consequence of the accumulation process of Giant planets, this process could have been responsible for accumulation of comets during the lifetime of solar system in the region of space, where one now finds the Oort cloud.

As can be seen from the above discussion, there are several hypotheses and suggestions that have been advanced to explain the origin of comets. In a field such as this where the data is meagre, on a sample of only one solar system and with our present limited knowledge there is bound to be room for multiple models and different explanations for the same observed phenomena. Only more and better data combined with ingenuity can help in narrowing down the existing explanations.

Problems

1. How can one say from the appearance of a comet in the sky whether it is a new comet coming from the first time or whether it is one which has already been seen?
2. What is Jupiter's family of comets? What is the most probable origin of these comets?
3. Is there any evidence that the Oort cloud was formed about 5×10^9 years ago? Can you think of some other process by which the Oort cloud be of more recent origin, formed some 10^6 to 10^7 years ago?
4. Is there any way to test observationally the existence of the Oort cloud?

References

The basic idea of the cloud of comets is from the classic paper
1. Oort, J. 1950, *Bull. Astr. Inst.* Netherlands **11**, 91.

Earlier references to the cometary clouds can be found in
2. Opik, E. 1932, *Proc. Am. Acad. Arts and Sciences*, **67**, 169.

The best book on the origin of Comets is the following:
3. Bailey, M. E., Clube, S. V. M. and Napier, W. M. 1990, *The origin of Comets*. Pergaman Press.

A good account of several aspects are given in the following papers:
4. Weissman, P. R. 1991. In *Comets in the Post-Halley* Era. eds. R. L. Newburn Jr. et al. *Kluwer Academic Publishers.* P. 463.
5. Chakrabarti, S. K. 1992, M. N. **259**, 37.
6. Fernandez, J. A. and Gallardo, T. 1994, *Astr. Ap.*, **281**, 911.

Searches carried out looking for objects in the trans-Neptunian region can be found in
7. Jewitt, D. C. and Luu, J. X. 1995, *AJ.*, **109**, 1867.
8. Weissman, P. R. 1995, *Ann. Rev. Astron. Ap.*, **33**, 327.
9. Cochran, A. L., Levison, H. F., Stern, S. A. and Duncan, M. J. 1995, *Bull. Am. Soc.*, **27**, 1122.

CHAPTER 13

RELATION TO OTHER SOLAR SYSTEM STUDIES

Comets appear to have a close relationship with other objects of the solar system like asteroids and meteorites. The meteor streams which often appear in the sky are also believed to come out of the cometary material. The study of the nature and composition of dust particles collected at high altitudes through rockets and satellites indicates that they could be of extraterrestrial origin and are generally associated with comets. In recent years comets have also attracted great attention in view of their possible impact on the life of our planet. This has come about basically from the observation that the complex molecules of various kinds appear to be present in the nuclei of comets. This in turn is related to the chemical composition of the primordial material of the solar system. We would like to elucidate some of these aspects here.

13.1. Asteroids

The minor planets of the solar system were classified as objects having dimensions less than Pluto. The Ceres which was discovered in 1801 was classified as a minor planet. This was followed by the discoveries of other minor planets called Pallas in 1802, Juno in 1804 and Vesta in 1807 and so on. However, these objects turned out to be far fainter than expected as their orbits are closer than Jupiter's distance. It was realized that these are of different kind of objects. Hence, minor planets are also called asteroids. The largest concentration of asteroids is in the range of 2 to 3.5 au, placing them between the orbits of Mars and Jupiter. This is generally called the asteroid belt. There are two other minor concentrations of asteroids known as Hildas and Trojans which are located at distances of around 4 and 5 au respectively. It is estimated that around 100,000 asteroids have been seen at least once. The computed orbits are accurately known for around 5000 asteroids. The period for most of the asteroids is in the range of around 3 to 6 years. They revolve round the Sun in the same direction as the principal planets. Most of them have orbits which lie nearly in the plane of the Earth's orbit and the average inclination is about $10°$. The average eccentricity of the orbits is around 0.15 with sharp cut offs on either side.

Observing the light curve, it is possible to determine the rotation period and shape of asteroids. The rotation period of most of the asteroids is of the order of a few hours. For example the rotation period of Eros $\sim 5^h\ 17^m$ and that of Juno $\sim 7^h\ 13^m$. The deduced mean geometrical albedo of asteroids is rather low, ~ 0.05 which implies that they are extremely dark objects. The diameter of asteroids can vary from the lower set limit of about 100 meters to about 1000 km as seen in the case of Ceres. Most of the asteroids are found to be extremely irregular in shape. This indicates that they were most probably produced out of the break up of some parent body. In addition, the collisions of asteroids within the belt is quite frequent and this leads to erosion and fragmentation.

Since 1965, space missions such as, the Mariner Series, the Vikings and the Hubble space telescope have been used for the study of asteroids. The two Martian satellites, Phobos and Deimos, two likely captured asteroids have been studied in great detail. They are found to be approximately elliptical in shape with radii of $13.5 \times 10.8 \times 9.4$ km and $7.5 \times 6.2 \times 5.4$ km respectively. The mass of Phobos derived from the perturbations of the spacecraft is around 1.3×10^{19} gm. The estimated average density of Phobos and Deimos is around 2.0 gm/cm^3 and the geometric albedo ~ 0.06. High resolution observations have shown many surface features, including large craters. In 1991, high spatial resolution observations around 200m on the surface of asteroid number 951 was obtained when the Galileo Spacecraft enroute to Jupiter passed through this asteroid at the closest distance of around 1600 km. The radii of this asteroid is around $10 \times 6 \times 5.5$ km and has sharp edges indicating that fragmentation must have taken place. The derived albedo is ~ 0.2 and the surface seems to be young ~ 3 to 5×10^8 yrs. The infrared observations show that it is composed of metal-rich silicate. The ultraviolet observations carried out with satellites on a large number of asteroids show that the geometrical albedo decreases towards shorter wavelengths. For example, the geometric albedo for Ceres at wavelengths 4300 and 2600 A is about 0.031 and 0.023 respectively. The extensive data obtained from the IRAS was used to detect asteroids by looking for moving objects. Several new asteroids were discovered.

Infrared observations have given information about the surface composition of asteroids through the detection of characteristic bands of minerals. The relative abundances of silicates like olivine, pyroxene, feldspar and so on has been determined. Most of them fall broadly into two types with respect to their chemical composition: the C type, the darkest asteroids due to the presence of carbonaceous material and the S type which has silicates like olivine and pyroxene. The C type asteroids are more numerous comprising around 60% of the observed asteroids.

The origin of asteroids is still not known but they could be of cometary origin. The short-period comets after many revolutions could become inactive as most of the volatiles would have vaporized. Such residual nuclei of comets are *indistinguishable* from those of asteroids. There are two other suggestions with regard to the origin of asteroids and are found in the belt itself. The irregular shape of the asteroids indicates that they were produced due to the break up of some parent body at its present location. The other more likely possibility is that the material that exists in the present location, is the material that could not have accreted into a planet due to some reason.

There are a number of asteroids with large orbital eccentricities and they are found to come regularly to the Earth. Their perihelion lies interior to Earth's aphelion and are called 'Apollo objects', named after the first object of this kind to be discovered. There is another class of asteroids which includes those which have somewhat larger orbits with perihelion of around 1.02 to 1.3 au and are called 'Amor objects'. In general these two populations are not fundamentally different from each other and are often called with a single population as Earth-approaching Apollo-Amor objects. The mean dynamical life time of Earth-crossing objects is $\sim 10^7$ to 10^8 yrs, which is short compared to the age of the solar system. Therefore there should be a constant source of these objects. Early studies carried out based on mechanism of perturbations as then understood, cannot modify the original orbits from the asteroidal belt into Earth Crossing asteroids in eccentric orbits to explain the observed number in a steady state. It was therefore suggested that the devolatilized nuclei of short period comets could be responsible for such asteroids. Later on, detailed dynamical calculations which takes into account secular resonances that could cause strong eccentric variations in asteroidal orbits along with periodic perturbations of planets like Jupiter and Mars have shown that the asteroidal belt could still explain the observed population of the near-Earth asteroids. The extinct comets could still contribute a fraction to the Earth-Crossing asteroid population. However, there is no way to differentiate between these two contributions. In fact, the traditional distinction between asteroids and comets is becoming nebulous at the present time, as some of the asteroids with comet like orbits could be comets which have temporarily exhausted their volatiles. Comets P/Neujmin 1 and P/Arend-Rigaux are believed to be nearing the end of their lifetime. There are some asteroids which appear to be parent objects to some meteor streams, for example the orbit of (3200) Phaeton coincides with that of the Geminid Stream. The other conflicting discovery came from the detection of a large cloud of gas and dust i.e. out-gassing, from the asteroid 2060 Chiron in 1988 and more recently in 1992 from (5145) Pholus, both moving in orbits beyond Saturn. Therefore, there is evidence for the evolution of active comets to inactive, near-Earth asteroids. These studies seem to indicate that during the final stages, whether the volatiles are present or not depend upon the dynamical life time and the life time for the volatile component to disappear from the cometary nucleus. In general, there is no necessary connection between these two lifetimes. For most of the comets, the dynamical lifetime is much shorter than the physical lifetime for the evaporation of volatiles. Therefore complete loss of volatiles would not have occurred. Hence, depending on the relative values of the two lifetimes, it is possible to have different situations.

13.2. Meteorites

The meteorites are extra-terrestrial bodies which survive the passage through Earth's atmosphere and fall to the ground. This is generally accompanied by a flash of light in the Earth's atmosphere like that of a meteor. Only the largest objects of some several tens of meters in diameter create large impact craters when they hit the ground. Since early times, meteorites have been studied extensively as they were shown to be extra-terrestrial in origin. Also, these are the only extra-terrestrial objects that were available for laboratory investigation for a long time. More than 3000 meteorites are known at the present time. Relatively more meteorites are found in Antartica as they are preserved over long periods of time. It is possible that due to the fragmentation of the body as it passes through the Earth's atmosphere, it may be scattered over hundreds of kilometers on the Earth.

A detailed analysis of the chemical and mineralogical components of several meteorites indicate that they are mainly of two kinds, namely the Stony (resembling terrestrial rocks) and iron meteorites. There are gradations between these two classes. A small fraction of the observed meteorites (about 4%) are called the carbonaceous chondrites which have high carbon content. Microscopic examination shows that there are some spherical inclusions of one millimeter or so in size, called *chondrules* that can be recognized as inserts in the matrix material. Among the meteorites that are seen, more than 90% are of the stony meteorite type. They are very much like terrestrial rocks.

Iron meteorites consist mostly of iron and nickel, generally with a nickel content of 5 to 10%. The carbonaceous chondrites contain chondrules consisting of mostly olivine and pyroxene. The Murchinson meteorite which fell in Australia in 1969 has been very well studied. A variety of around 400 types of organics have been identified in chondrites. They include substances like aliphatic and aromatic hydrocarbons, amino acids, amines and amides, carbolic acids and so on. Several of these are common with biological systems. It is generally believed that they are extraterrestrial in origin. The observed chemical composition can be explained from the equilibrium calculations condensing out of a high temperature material $(T \sim 1000°K)$. They can account quite well for the observed mineral morphology except for the carbon compounds. The chemical compositions of carbonaceous chondrites are very similar to each other and are also similar to the composition of the solar atmosphere, except for the volatiles H, C, N, O and rare gases which are not so abundant. This suggests that chondrites are the materials that formed at the same time as the Sun from the solar nebula and since then, they have undergone very minimal chemical modification. The idea of primitive character of chondritic meteorites is also strengthened by their radiometric ages, which is roughly the age of the solar system.

An important result that came out in 1973 is that several carbonaceous meteorites like Allende meteorite showed an anomalous isotopic composition of several elements like oxygen, neon, xenon etc. At present more than 10 elements have shown anomalies in many different meteorites. These anomalies have been seen mostly in their inclusions. Here, the isotopic anomalies refer to the values in comparison to that of terrestrial values. It is hard to explain these observed isotopic anomalies in terms of the local production mechanisms such as irradiation, mass fractionation process or known radioactive decay. Hence, these anomalies represent materials injected into the solar system from nucleosynthetic sites within a relatively short-time of their synthesis. It is highly unlikely that all the anomalies seen in meteorites have a common origin and therefore must have different nucleosynthetic origin. It is quite possible that they have formed in a supernova envelope and hence they may represent the true pre-solar material. Hence, a plausible scenario is that the pre-solar inclusions formed in a distant environment and could have been incorporated into the solar nebular material with normal isotopic composition, out of which the planetary bodies of the solar system were formed. The volatiles of the pre solar inclusions could have been lost during the process of accretion into the bodies. In fact, it is possible to isolate these inclusions from the meteoritic material and from the derived isotopic compositions attempts are being made to study the sites of nucleosynthesis, the region of condensation of these grains and so on, which can help in our understanding of the early stages in the evolution of the solar system. The mineralogical composition of meteorites seems to indicate that some are primitive and some have gone through thermal evolution in the sense of going through a high temperature phase, which can alter their mineral composition. The Comets can, in principle, explain the origin of primitive meteorites, but not the processed mateorites which have gone through thermal evolution. Hence asteroids may most probably be a better source of observed meteorites. In particular, the Earth-Crossing Apollo-Amor group of asteroids may be the source of meteorites. Spectral studies have also been used to see whether there is any similarity between asteroids and meteorites. They do show a great deal of similarity between the two. These studies indicate that C type asteroids could possibly be the source of carbonaceous chondrites while S type asteroids may be the sources of some stony meteorites.

13.3. Meteor Streams

Another interesting aspect, presumably associated with the cometary dust particles, is the meteor showers seen in the sky. The small particles comprising the meteor streams revolve round the Sun, producing meteor showers whenever their orbit crosses the Earth orbit. They are actually made visible as they burn away in the Earth's atmosphere. Many of the

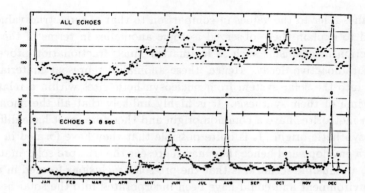

Fig. 13.1. A typical hourly rate of meteors seen throughout the year as observed by the Ottawa Meteor radar. The top figure represent total echo count and the bottom curve is for echos having duration $\geq$ 8 secs. The peaks represent showers corresponding to Q (Quadrantid), Y (Lyrid), E (η Aquarid), AZ (Arietid-Zeta-Perseid Complex), D (δ Aquarid), P (Perseid), O (Orionid), L (Leonid), G (Geminid) and U (Ursid) (McIntosh, B. A. 1991, *Comets in Post-Halley Era*, eds. R. L. Newburn, Jr., Kluwer Academic Publishers, p. 557).

meteor showers are seen regularly at specific times of the year. (Fig. 13.1). The most spectacular display in the sky ever recorded is that of the Leonid meteors which were seen in the month of November of 1833. In a span of a few hours, roughly about 20,000 meteors appear to have been seen. At the present time several hundred meteor streams exist in the solar system. What is of interest of course is the orbital characteristics of these meteors. From the observed direction and velocity, it is possible to get the orbit of the individual meteors in the Swarm.

It has been found that the elements of orbits of many of the meteor showers are very similar to those of the orbits of the known comets. In 1866, Schiaparelli showed that the Perseids shower had the same orbit as the Comet Swift-Tuttle. The showers Eta Aquarids and Orionids have the same orbit as Comet Halley and can be seen during the months of May and October. Of course, not all the meteor showers have been identified with the individual comets. Conversely, not all the known comets are associated with the corresponding meteor streams. Since the association between the two is well established only for a few cases out of a large number of known comets (Table 13.1) and meteor streams, at first sight these results might cast some doubt on the real association between the two. This could arise partly due to inaccuracies in the measured parameters of the orbit of the meteor

Table 13.1. Comets and their associated showers.

Comet	Associated shower	Maximum	Period of visibility
Halley (1910 II)	η Aquarids	May 4	May 2-7
	Orionids	Oct 21	Oct 17-24
Schwassmann-Wachmann 3 (1930 VI)	τ Herculids	June 9	
Pons-Winnecke (1951 VI)	June Bootids	June 23	
Encke (1977 XI)	Day time β Taurids	June 29	June 23-July 7
	Taurids	Nov 4	Oct 20-Nov 25
Switt-Tuttle (1862 III)	Perseids	Aug 12	July 29-Aug 18
Giacobini-Zinner (1946 V)	Draconids	Oct 10	Oct 10
Temple-Tuttle (1965 IV)	Leonids	Nov 16	Nov 14-19
Biela (1852 III)	Andromedids	Nov 20	Nov 15-Dec 6
Tuttle (1969 V)	Ursids	Dec 22	Dec 19-23
1861 I	Lyrids	Apr 21	Apr 20-22

stream. A more important reason could be that the dynamical evolution of the meteor stream could have modified the orbit such that it is different from that of the parent comet. Therefore a simple comparison of the two orbits may not be appropriate. On the other hand, the position of the perihelion of the comet and the meteor shower is a better indication of the real association as has been shown to be true for a large number of cases. Therefore, on a closer examination, the association between the comets and the meteor showers seems to be real. Let us first examine the formation of a Swarm of particles from a comet in a qualitative manner. The dust particles coming out of the nucleus are carried away by the gas. Since the flow of the dust from the nucleus is maximum near about the time of perihelion passage, it is also the place where most of the dust is introduced into the meteor stream. Each of these dust particles becomes independent and the orbit of these particles differs from that of a comet. The orbit of the dust particles depends upon its velocity and the effect of radiation pressure.

This varies with the size of the particles and their physical properties. Very small size particles are blown out of the system. The particles both lead and lag behind the comet and eventually lead to a continuous belt within a few revolutions of the period of the comet. The various particles in the meteor stream are subjected to various dispersive and degenerative effects, like gravitational perturbation by the planets, collisions and radiation pressure and so on. Therefore the orbit of these particles evolves into orbits which are difficult to predict. This may result in the orbit of particles in the meteor stream quite different from that of the orbit of the parent body. In fact the perturbations could disrupt the orbits of the particles such that it may make them sway away for crossing the Earth's orbit. This can possibly explain rather poor association between the comets and the meteor showers. But from the good correlation between the perihelion, it can be inferred that the meteor streams are of cometary origin.

Table 13.2. Computed and observed dates of Leonid Meteor Showers.

Computed (Based on Temple-Tuttle orbit)		Shower maximum time (observed)
902	Oct 12.7	Oct 13
1035	Oct 13.2	Oct 14
1202	Oct 18.7	Oct 18
1366	Oct 22.4	Oct 21, 22, 23
1538	Oct 25.0	Oct 26
1625	Nov 7.5	Nov 4, 5, 6
1799	Nov 12.5	Nov 11, 12
1866	Nov 13.7	Nov 14.1
1900	Nov 15.7	Nov 15-16
1932	Nov 16.5	Nov 16-17
1969	Nov 17.0	Nov 17.4
1997	Nov 17.4	

Yeomans, D. K. 1981 *Icarus* **47** 492.

A very interesting study pertains to Comet Temple-Tuttle and its associated Leonid meteor streams covering the meteor data over the period 902 to 1969. The comet and the meteor showers have roughly the same period of about 33 years. The comet itself does not appear to have been observed prior to 1366, although the Leonid meteors have been recorded even upto the year 902 A.D. From the presently known orbit of the Comet Temple-Tuttle, the orbit could be extrapolated backwards in time and the dates around which the meteor showers must have taken place can be calculated. A comparison of the calculated and the observed dates for the period between 902 to 1997 is shown in Table 13.2. The agreement between the two suggests that most of the particles of Leonid showers are from the Comet Temple-Tuttle.

13.4. Particles Collected at High Altitudes

Particle collection at high altitudes of the Earth's atmosphere is another way of studying the extraterrestrial dust particles and their possible origin. Various methods have been used for the collection of particles based on recoverable rockets, balloons and aircraft. These particles are generally called interplanetary dust particles (IDPs). The extensive collection of IDPs from the rarefied air in the stratosphere is carried out with NASA U2 aircraft, which flies at a height of about 20 km. The particles collected from these flights are subjected to a thorough laboratory investigation, not only to isolate the terrestrial contamination but also to study the morphological, structural and chemical properties of individual IDPs, in spite of their small size. Large number of the particles can be attributed either to rocket exhaust or to other known particle types. Many other particles cannot be explained as contamination from known causes and they could be extraterrestrial in origin. Roughly all these particles can be classified into three categories; 60% as chondritic, 30% as iron-sulfur-nickel and 10% as mafic silicate types. The chondritic particles are aggregates of small size grains of about 1000 Å. The size of the particles varies from 4 to 25μ. The grains are highly porous as well as compact. The elemental abundances of Fe, Mg, Si, C, S, Ca, Ni, Al, Cr, Mn and Ti in the first category of particles are in close agreement with the bulk compositions of chondritic meteorites. The abundances of C, S, Na and Mn indicate that the particles are volatile rich in composition and also have a black appearance because of the high content of carbon. The composition of iron-sulphur-nickel type of particles is similar to that of meteoritic triolite. The magic silicates are olivines and pyroxene. Evidence for the extraterrestrial origin of these dust particles comes from several sources. The direct proof for the extraterrestrial nature comes from the observations of the presence of solar noble gases, the deuterium enrichment (i.e. large D/H ratio) and the presence of solar flare tracks in mineral grains within the particles. Their extraterrestrial nature is also indicated by the presence of ^{10}Be, which is produced by cosmic ray bombardment.

The infrared transmission spectra between 2.5 and 25 μm of a large number of chondrites type of IDPs shows the presence of a strong 10μm silicate absorption feature and possibly the 3.4 μm feature. They can be classified into three groups as olivines, pyroxene and layer-lattice silicates. Particles in the olivine and pyroxene group are mostly crystalline in nature rather than amorphous. The carbonaceous material typically present is around 2-8% of the particle by mass. Carbonates are the important secondary material in layer-lattice-silicate IDPs, due to the presence of their characteristic feature at 6.8μm. The Raman spectra of several IDPs show double peaks at $\sim 1360 cm^{-1}$ and 1600 cm^{-1}, which are characteristic of aromatic molecular units smaller than 25A. The crystalline nature of 10 μm feature seen in Comet Halley and in other comets is in accord with the 10μm feature seen in IDPs. The overall properties pertaining to the physical and chemical nature of these particles strongly suggest that many of them are of cometary origin. Supporting evidence comes from the presence of meteor showers as well as the factual information that the vaporized material of the comet at each apparition is dispersed into the interplanetary medium, some of which may find its way on to the Earth.

13.5. Primordial Material

The material out of which the Sun and the planets were formed does not refer to the original material present at the time of formation of the universe about 15×10^9 years ago, but rather to the modified material which existed around 4.5×10^9 years ago. During all these times, the interstellar material was being constantly enriched with heavier elements through the element building in stars followed by ejection of this material into the interstellar medium. The same process has been happening since the formation of the solar system to the present time. It is not clear at present at what time in the history of the universe the Oort cloud was formed and hence its chemical composition is also not known. This will in turn be reflected on the original chemical composition of comets. For example, if the comets were formed along with other solar system bodies about 4.5×10^9 years ago, they would have the same composition as that of the solar system material. But on the other hand, if they were formed more recently they would have a different composition reflecting the contemporary interstellar abundances. One method to get information about the possible nature of the primordial cometary material and hence the time scale or the age is through the study of the isotopic ratios of various elements. As is well known, the relative abundances of various isotopes preserve the life history of the formation process and hence help in understanding the nature of the original material. Until recently, most of the measurements referred to the isotopic ratio $^{12}C/^{13}C$ in comets. The derived ratio of around 90 is the same as the solar system value. Other isotopic ratios were determined from *in situ* measurements of Comet Halley and their values are roughly in accordance with the solar value. More precise and accurate determination of these ratios are required for getting a better idea of the nature and time of formation of the Oort cloud and hence on the nature of the primordial cometary material.

13.6. Chemical Evolution

The origin of life on Earth has intrigued mankind since early times. In the standard scenario, the production of organics is the starting point of the whole complex process, as all life on Earth is composed of organic material. It is remarkable that relatively only a smaller number of organics appears to have been used in forming life system among a variety of organics that is possible. The compounds of major interest are those normally associated with water and organic chemistry in which carbon is bonded to itself and to other biogenic elements. The biogenic elements, H, C, N, O, S and P, are generally believed to be essential for all living systems. In 1953, Miller showed that when gaseous mixture of NH_3, CH_4 and H_2O is subjected to an electrical discharge, it produced various kinds of organic molecules including amino acids. This remarkable experiment showed for the first time the possibility of synthesis of organic molecules from a mixture of simple gases in the presence of an energy source. This led to the suggestion that a similar type of process could have taken place in the early stages of the earth leading to organics, the basic ingredient for the formation of life. However, studies have indicated that these simple molecules are unlikely to be present as major constituents in the early atmosphere as they are most probably photolyzed in a time scale of year or so. In addition, other studies indicate that the early Earth's atmosphere contained mostly CO_2, H_2O and N_2, which makes it difficult for the formation of organics. The existing observations show that the amount of organics in the solar system objects seems to increase with distance from the Sun. Therefore, it is intriguing that the organics necessary for chemical evolution is found in the outer solar system, whereas water, an essential ingredient for the life formation, is found in the inner solar system. This led to the suggestion that the organics might have been transferred from outer to inner regions of the solar system by some means, possibly through comets. Therefore the general conclusion is that it is highly unlikely that the biogenic could have formed on the Earth itself and therefore it had to be transported to the Earth through some means. The early chemical evolution was then believed to be followed by biological evolution, finally leading to life on the Earth. Even though this scenario is being taken as a working model, it is not at all known at the present time, how the whole process can take place in such a manner.

In recent years, the possible role of comets during the early chemical evolution on the Earth has been put forward. This is basically related to the fact that molecules of various complexities have been seen in comets (Chap. 4). In particular, observations of Comet Halley showed for the first time the presence of organics in comets through the detection of CHON particles. The composition of these grains showed that they are made up of highly complex organic molecules of various kinds. Comet Halley observations also indicated the detection of Phosphorus in the mineral core of the particles. Therefore, all the biogenic elements (H, C, N, O, P and S), essential for living system, have been detected in comets. There are several ways in which the organic material could have been incorporated into comets. It could have been incorporated directly at the time of formation from the material in the solar nebula which is a typical interstellar cloud. More than 80 molecules have been detected in interstellar clouds, most of which are organic in character. Therefore, organics in comets could be interstellar in origin. It could have also been formed in the nebula itself on surface layers of the cometary nuclei during its lifetime due to bombardment with energetic particles, UV radiation and so on. The energy source in the cometary case could be solar wind, solar UV radiation and cosmic rays.

There are several observations to show that the earth does seem to receive cometary material, such as meteor showers, dust particles collected at high altitudes, sedimentation at deep sea, cratering rates and so on. A rough estimate shows that $\sim 10^{23}$ to 10^{26} gm of cometary material should have entered the solar system in a time scale of about 4.5×10^9 yrs. There is also the suggestion that the Earth could have accreted a considerable amount of water during the early stages from comets. Therefore, the influx of cometary material could have injected large amounts of simple and complex organic molecules on the Earth. Particularly, the molecules accreted during the early times of planet formation could have triggered the chemical evolution cycle on the Earth. However, it is not clear whether these complex molecules could survive during the time of impact. This depends to some extent on the dynamics of the process.

At the present time there is evidence for the presence of some form of life on the Earth on or about 3.5×10^9 yrs ago. For times earlier than this, it is anyone's guess. Is it then possible that the organic molecules pouring from comets on to the Earth, particularly during the early times, might have started the whole cycle of chemical evolution, finally leading to the origin of life on the earth? This suggestion has been further extended to the possibility that some sort of basic primitive cells themselves could have come from the cometary material on to the Earth. In this connection it is interesting to see the extreme conditions under which microorganisms can exist. It is indeed remarkable that many microorganisms have been found to be able to survive under extreme environmental conditions, such as boiling water, freezing water, irradiation etc. They seem to adapt themselves to these environments and they are able to survive and grow. But one basic limitation is that they require liquid water. It has been suggested that ^{26}Al present in the early stages of solar system, which has a half life of only 0.7 million years, would have decayed within a few million years of formation of the solar system giving heat that could have melted the cometary material. Therefore it appears that comets *could have* played an important role in the process of chemical evolution, which finally led to life on Earth. However, many of the evidences brought forth to arrive at this conclusion have not yet been firmly established.

13.7. Overview

The nature of the original material of the solar nebula and its relation to the various solar system objects and their inter relationship is of great interest. Carbon chemistry or organics may be used as a probe for a qualitative discussion of this cosmic connection. Before coming to a discussion of this aspect, it may be worthwhile recapitulating some of the points regarding organics seen in various objects in the solar system and in interstellar space.

Evidences for the presence of organic molecules in interstellar clouds came from the observations made in the radio frequency region.

Table 13.3a. Interstellar molecules.

(a) Inorganic species (stable)

Diatomic		Tri-atomic	4-atom	5-atom
H_2[a]	HCl?	H_2O[a]	NH_3[a]	SiH_4[b]
CO[a]	PN	H_2S[a]		
CS[a]	NaCl[b]	SO_2[a]		
NO	AlCl[b]	OCS		
NS	KCl[b]			
SiO[a]	AlF[b]			
SiS[a]				

(b) Organic molecules (stable)

Alcohols		Aldehydes and ketones		Acids		Hydrocarbons	
CH_3OH	methanol	H_2CO	formaldehyde	HCN[a]	hydrocyanic	C_2H_2[b]	acetylene
EtOH	ethanol	CH_3CHO	acetaldehyde	HCOOH	formic	C_2H_4[b]	ethylene
		H_2CCO	ketene	HNCO	isocyanic	CH_4	methane
		$(CH_3)_2CO$?	acetone				

Amides		Esters and ethers		Organo-sulfur	
NH_2CHO	formamide	CH_3OCHO	methy formate	H_2CS	thioformaldehyde
NH_2CN	cyanamide	$(CH_3)_2O$	dimethy ether	HNCS	isothiocyanic acid
NH_2CH_3	melthylamine			CH_3SH	methyl mercaptan

Parafin derivatives		Acetylene derivatives		Other	
CH_3CN[a]	methyl cyanide	HC_3N[a]	cyanoacetylene	CH_2NH	methylenimine
EtCN	ethyl cyanide	CH_3C_2H	methylacetylene	CH_2CHCN[a]	vinyl cyanide

(c) Unstable molecules

Radicals		Ions	Rings	Carbon chains	Isomers
CH	C_3H[a,c]	CH^+	SiC_2[b]	C_3S[a]	HNC[a]
CN[a]	C_3N[a]	HCO^+	C_3H_2[a]	HC_5N[a]	CH_3NC
OH[a]	C_3N[a]	N_2H^+	C_3[a]	HC_7N[a]	
SO[a]	C_4H[a]	$HOCO^+$		HC_9N[a]	
HCO	C_5H	HCS^+		$HC_{11}N$[a]	
C_2[b]	C_6H	H_3O^+?		CH_3C_3N	
C_2H[a]	C_2S[a]	$HCNH^+$		CH_3C_4H	
C_3[b]	CH_2CN	H_2D^+?		CH_3C_5N?	

(Turner, B. E. 1989. *Space Sci. Rev.*, **51**, 235.)

[a] Seen in CSE as well as ISM sources.
[b] Seen only in CSEs.
[c] Both linear and cyclic forms.

Table 13.3b. Interstellar molecules (from 1989 to 1994).

Silicon carbide (1989)	SiC	Isocyanoacetylene	HCCNC
ptopynal	HCCCHO	Sulphur oxide ion	SO$^+$
—	SiC$_4$	Ethinylisocyanide	HNCCC
Phosphorous carbide	CP	Magnesium isocyanide	MgNC
Methylene	CH$_2$	Protonated HC$_3$N	HC$_3$NH$^+$
Propadienylidene	H$_2$CCC	—	NH$_2$
Butatrienylidene	H$_2$CCCC	Carbon monoxide ion	CO$^+$
Silicon nitride	SiN	Sodium cyanide	NaCN
Silylene	SiH$_2$?	—	CH$_2$D$^+$
—	HCCN	Nitrous oxide (1994)	N$_2$O
Carbon suboxide	C$_2$O	—	

(B. E. Turner, private communication 1995; The molecules are listed in order of their detection from 1989 to 1994).

A compilation of some of these molecules is given in Table 13.3. A close look at the Table reinforces the belief of the presence of numerous different organic compounds in interstellar clouds. The exact nature of the organic component of the dust is also not known. But it could be Polycyclic Aromatic Hydrocarbon (PAHs). A quantitative theory of the formation of these complex molecules has been developed.

The characteristic feature of the solar system is that the rocky planets (Mercury to Mars) form the inner region and icy planets form the outer regions (Jupiter to Neptune) and are separated by asteroids. The reducing atmosphere of the outer planets consisting of H, He, CH$_4$ and NH$_3$ is conducive to the formation of organic molecules. In fact several carbon compounds have been detected on Jupiter. Complex organic chemistry is operating on Titan, the Satellite of Saturn. The Earth has also organic material as the life on Earth goes back to about 3.5×10^9 years.

The new class of grains present in comets, called CHON particles, established that a large fraction of the comet is made up of very complex organics. They are shown in Table 13.4 . They provide evidence for the existence of many classes of cyclic and acyclic organic compounds. They include unsaturated hydrocarbons such as Pentyne, hexyne etc, nitrogen derivatives such as hydrogenic acid, aminoethylene etc., Heterocyclics with nitrogen such as purine and pyridine etc. and so on. HCN and H$_2$CO

Table 13.4. Types of organic molecules in Comet Halley dust.

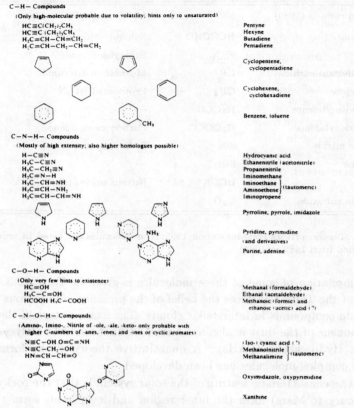

(Kissel, J. and Krueger, F. R. 1987. *Nature*, **326**, No. 6115, 755).

volatiles, allemine and pyrimidines and their derivatives which are biologically more significant has been seen in comets. The possible presence of Formaldehyde Polymers $(H_2CO)_n$ has also been suggested. It is quite possible that CHON particles contain many more molecules than identified so far due to the fact that the impact velocity of the dust at the dust mass spectrometer was very high ($\sim 70 km/sec$), which could possibly have destroyed many of them. The carbon content in cometary dust is close to 25% by weight. The character of the dust in comets i.e. silicate and carbon is similar to that of interstellar grains. Observations of comets reveal no evidence of thermal evolution that might have altered the material and hence must have kept the character of the pristine interstellar material more or less in tact.

The mineralogical composition of meteorites reflect differing degrees of thermal evolution indicated by Stony-iron and iron meteorites. Carbonaceous chondrites seems to be primitive in the sense that they have not altered much their state from the time of formation. The organic material found in carbonaceous chondrites is highly complex. There are two fractions present in them corresponding to soluble and insoluble portions. In Murchison meteorite, 74 different amino acids have been identified. Many of them occur on Earth biologically. A few of the different compounds seen are the following: Carboxylic acids, aliphatic and aromatic hydrocarbons, amines and amides, alcohols, ketones, purines, pyrimidines etc. The organic matter found in meteorites probably existed in the solar system a billion years before the appearance of life on the Earth. There are many similarities between the organic compounds in dust in Comet Halley and chondrites. The carbon content in carbonaceous chondrites is $\sim$ 3 to 5%, which is low by almost an order of magnitude compared to cometary dust. There is also an indication that they must have passed through a high temperature phase as well ($\sim$ 400K) in order to explain some of the compounds present in them. Therefore it is unlikely that comets are the source of meteorites as they remained at low temperature. Hence asteroids are most likely the source of meteorites. Observations show similarities between asteroids and meteorites. The isotopic anomaly seen in chondrules in meteorites is attributed to outside material survived and accreted and are therefore pre-solar grains.

Many of the IDPs which are thought to be of cometary origin is carbonaceous in character. The organics have not yet been analyzed in detail because of the small amount of material that is available. Besides terrestrial contamination may be present.

The various objects in the solar system are generally believed to have been formed as a by product of the formation of the Sun. The Sun was formed out of a typical interstellar cloud referred to as the primitive solar nebula. Model studies have shown that when contraction of the cloud takes place, the material will form a disk around a central core due to conservation of angular momentum. This material composed of gas and dust will accumulate in increasing numbers and form a thin disk of particles. This will then become unstable against gravitational forces and separate out from the system. These rings then clump together and form planetesimals and move in circular orbits around the centre of the nebula. These planetesimals act as nuclei for the formation of planets. This is also responsible for the nearly coplanar arrangement of planets and for a common direction of revolution of the planets around the Sun.

Condensation model calculations have shown that different types of material condense out of different temperature and densities, which imply different distances from the Sun. So there could be materials of high temperature phase, low temperature phase as well as a mixture of the two in the solar system objects. Carbon compounds can be vaporized upto around 2 to 3 au and water upto around 5 au. Therefore terrestrial planet zones will be rocky and the outer planet zones will be made up of water, methane, ammonia and CO_2-ice. This is in rough accordance with the gross properties of objects seen in the solar system. However, the detailed understanding of the whole problem is far from clear. The exact nature of compound produced depend upon various factors such as temperature, density, chemical composition of the constituents and various physical processes such as heating, cooling, annealing, selective evaporation, fractionation and so on. It is found from laboratory experiments that all the compounds seen in carbonaceous chondrites can be produced from CO, H_2 and NH_3 on a Fe_2O_3 or clay catalyst at a temperature of around 400 to 430K. Hence it is not surprising that there is a systematic change in composition going outwards from the Sun. Comets are produced in a cold temperature phase in the outskirt of the Solar System and so it has kept most of the molecules it had from the original material.

During the time interval of the formation of the Solar System, materials could have been injected into the solar system from outside (like from a supernova explosion). Some of this material may be preserved in the solar system objects as inclusions etc. In addition, comets colliding with Earth during the early stages of the formation could have deposited organic material, which might have started the chemical evolution on the Earth. Comets could also account for the biosphere on the Earth.

Therefore the organic material present in interstellar clouds and in solar system objects are the same material to start with but may look different in some of the solar system objects at the present time due to variable thermal evolution in these objects depending on their radial distance from the Sun (Cosmic Evolution).

So far we have been discussing the possible interrelationship between various solar system objects and comets. In certain cases the relationship to comets is more direct than in others. This is partly due to lack of good data as well as to the inadequacy in our present understanding of the nature of these objects. With further studies, it is hoped to understand the origin of the solar system itself and possibly even the origin of life on the Earth.

Problems

1. In the study of the origin of life, one deals mainly with the elements H, C, N and O. Explain.
2. Is there evidence of life beyond the Earth? If the answer is negative, discuss why?
3. Explain how the number of planetary systems in the Galaxy, with life, like on the Earth can be estimated
4. The meteor streams seen in the sky appear to diverge from a point in the sky. Explain.
5. What is the range of velocities at which meteors encounter the Earth? What is the explanation for the limiting values?
6. What has Titus-Bodes law to do with the question of asteroid belt?
7. Meteorites are said to act as beacons to the past. Explain.

References

The following reference gives a good account of asteroids and meteors
1. Milani, A., Di Martino, M. and Cellino, A. (eds.) 1994 *Asteroids, Comets, Meteors*. Dordrecht: Kluwer Academic Publishers.

The following reference gives a good description of meteorites:
2. Kerridge, J. F. and Matthews, M. S. (eds.) 1988. *Meteorites and the Early Solar System*. Tucson: University of Arizona Press.

The technique of particle collection at high altitudes and the results of such studies are given in the following review articles.
3. Brownlee, D. E. 1978 *In Cosmic Dust*. ed. McDonnell, J. A. M. New York: John Wiley and Sons. p. 295.
4. Sandford. S. A. 1987. *Fund. Cosmic Phys.* **12**, 1.

The early work on the various processes leading to element building in stars is discussed in the classic paper:
5. Burbidge, E. M., Burbidge, G. R., Fowler, W. A. and Hoyle, F. 1957, *Rev. Mod. Phys.* **29**, 547.

For recent discussion reference may be made to
6. Prantzos, N., Vangioni-Flam, E. and Casse, M. (eds.) 1993. *Origin and Evolution of the Elements*, Cambridge: Cambridge University Press.

The Oparin-Haldane hypothesis of the Origin of Life can be found in
7. Ponnamperuma, C. 1982 *Extraterrestrials: Where are they?* eds. Hart. M. H. and Zuckerman, B. New York: Pergamon Press, P. 87.

The other view point of cometary origin can be found in the following reference.
8. Hoyle, F. and Wickramasinghe, N. C. 1981. *Comets and the Origin of Life* ed. Ponnamperuma, C. Dordrecht: D. Reidel Publishing Company, p. 227.

The following book may be referred for solar system studies
9. Lewis, J. S. 1995, Physics and Chemistry of the Solar System, Academic Press.

CHAPTER 14

PROBLEMS AND PROSPECTS

14.1. Epilogue

So far we have discussed in some detail our present understanding with regard to various cometary phenomena which is summarized in Fig. 14.1. Although we have a reasonable working knowledge of the origin, nature and composition of comets, many of the aspects are still not well understood.

Over the years, our knowledge with regard to comets has improved enormously from the studies based on several bright Comets like Ikeya-Seki, Kohoutek, West, Bradfield and others. The Comet IRAS-Araki-Alcock, which came to within about 4-7 million kilometers from the Earth in May 1983, gave the opportunity to observe a comet at such a close approach. Comet Kohoutek gave a big boost to cometary science in 1974 through a concentrated and co-operative effort made by the scientists working in various fields. This venture actually created an interest among the physicists, chemists and biologists as well, in problems connected with cometary science. It also created an increased interest among the astronomers who normally work in other fields. At present we have beautiful techniques for both the ground based and space observations in different spectral regions, remarkable new instrumentation, (ground based equipment and computers), which are much better than what was available about 30 to 40 years ago. With these advances, phenomenal progress in the understanding of comets has been achieved.

However, there are still questions which cannot be answered, either from ground based or satellite observations. The only way to understand these problems is through probes or missions to comets which can pass close

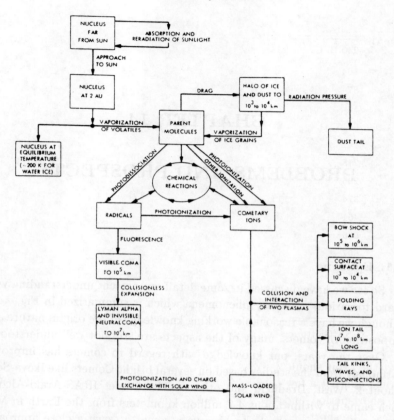

Fig. 14.1 The various processes arising out of the interaction of solar radiation and solar wind with a comet, which result in the observed features in a comet is shown in a block diagram (Report of the Science Working Group, The International Halley Watch, July 1980.)

to the nucleus. This was achieved with remarkable success for the first time during the apparition of Comet Halley in 1986. Comet Halley was chosen as a target for space missions as it satisfied most of the necessary requirements. First, the comet's orbit should be predicted accurately. This means that the arrival time of the comet can be predicted well in advance. This is very essential because of large time periods that are involved for developing experimental packages and the necessary preparations for making space flights. The comet should be bright, as well as exhibit as far as possible all the observed phenomena for the maximum scientific return. The condition

Table 14.1. Perihelion passages of Comet Halley.

240 B. C.	May 25 (probably observed)	912	Jul 19
164	Oct 13 (not observed)	989	Sep 6
87	Aug 6	1066	Mar 21
12	Oct 11	1145	Apr 19
66 A. D.	Jan 26	1222	Sep 29
141	Mar 22	1301	Oct 26
218	May 18	1378	Nov 11
295	Apr 20	1456	Jun 10
374	Feb 16	1531	Aug 26
451	Jun 28	1607	Oct 27
530	Sep 27	1682	Sep 15
607	Mar 15	1759	Mar 13
684	Oct 3	1835	Nov 16
760	May 21	1910	apr 20
837	Feb 28	1986	Feb 10

ESA Giotto mission pamphlet, 1981 (upto 1910).

that the orbit should be well known mean that the comet should have returned at least a few times and this eliminates new comets. It leaves only the short-period and the intermediate-period comets as potential candidates. The short-period comets on the other hand tend to decrease in gas production. Therefore they are also not suitable. Comet Halley which has a period of $\sim$ 76 years has a well-known orbit, as well as easily measurable production of gas and dust. Thus, Comet Halley satisfied most of the requirements and hence it was chosen as the first target, in addition to being the most famous of the known comets, which has been observed for the last 2000 years (Table 14.1). However, one disadvantage with Comet Halley has been that it being in a retrograde orbit, the encounter speed between the spacecraft and the comet was quite high, about 68 km/sec.

The *in situ* measurements of various kinds carried out in the inner coma by six spacecrafts, which passed through Comet Halley during March 1986 at distances ranging from 600 km to 10^6 km from the nucleus, when the comet was about its closest distance to the sun, was a great achievement and met with remarkable success. These *in situ* measurements, for the first time, gave a large number of unexpected results as well as showed the complexity of the physical processes occurring in the coma. These observations, combined with ground based and satellite observations, covered the entire range of the electromagnetic spectrum from the far ultraviolet to radio wavelengths providing the complete set of data available on any comet. These results have increased our knowledge about cometary science in a dramatic way. Some of the previously existing theories and hypothesis have been confirmed by these *in situ* measurements. The fly-by of ICE spacecraft through Comet Giacobini-Zinner and the Giotto spacecraft passing through P/Grigg-Skjerllerup which are short period comets compared to Comet Halley, have also given some data on these two comets, although not to the same extent as obtained for Comet Halley.

14.2. Future Studies

Although space missions to Comet Halley provided a remarkable insight into our understanding of the cometary phenomena, still there are several questions, which remain to be answered with regard to the origin and evolution of comets, structure and composition of the nucleus, parent molecules, the interaction of solar wind with Cometary Plasma, relation to chemical and biological evolution and so on. In addition, the space missions to Comets Halley and Giacobini-Zinner provided only snap shots of the physical and chemical conditions in these comets, as the observations could be carried out only for a very limited time period of a few hundred to a few thousand seconds or so. Hence one should have a cautious approach in extending the Halley data to other comets and in deriving general conclusions. Much more data on other comets, similar to that achieved for Comet Halley, is needed before arriving at general conclusions regarding the similarities and differences among comets of different types. It will also be of interest to probe the nucleus of a long period comet, coming directly from the Oort cloud, in order to examine the possible presence of the aging effect among comets. Also, since the encounter speed of Giotto spacecraft with Comet Halley was very high ~ 68 km/sec, it could have destroyed many of the heavy molecules as well as fragile dust grains. Hence, for these reasons more cometary missions of increasing complexities are needed.

The next logical step would be to achieve long duration close encounter of a short period comet to study the time evolution of the development of the coma as a function of the heliocentric distance. This can provide information about the nucleus, the elemental and molecular composition of the gas and the dust, solar wind interaction and so on. These studies will go a long way in comparison to those fly-bys to Comet Halley. This can be achieved by putting a satellite with instruments around a comet and tracking it as it moves with the comet. This should be followed by a probe to the nucleus with a penetrator for physical and chemical studies. The data could be radioed back to Earth. Of course, the best and a direct way to study the cometary material is to bring back some of this material to the Earth and subject it to various laboratory investigations. This is highly desirable, even though future instruments on-board the spacecrafts will undoubtedly become highly sophisticated. But, certainly, they cannot match the level of sophistication in instrumentation and analysis that can be carried out in the laboratory. This will provide the first opportunity to study the primordial material of our solar system formed around 4.8 billion years ago. Some of these ideas are being pursued and are in the planning stage of development. The space missions to comets, which have to be coupled with studies carried out with ground based and above the atmosphere observations, are also advancing at a rapid pace.

A new generation of powerful ground based optical/IR/Radio telescopes of large collecting area will be available in the near future. Combining this feature with advanced instrumentation and state of the art in optical and IR detectors will provide excellent opportunities for making very high resolution imaging, spectroscopy, polarimetry and so on. The extraordinary sensitivity combined with high spatial resolution should give exciting results. It will also be possible to extend the observations to faint objects. These will be supplemented with high quality observations carried out with rockets, satellites and KAO. The results of HST has already shown the vast potential of this telescope for the study of comets. The infrared space observatory (ISO) with a 60 cm telescope launched recently will provide wonderful opportunity for making imaging and spectral observations covering the spectral region from 3 to 200 μm with 10 to 15 times better sensitivity than IRAS. With these instruments, it should be possible to extend the observations to faint objects and also look for comet-like objects in the Kuiper belt. There is, thus, a bright future for cometary observations, with both the ground based and above the atmosphere instruments.

The laboratory simulation studies of the processes that could occur in a cometary environment is another important input necessary for an understanding the behaviour of comets. Of course, it is very difficult to create the actual cometary environment in the laboratory. In addition, it is very difficult to simulate a cometary target as we do not know the exact compositions and the physical state of a comet. However, efforts have been made to approximate the conditions of real comets depending on the problems of interest. With such set-ups, extensive studies have been carried out in the laboratory with regard to physical, thermodynamic and chemical properties of different types of ice systems, refractory organic and silicate compounds at very low temperatures. They have revealed the complex behaviour of the ice sublimation. Experiments have also been carried out on the physical and chemical changes induced by energetic ion and electron irradiation of material relevant to comets. These laboratory studies, though difficult, have already supplied some of the basic data needed for the development of more realistic comet models.

Future studies, based on combined effort of space missions, ground based and above the atmosphere observations and laboratory investigations should give new insights into the structure, evolution and origin of comets. Hopefully, through a series of such studies, it will be possible to resolve may of the basic and fundamental problems which have no answers at the present time. These studies should also give enormous information with regard to the early history of the solar system; physico-chemical, dynamical and thermodynamic conditions existing at that time; the information and relation to other solar system objects; the relation to interstellar molecules, chemical evolution, the origin of life and Exobiology, space plasma physics and so on.

14.3. Postscript

In contrast to the general experience so far that the appearance of a bright comet is a rare phenomenon, two exceptionally bright comets have been sighted recently. These have provided us with opportunities for verifying the physics and chemistry of comets discussed in the earlier chapters.

Comet Hyakutake (C/1996 B2) was discovered by the Japanese Amateur Astronomer Yuji Hyakutake on 30 January 1996. Within a few days of observations it was clear that this comet was going to be a bright comet visible to the naked eye around March and April 1996. This is the brightest comet discovered after Comet West in 1976. Comet Hyakutake appeared brightest at two positions in its orbit: once during its close encounter with the Earth (just $15 \times 10^6 km$ ($0.1 au$) away) in the months of March and April 1996 and again at its closest approach to the Sun in May 1996. The comet was visible to the naked eye in the northern hemisphere during March-April 1996, while it became a southern hemisphere object after its perihelion passage. It was an unusual comet as it passed through the perihelion passage (brightest) within a few months of its discovery and thus a rare event. The fortuitous circumstances of several spacecrafts and the Hubble Space Telescope in orbit covering a large part of the electromagnetic spectrum provided a unique opportunity for observing this comet very close to the Earth. Though, the detailed analysis of observations will take several months, the early startling results, something totally unexpected have emerged.

The first ever detection of X-rays from a comet was made with the ROSAT satellite on March 27, 1996. The strong X-ray intensity, primarily of energies $< 2 keV$ as well as its variation over a few hours was another surprise. The image made available through the internet shows that the X-rays appear to come from a crescent-shaped region of material on the sunward side of the comet located around 30,000 km from the nucleus. X-rays were also detected from the comet after its perihelion passage from June 22, 1996 and the image showed an extended source with a radial extent of at least 400,000 km. This shows the continuous behaviour of X-ray emission in the comet. X-ray emission from several comets (C/1990 K1(Levy); C/1990 N1 (Tsuchiya–Kiuchi); 45P/Honda-Mrkos-Pajdusakova) has also been seen in data obtained with the positive sensitive proportional counter of ROSAT during the all-sky survey. Therefore X-ray emission appears to be the general property of comets.

The possible physical processes for the production of X-rays could be 1) fluorescence process (discussed in Chap. 5) from the material in the coma, 2) from the interaction of the solar wind with the cometary material (Chap. 10) or 3) due to thermal bremsstrahlung. The last process is unlikely and the other two processes can in principle give rise to the observed crescent shape of the X-ray emission.

Simultaneously with the X-ray observations, ROSAT Wide Field camera (WFC) also detected for the first time the Extreme Ultraviolet (EUV) emission (band pass 140 to 60Å). The bright and diffuse crescent structure emission towards the direction of the Sun was seen to be roughly coincident with the X-ray emission. The EUV emission around the nucleus extended upto distances of $\sim 2 \times 10^5 km$ as was seen in the images provided in the internet. The WFC images also showed an arc of fainter diffuse emission extending to about 15'-40' from the center of the nucleus. The mechanism of origin of EUV emission is of great interest.

The spectroscopic observations carried out with the Hubble Space Telescope in the ultraviolet region (1297 to 3277Å) on April 1, 1996 showed clear detection of $(B-X)$ bands of S_2 molecule, a molecule of great interest as discussed in sections 4.1 and 6.6, in the wavelength region between 2850 and 3120Å. In addition, around 25 bands belonging to (A-X) system (Fourth positive system) and seven or more forbidden bands of the Cameron system of the CO molecule have been detected. This is much more than what was seen so far (Sec. 4.1). These bands could arise from the CO molecules in highly excited states created at the time of photodissociation of CO_2, which could in turn be used as tracer of CO_2. Such transitions generally called *prompt emission* have already been seen in the case of the OH molecule. Besides confirming several tentative or marginal detection of molecules in earlier comets which were discussed in the book earlier (Sec. 4.1), several newer transitions of molecules as well as new molecules have also been seen, in addition to many unidentified emissions being present in the spectra.

Another major surprise was the detection of ethane (C_2H_6) from the high resolution infrared observations in the 3 micron region (Mumma et al. 1996). Its abundance of about 1% of the methane (CH_4) shows the unusual chemistry of the cometary material as this molecule has not been detected in the interstellar material. The observations indicate the presence of chemical diversity of the material in comets introduced after the formation of the solar system from the primordial cloud. As the formation of C_2H_6 by gas-phase ion molecule reactions in the original cloud is difficult, processes such as formation on ice grain mantles through photolysis or chemical reactions due to solar UV radiation have been proposed. Therefore the existing idea of comets containing unaltered interstellar material might have to be replaced by a processed interstellar material model.

The photographs taken on 23 and 24 March 1996 show a solid bare nucleus and another with a tail clearly separated from each other. It appears to indicate the breakup of the nucleus of Comet Hyakutake as was seen in Comet West (Sec. 1.8).

Even before the dramatic visit of Comet Hyakutake, astronomers were eagerly looking forward to another bright comet, Comet Hale-Bopp (C/1995 O1). Even at the time of discovery in July 23, 1995, this comet was unusually bright as well as appeared to be remarkable in having an enormous coma $\sim 3 \times 10^6 km$ across, even at about $7au$ from the Sun.

The perihelion passage $(r \simeq 0.9au)$ should occur around April 1997. Based on early observations, it is optimistically predicted to be a bright spectacular naked-eye comet. But as discussed earlier in Sec. 1.7, the prediction of the brightness of a comet depends on several factors and therefore it should be taken with caution. The comet is observed continuously from the time of its discovery.

From the 1.3 mm observations, the CO production rate is deduced to be at $r \simeq 6.5au$ about 1300 kg/sec, using the procedure discussed in Sec. 6.2. This is unusually large compared to Comet Halley which produced roughly this amount when it was at $r \simeq 0.9au$ (Sec. 6.2; Jewitt et al 1996). The excessive brightness observed at $r \sim 6.5au$ does not appear to be due to the larger size of the nucleus but rather to the outgassing of highly volatile molecules like CO, as H_2O outgassing should be very weak at these distances. (Sec. 11.1; Sekanina 1996). However, water sublimation should take over from the CO-driven activity for $r \lesssim 3au$.

The optical images at $r \sim 6.5au$ show prominent features such as jets and spiral structure on three occasions (Sec. 7.3). The observed morphology of these features can be accounted for by a single active vent coming out of a rotating nucleus. The observed dust mass loss rate is found to be about 15 times larger than the CO mass rate (Sekanina 1996).

The burst of activity seen from Comet Hale-Bopp appears to be similar to the activities as seen in Comet Schwassmann-Wachmann 1 when far from the Sun (Sec. 11.2). Therefore, a systematic study of gas and dust release and morphology of Comet Hale-Bopp may give clues for a better understanding of the physical mechanism (i.e., the heat source) for the release of volatiles at larger distances and outbursts, complex state of rotation of the nucleus (Sec. 11.4) and so on. X-ray, EUV and other radiations will also be looked for in Comet Hale-Bopp. The confirmation or otherwise of these results with that of Comet Hyakutake is of great interest.

It is fortunate that astronomers have been given a unique opportunity for making observations on two bright comets of different characteristics coming so close to each other: Comet Hyakutake with $q \approx 0.23au$, time period $\sim 15,000$ yrs and Comet Hale-Bopp, a highly eccentric orbit with $q \approx 0.9au$, time period ~ 4000yrs. They have already posed challenging problems. With the advent of sensitive instruments, both ground based and space based, the apparitions of bright comets will bring in several data which will revolutionize our understanding of their origin and evolution. The necessary physics for such an understanding is dealt with in the earlier chapters.

Problems

1. What is the relation of comets to the origin of the Earth?

2. Can you suggest some way by which, short-period comets could be used as probes to explore other objects of the solar system?

References

1. IAU circular Nos: 6345 (March 21, 1996), 6353, 6364, 6366, 6373 (April 4, 1996; Discovery of X-rays), 6374, 6376, 6377, 6378, 6382, 6386, 6388, 6394 (May 9, 1996; Discovery of EUU radiation), 6404, 6408, 6413, NASA-Release: 96-66 (April 4, 1996; X-rays), 6433 (July 12, 1996).
2. Jewitt, D., Senay, M. and Matthews, H. 1996,Science **271**, 1110
3. Mumma, M. J., Disanti, M. A., Russo, N. D., Fomenkova, M., Magee-Sauer, K., Kaminski, C. D. and Xie, D. X. 1996, Science **272**, 1310 (31 May).
4. Sekanina, Z 1996, Astr. Ap. (Submitted) (Cometary Science Team Preprint No 160, Dec 95).

INDEX

aberration angle 272
abundances
    heavy elements 113, 151, 157, 311, 336
    isotopic 115, 118, 119
albedo 210, 262
anti-tail 20, 200, 251
asteroids 340
Atlas of Cometary Forms 288, 296
Atlas of Representative Cometary Spectra 64, 81
atomic spectroscopy 45
auroral lines 78, 106, 153

band sequence 49, 51
Bayeux tapestry 2
bend in the tail 291
biological extinction 332
black body radiation 42
black body temperature 42
blocking coefficients 86
Boltzmann distribution 57
bow shock 276, 279, 280
brightness profile 128, 233
brightness variations, heliocentric 11
Bruggeman rule 225

C_2 molecule 66, 95
    energy level diagram 97
    rotational structure 102, 112

		synthetic profile 102
		vibrational structure 102
chemical evolution 351
$C - H$ stretch 71, 249, 258
CHON particles 25, 77, 156, 312
clathrate hydrates 300
collisional excitation 93, 104, 150
column densities 171
cooling rate 180
co-ordinate systems 28
coma
	collisional effects 77, 93
	density variation 129
	expansion velocities 92, 177, 183
	parameters 23
	ultraviolet 19, 69, 157
	visible 64

Comets
	age 80, 314
	appearance 7
	beliefs 2
	brightness variations 11-13
	composition 178, 310, 356
	discovery 6
	extreme UV 367
	future studies 364
	Hale-Bopp 368
	Halley 1, 3, 4, 25, 76, 173, 362
	Hyakutake 366
	importance 10
	Jupiter's family 9-10
	mass 24
	new 8, 327
	nomenclature 6, 7
	observed characteristics 15, 288
	old 8
	orbit 8, 28, 363
	orbital statistics 8

origin 26, 324, 368
Shoemaker-Levy 9, 16
shower 331
spectra 64
splitting 15-17, 368
transient nature 7
X-rays 367
contact discontinuity 275-279
continuum 228, 263
comet-tail band 65
complex molecules 72, 149
Condon parabola 53
cretaceous-tertiary boundary 332

deuterium abundance 120, 349
dielectric constant 223
diatomic molecules 48-51
disconnection events 293
dissociative equilibrium 44
discrete dipole approximation 220
Doppler shift 45
dust particles, Cometary 228
dust feature 202
    chemical composition 356
    dust-to-gas ratio 79, 253
    multicomponent models 248
    production rates 236, 239, 250
    sizes 234, 237, 248
    spectral features 71, 72, 246, 254
    temperature 244, 247
    terminal speed 197
dust tail 20, 24, 193
    mechanism of formation 193
dynamical aberration 271
dynamics 193

efficiency factors for scattering, absorption 207, 215
    radiation pressure 194, 209
Einstein coefficients 54, 58

elemental abundances 113, 151, 157, 311, 336
electronic transitions 48, 64
electronic transition moment 55, 100
energy density 43
equation of state 44
equilibrium constant 44
evolution of orbits 333
excitation temperature 109, 114
extraterrestrial particles 257, 349
extreme UV 367

Fabry-Perot 78, 152, 165
flares 17, 303
Finson-Probstein theory 193
fluorescence efficiency factors 107, 124-126
fluorescence process 82
flyby missions 4, 173, 261, 280, 312, 363
fountain model 158, 233
forbidden lines 78
    oxygen lines 78, 140
forbidden transitions 78
Franck-Condon factors 52
Fraunhofer lines 83

g-factors 103, 105, 107, 124
gas-to-dust ratio 234, 253
gas-phase chemistry 167, 185
grain sizes 234, 237, 248
grain temperature 244, 247
Greenstein effect 92

Hα observations 140, 155, 165
Hale-Bopp 368
Halley's Comet 1, 3, 4, 25, 76, 173, 362
halos 17, 205
Haser's model 131
heating efficiency 180
heteronuclear molecule 48
Höln-London factors 55
homonuclear molecule 48

Hyakutake 366
hydrates 300
hydrodynamic flow 183
hydrogen cloud 20, 69
hydrogen lines 46

ice bands 260
icy-conglomerate model 23, 313
infrared lines 71
infrared band passes 243
infrared observations 242
in situ measurements 25, 173, 261, 280, 312, 363
interplanetary magnetic fields 282
    magnetic sector boundaries 61, 288, 293
interplanetary dust particles 257, 349, 357
interstellar molecules 354
ion-molecule reactions 167, 175
ions 146, 170, 173, 280
ion tail 20, 270
    acceleration 286
    interaction 275
    turbulence 285
    waves 283, 290
isotope ratio 115, 118, 119
isotope shift 52, 115
IUE Satellite 69

Kramers-Kroning relation 224
Kuiper belt 26, 334

lambda splitting 57, 104
lifetime of the molecule 127
long period comets 8, 26, 324
Lyman α of hydrogen 20, 157
    isophotes 160, 163, 165

magnitude, total 11
mass loss 69, 314, 319
mass spectrometer 76, 173
Maxwell-Garnett rule 225

meteorites 343
meteor showers 345
Mie theory 207
minor constituents 145, 190
molecular bands 64
molecular hydrogen 145
molecular polarization 107
molecular spectroscopy 48
Monte-Carlo approach 134, 163, 183, 328, 337
Morse potential 56

new comets 8, 327
non-gravitational force 316
nucleus 297
    albedo 304, 306
    composition 178, 310
    density 309
    expanding halos 17, 307
    icy model 23
    mass 24
    mass loss 314
    material strength 17, 315
    rotation 307
    size 304
    splitting 15-17, 368
    structure 314
    temperature 336
    undifferentiated 316

observations 64
    infrared 71, 368
    radio 72, 368
    ultraviolet 67, 367
    visible 65
observed species 23, 76
OH molecule 83, 136
    Λ-doubling 57, 104
    radio lines 58, 102
optical constants 223, 241, 248

orbital elements 28
    velocity 32
    calculation 37
Oort cloud 26, 324
origin 26, 324, 368
outbursts 17, 303
oxygen lines
    allowed 78, 106
    forbidden 78, 106, 153

Parent-daughter hypothesis 24, 130
parent molecules 24, 184, 189, 313
particle collection 349
phase function 217, 235
photochemistry 184
photodissociation rate 126
photoionization rate 126
photometric nucleus 66
photometric theory 14, 134
Planck's law 42
plasma tail 20, 25, 270
    structures 15, 288
polarization
    linear 212, 238
    circular 212
polyoxymethylene 175
primordial material 350
production rate 123
    calculations 128
    heliocentric variation 136, 143
    molecules 136
prompt emission 368

radial velocity 32, 45
radial velocity effect 83, 103, 138
radiation pressure 264
    to solar gravity 265, 271
radicals 76
radio transitions 58, 102

Rayleigh 161
r-Centroid 55
reddening 232
reflected solar radiation 11, 24, 65, 228
relative abundance 118, 144, 157, 178, 313
RKR potential 56
rocket effect 318
rotational structure 85
rotational temperature 109
rotational transition 73

sand-bank model 24, 306
scale length 130, 139, 142, 185
scattered intensity 211, 215
scattering functions 211, 217
scattering theories 207, 218
sector boundary 61, 288, 293
shock discontinuity 276
Shoemaker-Levy 9 16-17
short period comets 8, 324, 333
    origin 26
silicate features 72, 254
solar radiation 58, 83
    emission lines 59
    solar constant 59
    variations 60
    pressure 193
solar wind 61, 270
solar-wind interaction 270
    Parker's model 270
    physical conditions 276
    geomagnetic effect 292
space missions 4
species observed 64-77
    a compilation 76
    inferences 77
spectra, description 64
spectroscopy 45
    atomic 45

molecular 48
statistical equilibrium 86, 98
Stefan-Boltzmann law 43
strength of a line 54
surface brightness 128, 185, 233
sun-grazing comets 17, 65, 315
sunward spike 20, 200, 251
Swan band sequence 66, 98
Swan-like feature 22, 291
Swings effect 83, 103, 138
synchrone 195
syndyname (syndyne) 161, 195
synthetic spectra 89, 90, 102

tails, dust 20, 24, 76, 193
    composition 234, 235, 241, 254
    size 237, 241
    production 253
tails, ion 270
    filamentary structure 23, 288
    oscillations 23, 283, 290
    rays 23, 288
    streamers 233, 288
    structures 15, 288
temperature in the coma 177
tidal force 17, 315
time evolution 93

ultraviolet coma 19, 69, 157
ultraviolet observations 69
unsolved problems 364

vaporization theory 297
velocities in the coma 177
vibrational structure 88
vibrational temperature 95, 111
vibrational transitions 48, 64
vibrational transition probability 88

Wien's law 42
wind sock theory 275

X-rays 367